The Making of Copernicus

Intersections

INTERDISCIPLINARY STUDIES IN EARLY MODERN CULTURE

VOLUME 36 – 2015

The titles published in this series are listed at *brill.com/inte*

The Making of Copernicus

Early Modern Transformations of the Scientist and his Science

Edited by

Wolfgang Neuber, Thomas Rahn and Claus Zittel

BRILL

LEIDEN | BOSTON

Cover illustration: Roundel with a half-length figure of Nicolaus Copernicus in profile to left against a night sky, holding an astronomical instrument in his right hand; one of thirty-one fragments of a broadside on forty-eight famous astronomers. [*Hierinn findt man die namen der fürnämsten erfindern des himmels louffs, sampt iren contrafeitischen Instrumenten*] (Zurich, Eustachius Froschauer the Elder: 1545–1546); Woodcut and letterpress © The Trustees of the British Museum

Library of Congress Cataloging-in-Publication Data

The making of Copernicus : early modern transformations of the scientist and his science / edited by Wolfgang Neuber, Thomas Rahn, and Claus Zittel.
pages cm. — (Intersections, ISSN 1568-1181 ; volume 36)
Mostly in English; chapter 7 in German.
Includes bibliographical references and index.
ISBN 978-90-04-28110-3 (hardback : alk. paper) — ISBN 978-90-04-28112-7 (e-book) 1. Copernicus, Nicolaus, 1473–1543. 2. Astronomers—Poland—Biography. 3. Astronomy—History. 4. Astronomy—Miscellanea. 5. Planetology. I. Neuber, Wolfgang, editor. II. Rahn, Thomas, editor. III. Zittel, Claus, editor.

QB36.C8M25 2014
520.92—dc23
[B] 2014027278

This publication has been typeset in the multilingual 'Brill' typeface. With over 5,100 characters covering Latin, IPA, Greek, and Cyrillic, this typeface is especially suitable for use in the humanities.
For more information, please see brill.com/brill-typeface.

ISSN 1568-1181
ISBN 978-90-04-28110-3 (hardback)
ISBN 978-90-04-28112-7 (e-book)

This book is printed on acid-free paper.

Contents

PART 3
New Astronomers: Biographical Transformations

List of Illustrations

Figures 10.1–10.9 (accompanying the article of Claus Zittel)

Notes on the Editors

Wolfgang Neuber

has been Full Professor of German Philology/Early Modern German and Neo-Latin Literature in the European Context and Head of the 'Forschungsstelle für Mittlere Deutsche Literatur' at Freie Universität Berlin since 2000; he is currently on a five-year leave of absence to teach literature at New York University in Abu Dhabi. His Visiting Professorships include Marburg (Germany), Vrije Universiteit Amsterdam, Olsztyn (Poland), Wroclaw (Poland), Beida University (Beijing) and Korea University (Seoul). Neuber has published extensively on early modern travel accounts (including "Fremde Welt im europäischen Horizont", 1991) and is currently preparing a book on early modern European family books. His research also focuses on mnemonics, rhetoric and travel accounts. He is co-editor of the journals *Rhetorik, Daphnis, Chloe* and *Zeitsprünge* and inter alia of the volumes *Cognition and the Book* Intersections 4 (2005), and *Spirits Unseen* Intersections 9 (2007).

Thomas Rahn

teaches German literature at the Freie Universität Berlin. His primary research interests are in the areas of Early Modern German literature, court culture and the typographic dimension of texts. He is the author of *Festbeschreibung. Funktion und Topik einer Textsorte am Beispiel der Beschreibung höfischer Hochzeiten (1568–1794)* (Tübingen: 2006), editor of *Krieg und Rhetorik* (Tübingen: 2003), and co-editor (with J.J. Berns) of *Zeremoniell als höfische Ästhetik in Spätmittelalter und Früher Neuzeit* (Tübingen: 1995), (with M. Bauer) of *Die Grenze. Begriff und Inszenierung* (Berlin: 1997), (with B. Jahn and C. Schnitzer) of *Zeremoniell in der Krise. Störung und Nostalgie* (Marburg: 1998), (with W. Neuber) of *Theatralische Rhetorik* (Tübingen: 2008), (with R. Falk) of *Typographie und Literatur* (Frankfurt a. M. – Basel: 2014), (with J.J. Berns) of *Projektierte Himmel* (forthcoming).

Claus Zittel

teaches German literature and philosophy at the Universities of Stuttgart, Frankfurt am Main, and Olsztyn (Poland), and is deputy director of the Stuttgart Research Centre for Text Studies. His Visiting Professorships include Freie Universität Berlin (Germany), TU Darmstadt (Germany), Beida University (Beijing), University of Trieste (Italy) and the University of Zagreb

(Croatia). He is inter alia the author of *Das ästhetische Kalkül von Friedrich Nietzsches 'Also sprach Zarathustra'* (2000, 2012) and *Theatrum philosophicum. Descartes und die Rolle ästhetischer Formen in der Wissenschaft* (2009), editor of *René Descartes, Les Météores/Die Meteore* (2006); and co-editor (with Karl Enenkel) of *Die 'Vita' als Vermittlerin von Wissenschaft und Werk* (2013), (with Michael Thimann and Heiko Damm) of *The Artist as Reader* Intersections 27 (2013), (with Helen King and Manfred Horstmannshoff) of *Blood, Sweat and Tears. The Changing Concepts of Physiology from Antiquity into Early Modern Europe* Intersections 25 (2012); (with Sylwia Werner) of *Ludwik Fleck. Denkstile und Tatsachen* (2011); (with Moritz Epple) *Science as Cultural Practice Vol. 1: Cultures and Politics of Research from the Early Modern Period to the Age of Extremes* (2010); (with Gisela Engel and Romano Nanni) *Philosophies of Technology. Francis Bacon and his Contemporaries* Intersections 11 (2008); (with Wolfgang Detel) *Ideals and Cultures of Knowledge in Early Modern Europe* (2002).

Notes on the Contributors

Lucia Ayala
is an art historian specialised in astronomical images concerning the structure of the universe in Modern and Contemporary Europe. Her PhD thesis, accomplished in co-tutelle with the Humboldt-University of Berlin and the University of Granada, was focused on the popularisation of the images of the plurality of worlds through Fontenelle's *Entretiens sur la pluralité des mondes*, encompassing a wide period from 1686 until the present. In her current research project, started at UC Berkeley in 2012, she deals with the images of the universe in Modern cosmology. Ayala works at the intersection of art, science and technology, and has collaborated with artists and scientists in several curatorial and research projects. www.luayas.net

Tamás Demeter
is Senior Research Fellow at the Institute of Philosophy, Hungarian Academy of Sciences and teaches at the University of Pécs, Hungary. He received his PhD from the University of Cambridge and he has held various fellowships at the Universities of Bristol, Cambridge, Edinburgh, Singapore, the Netherlands Institute for Advanced Study, and the Max Planck Institute for the History of Science. He has published widely on early modern natural and moral philosophies in several collective volumes and journals.

Dana Jalobeanu
is lecturer in Philosophy at the Faculty of Philosophy, University of Bucharest. Her field is history and philosophy of science and her current research investigates the emergence of early modern experimental philosophy. She is co-editor of the *Journal of Early Modern Studies* and executive editor of *Society and Politics*. Recent publications include Dana Jalobeanu—Peter Anstey (eds.), *Vanishing matter and the laws of nature: Descartes and beyond* (2011).

Jörg Jungmayr
is research fellow for Early Modern Literature at Freie Universität Berlin. He studied German literature and history in Tübingen and Berlin, conferring a doctorate with the bilingual edition of the *Legenda Maior* ("Vita Catharinae Senensis") by Raymond of Capua (2004). His major fields of research are: early modern literature and its lexical reception, Klaus Mann (queer

literature) and editorial studies. He is co-editor of *Editionen in der Kritik* and *Im Dickicht der Texte: Editionswissenschaft als interdisziplinäre Grundlagenforschung* (with Gesa Dane and Marcus Schotte).

Stefan Kirschner
studied biology at the University of Munich. In 1997 he obtained his doctorate in history of science and became assistant at the Institute for the History of Science at the same University. He has been Professor of History of Science at the University of Hamburg since 2003. His current research focuses on medieval natural philosophy, especially Nicole Oresme. Further research interests concern the early debates on Copernicus's cosmology and astronomy in the second half of the 16th century.

Andreas Kühne
studied information technology, information/documentation, history of science and history of art; in 1981 he took his doctorate (Dr.-Ing.) at the Technical University of Ilmenau, Thuringia, East Germany; in 2000 his 'Habilitation' in the history of science at the University of Munich. Since 2001 he became honorary professor at the Academy of Fine Arts, Munich; since 2008 adjunct professor for history of science at the University of Munich; since 1991 co-worker and since 1999 co-editor of the "Nicolaus-Copernicus-Gesamtausgabe" (Complete Edition of the Works of Nicolaus Copernicus).

Sergius Kodera
is Dean of the Faculty of Design at *New Design University*, St. Pölten, Austria. Since he received his Doctorate in 1994 he has been teaching Renaissance philosophy at the Department of Philosophy at the University of Vienna. He has published on and/or is a translator of Renaissance authors such as Marsilio Ficino, Fernando de Rojas, Machiavelli, Leone Ebreo, Girolamo Cardano, Giovan Battista della Porta, and Giordano Bruno. Currently he is working on a book-length study on Della Porta in English. His last book is Giordano Bruno, *Candelaio. Translation, Edition, Introduction, Notes* (2013).

Steffen Schneider
is Akademischer Rat at the Department of Romance Literature at the University of Tübingen. His primary research interests are in the areas of Renaissance literature, theory and reception of myth, and Mediterranean literature studies. He is the author of *Archivpoetik. Zur Funktion des Wissens in Johann Wolfgang Goethes 'Faust II'* (2005), *Kosmos—Seele—Text. Formen der Partizipation und ihre literarische Vermittlung. Marsilio Ficino, Pierre de*

Ronsard, Giordano Bruno (2012). He is (co-)editor of *Die Kabbala des Pegasus. Zugänge zu einem Text Giordano Brunos* (2007), and of *Objektivität und literarische Objektivierung seit 1750*, Tübingen 2010. Currently, he is preparing two volumes: *Ethos und Form der Tragödie*, and *Aisthetik der Geister—Spiritustheorien in Kunst, Philosophie und Populärkultur der Frühen Neuzeit / Aesthetics of the Spirits—Theories of 'spiritus' in Early Modern Arts, Philosophy, and Popular Culture*.

Jonathan Schüz

is the author of *Johann Fischarts Dämonomanie. Übertragungs- und Argumentationsstrategien im dämonologischen Diskurs des späten 16. Jahrhunderts* (Dissertation FU Berlin, 2011). His research interests include demonology, rhetoric, and history of science.

Gereon Wolters

is professor emeritus of philosophy at the University of Konstanz. His primary research interests are in the areas of history and philosophy of science, science and religion, and Nazi philosophy. He is the author of *Basis und Deduktion. Studien zur Entstehung und Bedeutung der Theorie der axiomatischen Methode bei J.H. Lambert (1728–1777)* (1980), *Mach I, Mach II, Einstein und die Relativitätstheorie. Eine Fälschung und ihre Folgen* (1987), *Vertuschung, Anklage, Rechtfertigung. Impromptus zum Rückblick der deutschen Philosophie aufs 'Dritte Reich'* (2004), *Ambivalenz und Konflikt. Katholische Kirche und Evolutionstheorie* (2010). He is editor of *Franz Anton Mesmer und der Mesmerismus: Wissenschaft, Scharlatanerie, Poesie* (1988), (with Andreas Elepfandt) of *Denkmaschinen? Interdisziplinäre Perspektiven zum Thema Gehirn und Geist* (1993), (with Martin Carrier) of *Homo Sapiens und Homo Faber. Epistemische und technische Rationalität in Antike und Gegenwart—Festschrift für Jürgen Mittelstraß* (2005), of *Ernst Mach, Die Analyse der Empfindungen und das Verhältnis des Physischen zum Psychischen* (2008), (with Giora Hon) of *Ernst Mach, Die Mechanik in ihrer Entwicklung, historisch-kritisch dargestellt* (2012). He is (co-)editor of five volumes in the Konstanz-Pittsburgh Series in the History and Philosophy of Science (1994–2010).

Introduction: The Making of Copernicus

Wolfgang Neuber, Thomas Rahn and Claus Zittel

1

> Doch unter allen Entdeckungen und Überzeugungen möchte nichts eine größere Wirkung auf den menschlichen Geist hervorgebracht haben, als die Lehre des Copernikus. Kaum war die Welt als rund anerkannt und in sich selbst abgeschlossen, so sollte sie auf das ungeheure Vorrecht Verzicht thun, der Mittelpunct des Weltalls zu sein. Vielleicht ist noch nie eine größere Forderung an die Menschheit geschehen: denn was ging nicht alles durch diese Anerkennung in Dunst und Rauch auf: ein zweites Paradies, eine Welt der Unschuld, Dichtkunst und Frömmigkeit, das Zeugniß der Sinne, die Überzeugung eines poetisch-religiösen Glaubens; kein Wunder, daß man dieß alles nicht wollte fahren lassen, daß man sich auf alle Weise einer solchen Lehre entgegensetzte, die denjenigen, der sie annahm, zu einer bisher unbekannten, ja ungeahneten Denkfreiheit und Großheit der Gesinnungen berechtigte und aufforderte.[1]

Goethe's eulogy, typical in its central propositions and clear-sighted with regard to the consequences, bears witness to Copernicus's enormous posthumous reputation.[2] As is always the case when myths and legends entwine the image of a thinker, its original contours are hard to recognise. The historical patina can, however, not be easily scratched off, it rather comes across like an acid

---

1 'Yet among all discoveries and convictions none may have produced a greater effect on the human spirit than the doctrine of Copernicus. Hardly had the world been acknowledged as spherical and closed in itself when it should abandon the enormous prerogative to be the centre of the universe. Perhaps never a greater challenge has been imposed on mankind; for what did not dissolve by this acknowledgement into vapour and smoke, a second paradise, a world of innocence, poetry and piety, the testimony of the senses, the conviction of a poetic-religious belief; no wonder that they did not want to let go of all this, that they opposed such a doctrine in all manners, which entitled and summoned him, who accepted it, to a hitherto unknown, yea unimagined freedom of thought and greatness of views.' Goethe Johann Wolfgang von, *Geschichte der Farbenlehre*, "Zwischenbetrachtung", in *Goethes Werke, Weimarer Ausgabe* (Weimar: 1893) vol. II.3, 213.

2 Cf. Rosen E., *Copernicus and his Successors* (London: 1995); Zinner E., *Entstehung und Ausbreitung der copernicanischen Lehre*, 2., revised and enlarged edition, ed. Nobis H.M. – F. Schmeidler (Munich: 1988).

 | DOI 10.1163/9789004281127_002

that corrodes its object. At the same time, it generates for ever new effects, in Goethe's case the insistent idea that until shortly before Copernicus everyone believed that the world was a disc. If one compares more recent Copernicus legends with those notions that people of earlier times had about him, grave differences become apparent.[3] One sees with astonishment that Copernicus's work, strongly indebted to older astrological discussions [Fig. 1],[4] was hardly known in the 16th and 17th centuries. Rather the names of Brahe, Kepler and Galileo were in the 17th century associated with a revolution of astronomy[5] and Copernicus only successively rose to the position of a hero of science[6]—a position due to that he became during the early Enlightment even the target of satyrical science criticism [Fig. 2], but allowed him also to lay a claim to the role as the protagonist of numerous novels even in the 20th century.[7] Copernicus the artist, clergyman, physician[8] and scholar from the provinces became a doer, an active designer of a new world image.

For Goethe—as for many of us today[9]—the focal point about Copernicus was that his revolution was not only one of astronomy but one of thought

3 Cf. for example: Wolfschmidt G. (ed.), *Nicolaus Copernicus (1473–1543): Revolutionär wider Willen* (Stuttgart: 1994); Gingerich O., *The Book Nobody Read: Chasing the Revolutions of Nicolaus Copernicus* (New York: 2004).

4 Cf. Westman R., *The Copernican Question: Prognostication, Skepticism, and Celestial Order* (Berkeley, Calif.: 2011).

5 Cf. the article by C. Zittel in this volume.

6 Cf. Grant E., *In Defense of the Earth's Centrality and Immobility: Scholastic Reaction to Copernicanism in the Seventeenth Century* (Philadelphia: 1984); Rosen E., *Copernicus and the Scientific Revolution* (Malabar, Fla.: 1984).

7 Cf. the articles by W. Neuber and J. Jungmayr in this volume.

8 Cf. Buczkowski M., *Beitrag zum gegenwärtigen Stand der Forschung über die ärztliche Tätigkeit des Nicolaus Copernicus (1473–1543)* (Diss. Munich: 1989).

9 As also earlier for Gryphius. See:

Vber Nicolai Copernici Bildt.
DV dreymall weiser geist / du mehr den grosser Mann
Dem nicht die nacht der zeit / dem nicht der blinde wahn
Dem nicht der herbe neidt die sinnen hatt gebunden:
Die sinnen die den lauff der schnellen erden funden.
Der du der alten träum vnd dünckel widerlegt
Vnd vns recht dargethan was lebt und was sich regt.
Schaw' itzund blüht dein rhumb / den als auff einem wagen
Der kreiß auf dem wir sind mus vmb die Sonnen tragen.
Wen dis was irdisch ist wird mit der zeit vergehn;
Sol vnbewegt dein lob mitt seinem himmel stehn.

Gryphius Andreas, *Gesamtausgabe der deutschsprachigen Werke, vol. II: Oden und Epigramme*, ed. M. Szyrocki – H. Powell, (Tübingen: 1964) 152. Cf. Kühlmann W.,

FIGURE 1 Title-page to the second edition of George Hartgill, *Astronomicall TABLES* (London, Company of Stationers: 1656, first edition 1594), Etching, © The Trustees of the British Museum. Hartgill was an Elizabethan astrologer and astronomer. An accompanying poem explains the frontispice: 'The Learned Artist works his brain / To make his Observation plain, / Hee, by his Instrument can tell / Thee where the *Stars* and *Planets* dwell. /*Philosophers*, do Knowledge gaine by books,/ *Astrologers*, by Heavens various looks.'

generally,[10] opening new possibilities also for the arts; this view, however, was disputed and even ignored by numerous anti-Copernicans amongst whom

"Neuzeitliche Wissenschaft in der Lyrik des 17. Jahrhunderts. Die Kopernikus-Gedichte des Andreas Gryphius und Caspar Barlaeus im Argumentationszusammenhang des frühbarocken Modernismus", *Jahrbuch der Deutschen Schillergesellschaft* 23 (1979), 124–153.

10 Cf. Kaulbach F., "Die Copernicanische Denkfigur bei Kant", *Kant-Studien* 64 (1973), 30–48.

FIGURE 2 Frontispice to Johann Gottfried Zeidler, *Die Hoch-Edle/ Veste Hochgelahrte und Hocherfahrene PHYSICA, Oder Natur-Lehre/ Als Hoch-betraute Ehestands-Räthin/ Ober-Kindmutter/ Inspectorin Uteri, Urin und Milch-Kosterin/ wie auch General Bratenwenderin des Sonnenlauffs am Königl. Hofe der grossen lateinischen Kunst-Göttin MINERVEN. Allen deutschen Micheln zur Verwunderung und Anbetung aus der Lateinischen Grund-Sprache in unsere Hochdeutsche Fr. Mutter-Sprache getreulich übersetzet/ und in eine Kirsche/ nicht grösser als ein Sperlings-Ey gefasset. Mit einer Vorrede FACULTATIS PHILOSOPHICAE der hoch-löbl. uralten Unversteht zu Abel* (Abel, Griller: 1700). © Württembergische Landesbibliothek Stuttgart.

were not only representatives of the Vatican but also philosophers like Francis Bacon.[11] Some only adapted what they could use and concealed the rest,[12] for others the astronomical revolutions played a merely subordinate role.[13] E.g.,

11 Cf. the articles by D. Jalobeanu and G. Wolters in this volume.

12 Cf. the articles by S. Kodera, S. Schneider and S. Kirschner – A. Kühne in this volume.

13 Cf. Stimson D., *The Gradual Acceptance of the Copernican Theory of the Universe* (Gloucester, Mass.: 1972).

after Johann Gottfried Herder in an essay 'Something of Nicolaus Copernicus's Life, towards His Image'[14] had presented the astronomer as a Sarmatian, i.e. as Polish, a century-long bold dispute arose on the question if Copernicus had been a German or a Pole.[15] This dispute, by the way, is still of the utmost importance in Poland as we could see on the occasion of our workshop that we held in the Copernicus city of Olsztyn in preparation of this volume.

These findings prompted us to gather in a volume articles that should make it possible to understand exemplarily how some of the Copernicus myths came about and if they could hold their ground or have vanished again. Hence, how and why did the notions change about Copernicus and the revolution that was named after him? Are there links between a factual or postulated transformation of world images and the application of certain scientific metaphors, especially the metaphor of a revolution? Were there interactions and amalgamations of the literary and scientific enthronement[16] or outlawry[17] of Copernicus and if so, how did they take place? Are present-day notions in science theory of a change of paradigm the effect of a rhetorical stage-management in earlier centuries? On the other hand, are there repercussions of the scientific-historical reconstructions and hagiographies on the literary image of Copernicus as sketched by novelists even in the 20th century?

The essays gathered in this volume try to give an answer to these central questions.

2

The history of the reception of Copernicus in this volume shall not be dominantly dealt with from the point of view of a *factual* affirmation and rejection of the astronomer and his doctrine but rather as accomplishments of transformation respectively. We consciously chose the term 'transformation' because it

14 Herder Johann Gottfried, "Etwas von Nikolaus Kopernikus Leben, zu seinem Bilde", *Der Teutsche Merkur*, No. 4 (1776), 169–179, reprinted in Kühne A. – Kirschner S. (eds.), *Biographia Copernicana. Die Copernicus-Biographien des 16. bis 18. Jahrhunderts* (Berlin: 2004) 292–299. Cf. the article by W. Neuber in this volume esp. 301-303.

15 Cf. on this Schröder R. "War die copernicanische Reform der Astronomie ein Weltbildwandel?", in Markschies C. – Zachhuber, J. (eds.); *Die Welt als Bild. Interdisziplinäre Beiträge zur Visualität von Weltbildern*, Arbeiten zur Kirchengeschichte 107 (Berlin: 2008), 91–112.

16 Exemplarily in Blumenberg H., *Die kopernikanische Wende* (Frankfurt am Main: 1965); Idem, *Die Genesis der kopernikanischen Welt* (Frankfurt am Main: 1975).

17 Koestler A., *Die Nachtwandler. Das Bild des Universums im Wandel der Zeit* (Bern – Stuttgart – Wien: 1959).

signifies a modification by which something is still preserved *in another state*.[18] Thus, instead of contrasting (within the framework of a history of the reception of Copernicus) affirmation as the identical and continued existence of something with rejection as the total sublation of something, the articles in this volume investigate *transformations*: methodological, institutional, textual and visual transformations of the Copernican doctrine and the topical, rhetorical and literary transformations of the historical person of Copernicus respectively.

The first part of the volume is dedicated to methodological transformations. Stefan Kirschner and Andreas Kühne offer an explanatory approach to the fact that in the second half of the 16th century at Wittenberg university, where astronomy and astrology formed focal points, the empirical data from Copernicus's research were taken into consideration but his theory of a heliocentric world system was largely ignored; while the specific structure of debate, derived from the scholastic practice of disputation and the medieval commentation of Aristotle respectively, allowed for the discussion even of unorthodox doctrines like the rotation of the earth, the humanist reform of the university induced with its teaching orientation towards pragmatic questions and rhetorical skills an eclipse of the question of the claim for truth within the Copernican hypothesis. In the course of the decay of the textual genre 'questio disputata' and of the disputation system at Wittenberg since around 1520 an epistemological situation arose where the Copernican hypothesis could no longer be of relevance.

Gereon Wolters examines a related question: what caused the Inquisition to turn against Copernicus's cosmology, which it had ignored for decades? In order to give an answer, the article focuses on Cardinal Bellarmin's efforts to take seriously the older doctrine of the superiority of theology over all other forms of knowledge and to enforce the counter-reformationist monopolisation of biblical exegesis as it had been agreed upon at the Council of Trent. Thus, Copernicus's hypothesis as a deviation from biblical statements and as an 'unorthodox' position only became virulent when the Catholic Church occupied an epistemic monopoly (and it may well be said that thus only from then on the church exposed something as 'revolutionary' what had before lain in the shadow of a special discourse).

18 Cf. the considerations regarding the term 'transformation' in: Leinkauf T., "Überlegungen zur Transformation des antik-scholastischen Methoden- und Wissensbegriffs in der Frühen Neuzeit: Autopsie, Experiment, Induktion", in Toepfer G. – Böhme H. (eds.), *Transformationen antiker Wissenschaften*, Transformationen der Antike 15 (Berlin – New York: 2010), 215–241, here 215f.

While Kirschner/Kühne's and Wolters' contributions deal with methodological transformations within an institutional framework, i.e. with modes of the Copernicus reception that are conditioned by general didactical and epistemological changes of attitudes in churches and universities, the following two articles in this section focus on Bacon and Hume's *specific* examinations of Copernicus's method. Bacon's anti-Copernicanism must be located, as Dana Jalobeanu demonstrates, within the framework of a critique of 'idolatrous' projections of mathematical ideals of order into the sky, which Bacon criticises generally about the old and common astronomy. Bacon sees Copernicus's attempt to simplify astronomical theory as a part of this geometrical 'fictionalisation'. By contrast, designing the project of a 'natural history of the heavens', Bacon envisages the ideal of a renewed 'living astronomy' that asks for the physical causes of phenomena and accomplishes a continuous mutual control of the results of mathematical astronomy and natural philosophy.

Tamás Demeter investigates an *inter-discursive* transformation of methods in Hume between the astronomical research of Copernicus and the moral philosophy and science of man respectively. While the reception of Copernicus is generally related to the new world model in Scottish enlightenment, Hume refers to Copernicus's achievement as an early example of and model for an *explanatory reductionism* which was to be employed in the search for the principles of human nature and behaviour. Hume's example makes it clear that the Copernican revolution was also perceived as a *methodological revolution* of science.—Contrastingly, Jalobeanu and Demeter's contributions show how one and the same approach—the simplification of theory—can be assessed as a 'mathematical-neurotic' eschewing of the phenomena on the one hand or as a condition for the understanding of the phenomena on the other—according to the respective methodological perspective and goal.

The second part of the volume assembles contributions on textual and graphical transformations of the Copernican doctrine. Here, the playful character of literature in its negotiations of knowledge is illuminated as well as the textual constitution of scientific and philosophical texts; it has become a commonplace even in the historiography of science to consider the rhetoricity of science[19] or even the 'poetic' modelling of scientific knowledge.[20]

Jonathan Schüz examines Jean Bodin's anti-Copernicanism predominantly on the level of rhetoric and modes of argumentation in Copernicus's *De Revolutionibus* and Bodin's *Universae naturae theatrum*. He demonstrates

19 Cf. Moss J.D., *Novelties in the Heavens. Rhetoric and Science in the Copernican Controversy* (Chicago – London: 1993).

20 Cf. Hallyn F., *The Poetic Structure of the World. Copernicus and Kepler* (New York: 1990).

the methodological and rhetorical contrast of both texts—logical vs. dialectical syllogism, mathematical argumentation vs. the topics of common sense, a focus on detail vs. a general view of reality—whereby the technical text appears to be in a defensive position (at least from the point of view of the 16th century). The rhetorical efficiency of the anti-Copernican argumentation is based upon the transformation of the topical frame within which theory is negotiated.

The link between epistemology and poetics is the focus of Steffen Schneider's article. Using the example of the *Ash Wednesday Supper*, the epistemological level of Giordano Bruno's Copernicus reception is addressed. Bruno's adherence to geocentric thinking can be explained on the one hand with the restrictions by the naked eye of celestial observations, on the other hand with a misguided imagination. The article exposes Bruno's epistemology and theory of reason and examines, combined with it, the dialogues' specific literary calculation: their aim is to oppose the imagination of a chimeric sky.

Another imaginary sky (also not completely free of chimeras) is employed in space-travel narratives of the 17th century in order to confirm Copernicus's world model. Thomas Rahn examines Kepler's *Somnium*, Godwin's *Man in the Moone* und Cyrano de Bergerac's *L'Autre Monde* with a thematic focus on the production of rhetorical evidence of the planetary movements and the transformation of methodological problems into games of inversion.

Lucia Ayala's contribution deals with the *media* transformation of a *cosmological* transformation. In the course of the replacement of the Copernican model of *one* world by the notion of a plurality of worlds, the graphical representation of the sky is changed—this iconographic change, predominantly explained by using the example of the illustrations of Fontenelle's *Entretiens sur la Pluralité des Mondes*, not only marginalises Copernicanism but also subverts the iconography of Louis XIV as the sovereign sun king.

The volume's third part turns to Copernicus the scientist himself but also to historical figures and novels' protagonists that appear as proxies of Copernicus. Sergius Kodera examines literary strategies of self-fashioning within the framework of Giordano Bruno's reception and assimilation of the Copernican theory. In his texts, Bruno stages himself as the more accomplished cosmologist than his paragon, i.e. not as an adept but as a prophet of the traditional doctrine who has assumed the role of a scientific leader.

Claus Zittel analyses, based upon texts and images of the life of Copernicus by Gassendi, the example of a scientist's vita that addresses with its doxographical emphasis first and foremost a professional astronomical audience. With its historical-genetical reconstruction of the theory construction, the *Vita* betrays a proximity to modern modes of the historiography of science. At the same time, however, the text contains a subversive, ironical level that allows for a reading as a 'crypto-apologia' for Galileo.

The heroisation of Copernicus since the 18th century to be found in German literature stands in contrast to this. Wolfgang Neuber traces into the 20th century the topic of the science hero—be that as a philosopher (as in Gottsched) or as a 'man of power' (as in Herder), but always a lone hero. The Copernican turn as manifested in Kant, Nietzsche and Freud eventually evolves as the model per se for a theory construction that sees itself as a revolutionary fact-act (to use Fichte's term 'Tathandlung').

Jörg Jungmayr's contribution is dedicated to Max Brod's novel *Tycho Brahes Weg zu Gott* (1915). There, the controversy around Copernicanism is dramatised and psychologised as a pupil-teacher relationship between Brahe and Kepler which is eventually inverted. Copernicus is present as a relic and plot-ferment in the shape of a *Commentariolus* manuscript, which in the end changes from Brahe to Kepler. If Kepler, at the same time, according to Brod is modelled on Einstein (very much alive in 1915) then a transformative line can be drawn from the 16th into the 20th century.

To sum it all up: we find Bruno as a better Copernicus; Copernicus's life as a subject for a covert partisanship for Galileo; Kant, Nietzsche and Freud as Copernican revolutionaries; Copernicus as Kepler as Einstein—some of the biographical Copernicus transformations covered in the last part of this book are figures of *translatio*, however not only of the transfer of a certain knowledge that is considered to be true but also of the transfer of a scientific ethos aimed at unconditional cognition.

Acknowledgements

For help with particular issues concerning the editorial work, we would like to thank Martin Bisse.

Selective Bibliography

Blumenberg H., *Die kopernikanische Wende* (Frankfurt am Main: 1965).

Blumenberg H., *Die Genesis der kopernikanischen Welt* (Frankfurt am Main: 1975).

Buczkowski M., *Beitrag zum gegenwärtigen Stand der Forschung über die ärztliche Tätigkeit des Nicolaus Copernicus (1473–1543)* (Diss. Munich: 1989).

Gingerich O., *The Book Nobody Read: Chasing the Revolutions of Nicolaus Copernicus* (New York: 2004).

Goethe Johann Wolfgang von, *Geschichte der Farbenlehre*, "Zwischenbetrachtung", in *Goethes Werke, Weimarer Ausgabe* (Weimar: 1893) vol. II.3.

Grant E., *In Defense of the Earth's Centrality and Immobility: Scholastic Reaction to Copernicanism in the Seventeenth Century* (Philadelphia: 1984).

Gryphius Andreas, *Gesamtausgabe der deutschsprachigen Werke, vol. II: Oden und Epigramme*, ed. M. Szyrocki – H. Powell (Tübingen: 1964).

Hallyn F., *The Poetic Structure of the World. Copernicus and Kepler* (New York: 1990).

Herder Johann Gottfried, "Etwas von Nikolaus Kopernikus Leben, zu seinem Bilde", *Der Teutsche Merkur* 4 (1776) 169–179.

Kaulbach F., "Die Copernicanische Denkfigur bei Kant", *Kant-Studien* 64 (1973) 30–48.

Koestler A., *Die Nachtwandler. Das Bild des Universums im Wandel der Zeit* (Bern – Stuttgart – Wien: 1959).

Kühlmann W., "Neuzeitliche Wissenschaft in der Lyrik des 17. Jahrhunderts. Die Kopernikus-Gedichte des Andreas Gryphius und Caspar Barlaeus im Argumentationszusammenhang des frühbarocken Modernismus", *Jahrbuch der Deutschen Schillergesellschaft* 23 (1979) 124–153.

Kühne A. – Kirschner S. (eds.), *Biographia Copernicana. Die Copernicus-Biographien des 16. bis 18. Jahrhunderts* (Berlin: 2004).

Leinkauf T., "Überlegungen zur Transformation des antik-scholastischen Methoden- und Wissensbegriffs in der Frühen Neuzeit: Autopsie, Experiment, Induktion", in Toepfer G. – Böhme H. (eds.), *Transformationen antiker Wissenschaften*, Transformationen der Antike 15 (Berlin – New York: 2010) 215–241.

Moss J.D., *Novelties in the Heavens. Rhetoric and Science in the Copernican Controversy* (Chicago – London: 1993).

Rosen E., *Copernicus and his Successors* (London: 1995).

Rosen E., *Copernicus and the Scientific Revolution* (Malabar, Fla.: 1984).

Schröder R. "War die copernicanische Reform der Astronomie ein Weltbildwandel?", in Markschies C. – Zachhuber, J. (eds.), *Die Welt als Bild. Interdisziplinäre Beiträge zur Visualität von Weltbildern*, Arbeiten zur Kirchengeschichte 107 (Berlin: 2008) 91–112.

Stimson D., *The Gradual Acceptance of the Copernican Theory of the Universe* (Gloucester, Mass.: 1972).

Westman R., *The Copernican Question: Prognostication, Skepticism, and Celestial Order* (Berkeley, Calif.: 2011).

Wolfschmidt G. (ed.), *Nicolaus Copernicus (1473–1543): Revolutionär wider Willen* (Stuttgart: 1994).

Zinner E., *Entstehung und Ausbreitung der copernicanischen Lehre*, 2., revised and enlarged edition, ed. H.M. Nobis – F. Schmeidler (Munich: 1988).

PART 1

The Copernican Turn: Methodological Transformations and Rejections

CHAPTER 1

The Decline of Medieval Disputation Culture and the 'Wittenberg Interpretation of the Copernican Theory'

Stefan Kirschner and Andreas Kühne

Abstract

In his first book of *De revolutionibus* Copernicus offered an alternative to Aristotle's account of gravity, levity and natural elemental motion. However, Copernicus's new world system, in particular his cosmological and physical theses, received almost no deeper discussion by scholars at the University of Wittenberg in the second half of the sixteenth century. This is not only true of the astronomical textbooks that Wittenberg scholars composed but applies also to their lectures. In fact, the Wittenberg scholars confined themselves to extracting certain astronomical data from Copernicus's work and to considering it a mere mathematical model that enabled them to render the celestial motions in a more accurate way. Westman has called this phenomenon the 'Wittenberg Interpretation of the Copernican Theory'. Erasmus Reinhold and Philipp Melanchthon played a pivotal role in establishing this tradition.

Adequate as the term 'Wittenberg Interpretation' is to describe the reception of Copernicus's theory in the second half of the sixteenth century, it contains a major shortcoming, as it is rather the description of an effect than of a cause. The question that must be raised is *why* the Wittenberg astronomers and natural philosophers were so reluctant to discuss Copernicus's cosmological and physical theses. The thesis we want to present in this paper is that the decline of the medieval culture of disputation was one of the main reasons, if not the decisive one, that prevented an adequate and vivid discussion of Copernicus's cosmological theses within university circles.

1 Introduction

The University of Wittenberg, founded in 1502, played a major role in the controversies over Copernicus's heliocentric system during the second half of the sixteenth century. Georg Joachim Rheticus (1514–1574), the young Wittenberg professor of 'lower' mathematics (*professor mathematum inferiorum*), who had visited Copernicus at Frauenburg and had become an enthusiastic adherent

 | DOI 10.1163/9789004281127_003

of heliocentrism, did not succeed in convincing his colleagues and students of the advantages of Copernican astronomy when he returned to Wittenberg in October 1541. Moreover, in October 1542 he left Wittenberg to assume the chair of 'higher mathematics' (*professor mathematum superiorum*) at Leipzig University. Thus it was Melanchthon and Reinhold who in the 1540s and the early 1550s shaped the later reception of Copernican astronomy at Wittenberg University.

Philipp Melanchthon (1497–1560), professor of Greek at Wittenberg University from 1518 and the intellectual head of the educational reform movement at Protestant universities, adopted a critical and negative attitude towards the Copernican system.[1] However, he was prepared to adopt empirical data and theoretical concepts from Copernicus's work, since many of them could be translated into the framework of a geocentric system.[2] It was probably under Melanchthon's influence that such an attitude became typical of a large group of scholars, and this led Westman to speak of a 'Melanchthon circle' and the 'Wittenberg Interpretation of the Copernican Theory'.[3]

Erasmus Reinhold's (1511–1553) *Prutenicae Tabulae coelestium motuum* (first edition Tübingen 1551) are a striking and influential example of this Wittenberg interpretation. In computing his tables, Reinhold drew upon Copernicus's

1 See Lerner M.-P., "Aux origines de la polémique anticopernicienne (II)", *Revue des sciences philosophiques et théologiques* 90 (2006) 409–452, here 433–447.

2 Westman R.S., "The Melanchthon Circle, Rheticus, and the Wittenberg Interpretation of the Copernican Theory", *Isis* 66 (1975) 164–193, here 172–174. By contrast, Ferry argues that primarily under the influence of Melanchthon 'the University of Wittenberg and its faculty helped shape an intellectual milieu that proved to be helpful to the expansion of Copernican teaching' (Ferry P.T., "The Guiding Lights of the University of Wittenberg and the Emergence of Copernican Astronomy", *Concordia Theological Quarterly* 57 [1993] 265–291, here 266) and 'did more to enhance than impede the progress of the new scientific astronomy' (ibidem 287). In our view Ferry has made the mistake of not sufficiently distinguishing between the reception of mathematical elements of the Copernican theory and its cosmological and physical contents. A similar dissent among historians of science concerns Luther's attitude towards Copernican astronomy; cf. Ferry, "The Guiding Lights"; Kleinert A., "'Eine handgreifliche Geschichtslüge'. Wie Martin Luther zum Gegner des copernicanischen Weltsystems gemacht wurde", *Berichte zur Wissenschaftsgeschichte* 26 (2003) 101–111.

3 Westman, "The Melanchthon Circle"; Westman R.S., *The Copernican Question. Prognostication, Skepticism, and Celestial Order* (Berkeley, Calif. et al.: 2011) 141–170. As Westman himself emphasises ("The Melanchthon Circle" 167, n. 8), the term 'Melanchthon circle' was coined by Lynn Thorndike. Brosseder criticises the concept of a 'Melanchthon circle' as 'myth' (Brosseder C., *Im Bann der Sterne. Caspar Peucer, Philipp Melanchthon und andere Wittenberger Astrologen* [Berlin: 2004] 11–17).

observations and planetary mechanisms, but he did not discuss heliocentrism.[4] Similarly, in his second edition of Peurbach's *Theoricae Novae Planetarum* (Wittenberg 1553) Reinhold mentions Copernicus several times and praises him as a second Ptolemy,[5] but again remains silent on his own attitude towards Copernicus's cosmology. His personal copy of *De revolutionibus* contains 'no annotations in the preface and few in Book I', while there are extensive mathematical and astronomical annotations on other parts of the work.[6] In one of his annotations Reinhold explicitly argued that the idea of the Earth's motion was absurd.[7] There is an indication that Reinhold composed a commentary on the whole of *De revolutionibus*,[8] but what has survived in manuscript form is a commentary only on books 3, 4, and 5.[9] It is unclear whether Reinhold ever treated the other books. Reinhold's attitude towards heliocentrism is generally difficult to assess. Two remarks in his commentary indicate that he was no supporter of heliocentrism.[10] In this context he mentions briefly his own, new

4 See Gingerich O., "The Role of Erasmus Reinhold and the Prutenic Tables in the Dissemination of Copernican Theory", in *Colloquia Copernicana II. Études sur l'audience de la théorie héliocentrique. Conférences du Symposium de l'UIHPS Toruń 1973* (Wrocław et al.: 1973) 43–62, 123–125.

5 Reinhold Erasmus, *Theoricae Novae Planetarvm Georgii Pvrbachii Germani, ab Erasmo Reinholdo Salueldensi pluribus figuris auctae, & illustratae scholijs* [...]. *Recens aeditae & auctae nouis scholijs in Theoria Solis ab ipso autore.* [...] (Wittenberg, Ioannes Lufft: 1553) for instance f. 23r, 25v, 31r ('Copernicum, quem iure dixeris alterum Ptolemaeum'), 31v.

6 See Gingerich O., *An Annotated Census of Copernicus'* De revolutionibus (*Nuremberg, 1543 and Basel, 1566*) (Leiden et al.: 2002) 268–278, here 269.

7 Westman, *The Copernican Question* 152.

8 The privilege issued by King Ferdinand for Reinhold ("Diploma caesarevm concessum Erasmo Rheinholt Salueldensi") that precedes Reinhold's *Prutenicae Tabulae Coelestium Motuum* explicitly mentions an "Eruditus Commentarius in totum opus Reuolutionum Nicolai Copernici"; see Reinhold Erasmus, *Prutenicae Tabulae Coelestium Motuum* (Tübingen, Ulrich Morhard: 1551) f. α2r. Cf. the annotation on the title page of one of the copies of the first edition of *De revolutionibus* kept at the Royal Library of Copenhagen (Gingerich, *An Annotated Census* 33).

9 Reinhold Erasmus, "Commentarius in opus revolutionum Copernici", ed. H.M. Nobis – A.M. Pastori, in Nobis H.M. et al. (eds.), *Receptio copernicana. Texte zur Aufnahme der copernicanischen Theorie, Nicolaus Copernicus Gesamtausgabe* VIII/1 (Berlin: 2002) 189–358.

10 Ibidem 295, ll. 26–27; 296, ll. 27–28; 296, ll. 33–35. Cf. Schofield C., "The Geoheliocentric Mathematical Hypothesis in Sixteenth-Century Planetary Theory", *The British Journal for the History of Science* 2 (1965) 290–296, here 292–293; Gingerich, "The Role of Erasmus Reinhold" 58–59; Henderson J.A., *On the distances between sun, moon and earth according to Ptolemy, Copernicus and Reinhold* (Leiden et al.: 1991) 10.

hypotheses, which however he omits to illustrate. Birkenmajer interpreted Reinhold's remarks as an anticipation of Tycho Brahe's geoheliocentric world system published in 1588.[11]

In an earlier paper we have explored the role that Copernicus's world system played in the astronomical lectures at Wittenberg University.[12] To answer the question of whether and in what way Copernicus's theory was presented and in which way it was criticised or even attacked, we have examined several surviving manuscripts of astronomical lectures at the University of Wittenberg during the second half of the sixteenth century. The result shows that the astronomical textbooks published by Wittenberg scholars must not be taken as the only basis of judgement on the status of Copernicus's world system in Wittenberg. In the lectures, Copernican heliocentrism was presented in a fair and unprejudiced yet very rudimentary way. Several references to passages in *De revolutionibus* leave the impression that students should be encouraged to occupy themselves with Copernicus's work rather than be deterred from it. However, it is also clear that this statement refers to the mathematical content of Copernicus's work and that there was no real interest in launching into an examination of the pros and cons of the Copernican system in natural philosophical and cosmological respects.

Adequate as the term 'Wittenberg interpretation' is to describe the reception of Copernicus's theory in the second half of the sixteenth century, it has a major shortcoming, as it is rather the description of an effect than of a cause. The question that must be raised is *why* the Wittenberg astronomers and natural philosophers were so reluctant to discuss Copernicus's cosmological and physical theses as exposed in the first book of his *De revolutionibus*.

The thesis we want to present in this paper is that the decline of the medieval culture of disputation was one of the main reasons, if not the decisive reason which prevented an adequate and vivid discussion of Copernicus's cosmological theses within university circles.

11 Birkenmajer A., "Le commentaire inédit d'Erasme Reinhold sur le *De Revolutionibus* de Nicolas Copernic", in *La science au seizième siècle. Colloque international de Royaumont 1–4 juillet 1957*, Histoire de la pensée 2 (Paris: 1960) 169–177, here 176–177.

12 See Kirschner S. – Kühne A., "Zur Copernicus-Rezeption an der Universität Wittenberg in der Frühen Neuzeit", in Dauben J.W. et al. (eds.), *Mathematics Celestial and Terrestrial. Festschrift für Menso Folkerts zum 65. Geburtstag* (Halle [Saale]: 2008) 507–522.

2 Copernicus's Cosmological and Physical Theses

Whenever we speak in this paper of Copernicus's cosmological and physical theses or use related formulations we mean much more than merely heliocentrism and the several motions of the Earth, which Copernicus described in his *De revolutionibus*. The term 'Copernicus's cosmological and physical theses' refers to the whole range of physical and cosmological arguments by which Copernicus wanted to refute Aristotle's reasoning for geocentrism.

Copernicus advances the following arguments against Aristotle's doctrine of an immobile Earth:

(1) Since it is the heavens that contain everything and are the common place of all things (*communis universorum locus*), it seems more adequate to ascribe motion to that which is contained and located (*locato*) than to that which contains and locates (*locanti*).[13] Since Copernicus partly uses Aristotelian terminology here, it is worthwhile to take a short look at Aristotle's definition of place and its medieval derivatives. Strictly rejecting the alternative view that the place of a body is the space filled by the body, Aristotle had defined the place of a body as the 'innermost motionless boundary of what contains it'.[14] There are many shortcomings of the Aristotelian definition of place and these were intensively discussed in the Middle Ages.[15] It was William of Ockham (c. 1285–1348) who held that place is not the innermost boundary of the containing body but the whole surrounding body.[16] Therefore, strictly speaking, Copernicus's argument addresses Ockham's view of place rather than Aristotle's. Nevertheless, we should not conclude that Copernicus knew Ockham's view on the nature of place. It is more probable that this correspondence is accidental, since Copernicus's statement that the heavens contain everything is quite generic and reminiscent of what Pliny writes in the second book of his *Historia naturalis*, namely that the world—i.e. the heavens—'externally and internally

13 Copernicus Nicolaus, *De revolutionibus libri sex*, ed. H.M. Nobis – B. Sticker, *Nicolaus Copernicus Gesamtausgabe II* (Hildesheim: 1984) bk. 1, ch. 5, 11, ll. 5–8; see also bk. 1, ch. 8, 16, ll. 22–24.

14 Aristotle, *Physica*, IV.4.212a.20–21 (the English translation is taken from Aristotle, *The Complete Works of Aristotle. The Revised Oxford Translation*, ed. J. Barnes, 2 vols. [New Jersey – Oxford: 1991] vol. 1, 361); see also IV.4.212a.5–6 and 4.211b.11–14.

15 See Grant E., "The Medieval Doctrine of Place: Some Fundamental Problems and Solutions", in Maierú A. – Paravicini-Bagliani A. (eds.), *Filosofia e scienze nella tarda scolastica: Studi sul XIV secolo in memoria di Anneliese Maier* (Rome: 1981) 57–79.

16 See ibidem 69; Shapiro H., *Motion, Time and Place according to William Ockham* (Louvain – Paderborn: 1957) 112–128.

embraces all things in itself'.[17] It is all the more probable that it was Pliny on whom Copernicus drew here, since Pliny's influence on Copernicus is manifest also in other places.[18]

(2) As we do not know for sure whether the world is finite or infinite, that is, whether the heavens are finite or extend into infinity, it is better to ascribe the daily rotation to the Earth than to the heavens, because what is infinite cannot be moved.[19] Copernicus's argument, which we have summarised here, starts from the uncertainty of whether the world is finite or infinite. At first sight one might suppose that this is an allusion to the extensive and controversial scholastic discussions on the question of whether there is an infinite void space beyond the world.[20] What speaks against such an assumption is that Copernicus obviously means an infinite material extension of the heavens themselves. As Knox has convincingly shown,[21] it was Pliny on whom Copernicus is drawing here. However, despite Pliny's obvious influence on Copernicus, which is also manifest in other places,[22] one must not forget that Copernicus advances an additional argument for the possible infinity of the heavens that sounds rather original and again refers to his idea of the heavens as the container of all other things. Having cited Aristotle's doctrine[23] that there is neither body nor place nor a vacuum, but absolutely nothing outside the heavens, Copernicus briefly and indirectly touches on the old problem of what constitutes—in an Aristotelian sense—the place of the heavens, if the heavens are surrounded by nothing, since it sounds strange that nothing shall contain something. Omitting Aristotle's solution to this problem,[24]

---

17 Pliny, *C. Plini Secundi naturalis historiae libri XXXVII*, ed. L. Ian, 5 vols. (Leipzig: 1854–1860) bk. II, I.2: 'extra intra cuncta complexus in se'. The English translation is from Pliny, *The Natural History of Pliny*, ed. J. Bostock (London: 1855–1857).

18 See Knox D., "Copernicus's Doctrine of Gravity and the Natural Circular Motion of the Elements", *Journal of the Warburg and Courtauld Institutes* 68 (2005) 157–211, here 189–190; Goddu A., *Copernicus and the Aristotelian Tradition* (Leiden – Boston: 2010) 237–238.

19 Copernicus, *De revolutionibus* bk. 1, ch. 8, 14, ll. 31–15, ll. 13. For the principle that what is infinite cannot be moved see Aristotle, *De caelo* I.5.271b.26–273a.6; I.7.274b.27–32; I.7.275b.12–29.

20 Cf. Grant E., *Much Ado About Nothing. Theories of Space and Vacuum from the Middle Ages to the Scientific Revolution* (Cambridge et al.: 1981) 101–147; Grant E., *Planets, Stars, and Orbs. The Medieval Cosmos, 1200–1687* (New York: 1996) 169–185.

21 Knox, "Doctrine of Gravity" 190.

22 See above, n. 18.

23 Aristotle, *De caelo* I.9.278b.21–279a.18.

24 For Aristotle the heavens and the world as a whole are in no place per se (Aristotle, *Physica*, IV.5.212b.8–10, 14–17) because only that which is contained by another body is

which incidentally never gained much acceptance among the scholastic commentators,[25] Copernicus propounds his own solution: 'If the heavens are infinite, however, and finite at their inner concavity only, there will perhaps be more reason to believe that beyond the heavens there is nothing. For, every single thing, no matter what size it attains, will be inside them, but the heavens will abide motionless. For, the chief argument by which it is sought to prove that the universe is finite is its motion.'[26]

(3) Necessarily, Copernicus launches his main attack on Aristotle's account of gravity, levity and natural elemental motion. It is primarily the following basic physical tenets of Aristotelian natural philosophy that Copernicus had to refute or at least to neutralise:

- There are three sorts of simple local motion, namely to the centre of the cosmos, from the centre of the cosmos, and around the centre of the cosmos.[27]
- Each simple body, that is, each element, has only one simple motion, which is its natural motion.[28]
- The natural motions of the four elements are rectilinear. The element earth, which is simply heavy, moves rectilinearly to the centre of the cosmos, which is its natural place, whereas fire, which is simply light, moves upwards on a rectilinear path away from the cosmos' centre.[29]
- Having reached its natural place, each element has no longer any tendency to move, but rests.[30] Therefore the Earth as a whole cannot move from the cosmos' centre, but rests therein.[31]

---

in place (IV.5.212a.31–32). However, outside the last sphere there is nothing (212b.16–17). Therefore the heavens are only per accidens 'in place, for all its parts are: for on the orb one part contains another' (212b.11–13) (English translation from: Aristotle, *The Works of Aristotle. Translated into English under the editorship of W.D. Ross*. Vol. II. *Physica. De caelo. De generatione et corruptione* [Oxford: 1930]).

25 See Grant E., "Cosmology", in Lindberg D.C. (ed.), *Science in the Middle Ages* (Chicago – London: 1978) 265–302, here 271–275; Grant, "The Medieval Doctrine of Place" 72–79.

26 Copernicus, *De revolutionibus* bk. 1, ch. 8, 15, ll. 3–6. The English translation is taken from Copernicus Nicolaus, *Nicholas Copernicus on the Revolutions*, ed. J. Dobrzycki, translation and commentary by E. Rosen (London: 1978) 16.

27 See Aristotle, *De caelo* I.2.268b.14–26.

28 Ibidem I.2.269a.8–9.

29 Ibidem IV.4.311a.16–21, IV.4.311b.13–29.

30 Ibidem I.9.279b.1–2.

31 Ibidem II.13.295b.19–23, II.14.296b.25–297a.2; *Physica*, III.5.205b.12–13.

- A rotation of the Earth would be a violent motion, because the Earth's natural motion is rectilinear towards the centre of the cosmos. Since a violent motion cannot be eternal, whereas the order of the cosmos is eternal, a rotation of the Earth is impossible.[32]

In sharp contrast to Aristotle, Copernicus advances the following propositions:[33]

1) Circular motion is the only motion that is truly uniform, simple and even, whereas the motion of falling bodies is accelerated and that of rising bodies is decelerated and thus cannot be regarded as even.[34]
2) Bodies in their natural place and persistent in their unity show circular motion,[35] since 'motion in place is none other than circular motion, which abides in itself, the whole being similar to something at rest.'[36] Rectilinear motion occurs additionally in those bodies that are outside their natural place, that is, in bodies that 'are separated from their whole and have abandoned their unity.'[37] 'Since, therefore, circular motion appertains to wholes, whereas rectilinear motion appertains additionally to parts, we can say that circular motion abides rectilinear motion in the same way as an animal abides a sickness.'[38] Accordingly, 'Aristotle's distinction of simple motion into three kinds—from a centre, to a centre and around a centre—should be considered merely a theoretical abstraction.'[39]
3) As the Earth rotates, so does the water that is contiguous to it.[40] Concerning the ambient air Copernicus states that it follows the Earth's rotation either because it is mixed with earthy or watery matter and thus obeys the same nature as earth or because the Earth's rotation is imparted to the air due to the latter's being contiguous to the rotating Earth and the resistance resulting therefrom.[41] Accordingly, due to its great distance from the Earth, the uppermost region of air where the comets are said to

32 Aristotle, *De caelo* II.14.296a.24–34.
33 Cf. Knox, "Doctrine of Gravity" 164–179.
34 Copernicus, *De revolutionibus* bk. 1, ch. 8, 16, ll. 8–15.
35 Ibidem bk. 1, ch. 8, 16, ll. 1–2.
36 Ibidem bk. 1, ch. 8, 16, ll. 2–4; English translation by Knox, "Doctrine of Gravity" 171.
37 Ibidem bk. 1, ch. 8, 16, ll. 6–8; Knox, "Doctrine of Gravity" 171.
38 Ibidem bk. 1, ch. 8, 16, ll. 15–17; Knox, "Doctrine of Gravity" 172.
39 Ibidem bk. 1, ch. 8, 16, ll. 17–19; Knox, "Doctrine of Gravity" 172.
40 Ibidem bk. 1, ch. 8, 15, ll. 19–20.
41 Ibidem bk. 1, ch. 8, 15, ll. 18–24.

be generated[42] is free from any motion.[43] There are several indications that Copernicus did not assume the existence of a sublunary fire sphere.[44]

4) The motion of rising and falling bodies is a double one in relation to the universe and composed of a rectilinear and a circular motion because things that sink of their own weight, being predominantly earthy, undoubtedly retain the same nature as the whole of which they are parts. Nor is the explanation different in the case of those things which, being fiery, are driven forcibly upward. For also fire here on the Earth feeds mainly on earthy matter, and flame is defined as nothing but blazing smoke.[45]

Evidently Copernicus does not attribute the circular part of the composed motion of falling bodies to their being dragged along by the rotating air but refers to the common nature of parts and wholes, by which he means their natural tendency to circular motion.

With the above-mentioned propositions Copernicus expounded an alternative to Aristotle's account of gravity, levity and natural elemental motion.[46] His own idea of gravity is completely different from and absolutely incompatible with Aristotle's. Having made the Earth a planet circling on a path around the sun, Copernicus had to explain why heavy bodies fall vertically onto the surface of the Earth on a path radial to the Earth's centre although the Earth is no longer in the centre of the cosmos. Setting forth his own definition of gravity he declares:

42 Copernicus refers to the view, current at his time, of the formation of comets that traces back to Aristotle (cf. Aristotle, *Meteorologica*, 1.7.344a.8–b.17). Strictly speaking, Aristotle held that comets are formed within the region of warm and dry exhalation called 'fire'. The upper part of the air participates in this process inasmuch as it is—together with the region of fire—carried around by the circular motion of the heavens.

43 Copernicus, *De revolutionibus* bk. 1, ch. 8, 15, ll. 24–28.

44 See Knox, "Doctrine of Gravity" 167–168.

45 Copernicus, *De revolutionibus* bk. 1, ch. 8, 15, ll. 31–36; The English translation is taken from Copernicus, *Nicholas Copernicus on the Revolutions* 16.

46 Cf. Knox, "Doctrine of Gravity". Goddu takes a different view, emphasising that 'Copernicus's doctrine of motion was undeveloped' and arguing that Copernicus 'was less interested in developing a complete and coherent theory than in raising doubts about Aristotle's theory' (Goddu, *Tradition* 336–355, here 344).

> I believe for my part that gravity is nothing other than a certain natural appetite implanted in parts by the divine providence of the creator of all things so that they would join themselves into unified wholes, collecting together in the form of a globe.[47]

Copernicus's cosmological and physical theses did not raise much interest in the second half of the sixteenth century. Even the few who sided with Copernicus's heliocentrism did not completely seize upon the other theses. In his *Narratio prima*, Rheticus shares Copernicus's line of reasoning that the sphere of the fixed stars embraces all things and should therefore be immobile.[48] On the other hand, Rheticus keeps closer to Pliny's agnostic view that it is useless to inquire into what lies beyond the world. Unlike Copernicus he does not transform the possible infinity of the world into an argument for the heavens' immobility.[49] Rheticus wastes no words in his *Narratio prima* on Copernicus's physical ideas as an alternative to Aristotle's account of gravity, levity and natural elemental motion. Only in his anonymous treatise *De terrae motu et scriptura sancta* does Rheticus go into this subject in great detail.[50]

Achilles Pirmin Gasser (1505–1577), Rheticus's friend and fellow adherent to Copernican theory from the very time he had received Rheticus's *Narratio prima*, tried to explain the Earth's rotation by magnetic forces, drawing on Petrus Peregrinus's (13th century) suggestion[51] that a spherical magnet would constantly rotate in accordance with the heavens' rotation if mounted in such a way that the axis through its poles points to the north and south celestial

47 Knox, "Doctrine of Gravity" 172. Cf. Copernicus, *De revolutionibus* bk. 1, ch. 9, 17, ll. 3–6. For the possible neoplatonic and other sources for Copernicus's view on gravity and his physical doctrines see Knox, "Doctrine of Gravity" 182–208; Goddu, *Tradition* 348–353.

48 See Rheticus Georg Joachim, "Narratio prima", ed. H.M. Nobis – A.M. Pastori, in Nobis H.M. et al. (eds.), *Receptio copernicana. Texte zur Aufnahme der copernicanischen Theorie, Nicolaus Copernicus Gesamtausgabe* VIII/1 (Berlin: 2002) 3–48, here 22, ll. 22–24.

49 See Rheticus, "Narratio prima" 23, ll. 10–20.

50 See Rheticus Georg Joachim, "De terrae motu et scriptura sancta", ed. H.M. Nobis – A.M. Pastori, in Nobis H.M. et al. (eds.), *Receptio copernicana. Texte zur Aufnahme der copernicanischen Theorie, Nicolaus Copernicus Gesamtausgabe* VIII/1 (Berlin: 2002) 57–73, here 69, ll. 4–21.

51 De Maricourt Petrus Peregrinus, "Epistola Petri Peregrini de Maricourt ad Sygerum de Foucaucourt militem de magnete", in Hellmann G. (ed.), *Rara Magnetica* (Berlin: 1898) 27–40, ch. X (35–36); de Maricourt Petrus Peregrinus, *Opera. Epistula de magnete. Nova compositio astrolabii particularis*, ed. L. Sturlese – R.B. Thomson (Pisa: 1995) 79–80.

pole.[52] Such ideas are not to be found in *De revolutionibus*. Moreover, the 'infrequent underlinings and marginal comments' that Gasser made in his personal copy of *De revolutionibus* 'generally call attention to, but do not evaluate, Copernicus's cosmological hypotheses'.[53]

A peculiar step towards popularising Copernicus's physical ideas was taken by the English Copernican Thomas Digges (1546–1595).[54] His *A Perfit Description of the Caelestiall Orbes according to the most aunciente doctrine of the Pythagoreans, latelye reuiued by Copernicvs and by Geometricall Demonstrations approued*, printed in 1576 as an addition to his late father's, Leonard Digges's (1520–1559) *Prognostication euerlastinge*, consists mainly of an English translation of the chapters 7 to 10 of Copernicus's *De revolutionibus* in which Copernicus defended his new world system with cosmological and physical arguments. Digges was convinced of the infinite extension of the sphere of the fixed stars,[55] while Copernicus had refrained from a definite decision on this question (see above). It has long been assumed that Palingenius's metaphysical speculations in his *Zodiacus Vitae*[56] gave Digges the idea of an infinite world.[57] However, despite Palingenius's indisputable influence on Digges, it is noteworthy that the *Zodiacus Vitae* contains the assumption of an infinite space beyond the world[58] but has nothing to say about an infinite extension of the last sphere. In our view it is probable that Digges seized on Copernicus's idea that the heavens might be infinite. What Copernicus regarded as possible, Digges regarded as certain. The assumption of an infinite world was much too attractive and tempting to be resisted, for an infinite extension of the sphere of the fixed stars excluded any circular motion,

52 Kremer R.L., "Copernicus among the Astrologers: A Preliminary Study", in Folkerts M. – Kühne A. (eds.), *Astronomy as a Model for the Sciences in Early Modern Times* (Augsburg: 2006) 225–252, here 232–233.

53 Ibidem 236 n. 26.

54 See Westman, *The Copernican Question* 268–280.

55 Digges Thomas, *A Perfit Description of the Caelestiall Orbes according to the most aunciente doctrine of the Pythagoreans, latelye reuiued by Copernicvs and by Geometricall Demonstrations approued* (London, Thomas Marsh: 1576) fol. N4r.

56 Palingenius Stellatus Marcellus, *Marcelli Palingenii Stellati Poetae doctissimi Zodiacus vitae hoc est, De hominis vita, studio, ac moribus optime instituendis, libri XII. Cum indice locupletissimo* (London, Thomas Marsh: 1569).

57 Cf. Johnson F.R. – Larkey S.V., "Thomas Digges, the Copernican System, and the Idea of the Infinity of the Universe in 1576", *The Huntington Library Bulletin* No. 5 (April 1934) 69–117.

58 Ibidem 103.

as Copernicus had pointed out, drawing indirectly on Aristotle's own proofs of the immobility of infinite bodies.

Michael Maestlin's (1550–1631) advocacy of Copernicus was grounded primarily in astronomical and mathematical reasoning. As early as 1573, in his comment on the *stella nova* of 1572, Maestlin acknowledged Copernicus, 'the prince of astronomers after Ptolemy', for having demonstrated the 'certain distances of the planetary orbs from the center of the world'.[59] Moreover, in his treatise on the comet of 1577–1578, Maestlin made explicit use of Copernicus's heliocentric theory to calculate its motion.[60] However, it was not until 1596, when he published a third edition of Rheticus's *Narratio prima* together with Kepler's (1571–1630) *Mysterium cosmographicum*, that he thoroughly subscribed himself to Copernicus's theory including its natural philosophical foundations as is evident from his foreword to the reader.[61] As Westman has pointed out, Maestlin's annotation in his personal copy of *De revolutionibus* that Copernicus wrote his book 'as an astronomer, not as a physicist' may be understood as self-referential.[62] Maestlin, 'a slavish admirer of the physical arguments set forth in Book I'[63] of *De revolutionibus*, as his annotations reveal, does not seem to have had a deep interest in dealing with the natural philosophical tenets of *De revolutionibus*. Maestlin would have been the ideal candidate for writing a commentary on *De revolutionibus* from a favourable point of view. Instead, he preferred to re-edit Rheticus's *Narratio prima* and *Encomium Borussiae*, which two treatises he wanted to be regarded as a short commentary on Copernicus's work and as a guide for understanding Kepler's *Prodromus*.[64]

---

59 Westman, *The Copernican Question* 261; Methuen C., "Maestlin's Teaching of Copernicus", *Isis* 87 (1996) 230–247, here 232.

60 Methuen, "Maestlin's Teaching" 232–233.

61 See Westman R.S., "Three responses to the Copernican theory: Johannes Praetorius, Tycho Brahe, and Michael Maestlin", in idem (ed.), *The Copernican Achievment* (Berkeley et al.: 1975) 285–345, here 332–333.

62 Westman, *The Copernican Question* 265, 267

63 Westman, "Three responses" 332.

64 Maestlin, preface to the third edition of Rheticus's *Narratio prima*: M.<AGISTER> MICHAEL MAESTLIN GOEPPINGENSIS, Candido Lectori S.<alutem>), in Kepler Johannes, *Prodromus Dissertationvm Cosmographicarvm, Continens Mysterivm Cosmographicvm* [...]. *Addita est erudita Narratio M. Georgii Ioachimi Rhetici, de Libris Reuolutionum* [...] *D. Nicolai Copernici* (Tübingen, Georg Gruppenbach: 1596) 86–90, here 89 (see also *Nicolaus Copernicus Gesamtausgabe*, vol. VIII/1 [Berlin: 2002] 448–452, here 451, ll. 26–35).

In Christoph Rothmann (c. 1550/60–c. 1600) we finally have a faithful defender of Copernicus's alternative to Aristotle's account of gravity, levity and natural elemental motion, as his correspondence with Tycho Brahe (1546–1601) demonstrates.[65] Not surprisingly, clear influences of Copernicus's doctrine are also manifest in the works of Giordano Bruno (1548–1600).[66]

3 The Lack of University Debates on Copernicus's Cosmological Theses: Causes and Consequences

There are no indications that Copernicus's cosmological theses had become the subject of extensive university debates in the second half of the sixteenth century. Even Maestlin, who taught mathematics from 1580 to 1584 at the University of Heidelberg and at the University of Tübingen where he became Kepler's teacher from 1584 until his death, does not seem to have encouraged public discussions of Copernicus's world system, to say nothing of its natural philosophical foundations. At least this is the impression one gets from the—admittedly very few—surviving astronomical disputations over which Maestlin presided. Only in a disputation in 1606 does the Copernican system become a major topic.[67]

This finding correlates very well with Maestlin's statement in his preface to the third edition of Rheticus's *Narratio prima* (Tübingen 1596) on the role

65 See Rothmann's letters to Brahe from 19 September 1588 (Brahe Tycho, *Tychonis Brahe Dani opera omnia*, ed. I.L.E. Dreyer, 15 vols. [Copenhagen: 1913–1929] vol. VI, 149–161, here 159) and 18 April 1590 (ibidem 214–217, here 215–216). For Brahe's counter-arguments see ibidem 218–223. Brahe claimed that he had convinced Rothmann during the latter's stay with him of the untenability of the Copernican system (ibidem 218, 222). On the controversy between Brahe and Rothmann see Granada M.A., "Christoph Rothmann und die Kosmologie in Kassel zur Zeit Wilhelms IV.", in Gaulke K. (ed.), *Der Ptolemäus von Kassel. Landgraf Wilhelm IV. von Hessen-Kassel und die Astronomie* (Kassel: 2007) 63–75; Seidengart J., "Le copernicianisme de Christoph Rothmann dans sa controverse avec Tycho Brahe et son prétendu 'retrait'", in Granada M.A. – Mehl E. (eds.), *Nouveau ciel. Nouvelle terre. La revolution copernicienne dans l'Allemagne de la Réforme (1530–1630)* (Paris: 2009) 179–205.

66 See Knox D., "Copernicus and Bruno on the Motion of the Earth", *Bruniana & Campanelliana* 5 (1999) 333–366; Knox D., "Bruno's Doctrine of Gravity, Levity and Natural Circular Motion", *Physis* 38 (2001) 171–209, here 198–208.

67 Cf. Methuen, "Maestlin's Teaching".

of the geocentric world system in teaching astronomy and in disputations.[68] Thus he believes it to be more appropriate and more advisable to propose to the public the common hypotheses, that is, the geocentric ones, and to teach these hypotheses to beginners at first, as they are generally known and therefore easier to understand. For the same reason, Maestlin continues, it is mostly better to retain the geocentric hypotheses even in the other disputations except when it comes to entering the inner chambers of astronomy.

Maestlin's remarks are as revealing as they are astonishing, since Copernicus's cosmological and physical tenets would have fitted very well into a lively discussion of the kind that Nicole Oresme (c. 1320–1382) and Jean Buridan (c. 1305–c. 1360) held at the University of Paris in the fourteenth century on the question of the possible rotation of the Earth. It is hard to believe that theses like the ones proposed by Copernicus as alternatives to Aristotelian tenets would have been considered unworthy of discussion in the intellectual atmosphere of medieval Paris. By contrast, in the second half of the sixteenth century obviously nobody at the universities felt the intellectual need or the interest to make Copernicus's cosmological theses the subject of extensive and open-minded disputations the aim of which is the investigation of truth and in which the opposing view, even if it is finally rejected, is given the same attention and fair treatment as one's own. The obvious decline of the medieval scholastic culture of debate at the universities constitutes an intellectual impoverishment the consequences of which have not yet been investigated in even a rudimentary way.

The intellectual potential of scholastic natural philosophical debates becomes apparent when we look at fourteenth-century commentaries on Aristotle's *De caelo* and the extent to which medieval natural philosophers like Buridan or Oresme dealt with the question of a possible rotational movement of the Earth around its axis.[69] In his famous *Livre du ciel et du monde* Oresme

68 Maestlin, preface to the third edition of Rheticus's *Narratio prima*: M.<AGISTER> MICHAEL MAESTLIN GOEPPINGENSIS, Candido Lectori S.<alutem>), in Kepler Johannes, *Prodromus Dissertationvm Cosmographicarvm, Continens Mysterivm Cosmographicvm* [...]. *Addita est erudita Narratio M. Georgii Ioachimi Rhetici, de Libris Reuolutionum* [...] *D. Nicolai Copernici* (Tübingen, Georg Gruppenbach: 1596) 86–90, here 87 (see also *Nicolaus Copernicus Gesamtausgabe*, vol. VIII/1 [Berlin: 2002] 448–452, here 449, ll. 19–23).

69 Cf. Oresme Nicole, *Questiones de spera*, ed. G. Droppers (Ph.D. Dissertation, University of Wisconsin: 1966) 390–410 (commentary by Droppers on Qu. 8); Fellmann F., *Scholastik und kosmologische Reform* (Münster: 1971); Hugonnard-Roche H., "Le problem de la rotation de la terre au XIVe siècle", in *Avant, avec, après Copernic: la représentation de l'univers et ses conséquences épistémologiques. XXXIe semaine de synthèse, 1–7 juin 1973* (Paris: 1975)

demonstrated that every Aristotelian or other traditional, conventional argument for the immobility of the Earth can be refuted or neutralised and that there is neither a sensual experience nor a rational argument that would be apt to prove the Earth's immobility at the centre of the cosmos.[70] Furthermore, he advances a long list of arguments in favour of a daily rotation of the Earth,[71] even if he finally decides to vote for its immobility.[72]

Oresme and other medieval natural philosophers were accustomed to, and well trained in, raising doubts even against common opinions and seemingly well-grounded Aristotelian tenets. Therefore our hypothesis is that, had there still been a living tradition of scholastic disputation at the universities in the second half of the sixteenth century comparable to what we observe in the fourteenth century, Copernicus's cosmological and natural philosophical ideas would have been given the attention they actually deserved. Attention does not necessarily mean acceptance, it is true, but we have here much more than the intellectual lack of interest with which Copernicus's theory found itself confronted among the Wittenberg scholars.

We are aware that it is one thing to be open-minded enough to fairly discuss unusual and unorthodox opinions, but another to propound a completely new theory. Even Oresme did not extend his detailed analysis of the problem of a possible movement of the Earth beyond the question of whether the Earth might rotate around its axis. Despite medieval scholastics' delight in discussing unorthodox views and despite fourteenth-century plurality of opinions, it was not in the fourteenth century but at the beginning of the sixteenth century that somebody stood up to propose a heliocentric world system.

Doubtless, the transition from the Middle Ages to the Early Modern Period was conducive to the generation of completely new and alternative views and methods. However, as is well known, any progress implies some regression. In our case the decline, happening at the same time, of a living university disputation culture might have been the cause that prevented extensive debates on the cosmological theses of Copernicus.

62–65; Souffrin P., "Oresme, Buridan, et le mouvement de rotation diurne de la terre ou des cieux", in Ribémont B. (ed.), *Terres médiévales* (Paris: 1993) 277–303; Grant, *Planets, Stars, and Orbs* 639–647.

70 See Oresme Nicole, *Le livre du ciel et du monde*, ed. A.D. Menut – A.J. Denomy, translated with an introduction by A.D. Menut (Madison et al.: 1968) bk. II, ch. 25, ll. 33–242 (520–530).

71 Ibidem bk. II, ch. 25, ll. 243–344 (530–536).

72 Ibidem bk. II, ch. 25, ll. 354–364 (536–538).

We shall now briefly sketch the factors that provided the basis for a fruitful disputation tradition at medieval universities with the *questio disputata* as its primary representative, and how this situation changed during the Renaissance.

(1) When it came to discussing cosmological issues, Aristotle's *De caelo* was considered the ideal starting point. Plato's *Timaeus* could not compete with Aristotle's *De caelo*. Whoever wanted to deal with cosmology on a higher scholarly level and possibly present new ideas was well advised to do this in the framework of a commentary on Aristotle's *De caelo*. Another candidate was Johannes de Sacrobosco's (c. 1195–c. 1256) *Sphaera*, which served as an introduction to astronomy at medieval universities. For instance, besides his Latin and French commentaries on Aristotle's *De caelo*, Oresme composed a Latin and a French commentary on Sacrobosco's textbook. In his Latin commentary, which is composed in the form of *questiones*, he proposes many original ideas concerning the question of whether the Earth might rotate around its axis.[73] Nevertheless, generally speaking, commentaries on Sacrobosco's *Sphaera* were mostly restricted to elucidating basic astronomical and cosmological concepts and commonly accepted tenets, and that is in accordance with its role as an isagogic work.

At first sight, the concentration on a single work, i.e. Aristotle's *De caelo*, appears to be a serious drawback. Nevertheless, to be forced to present one's own cosmological ideas in the framework of a commentary on Aristotle's *De caelo* did not only have disadvantages. Numerous factors decided the fate of a medieval piece of scholarship, especially before the invention of printing with movable type. Since lectures on Aristotle's *De caelo* were quite common and usually obligatory at medieval universities, the author of a commentary on Aristotle's treatise probably had a greater chance that his ideas exerted a broader impact on contemporary and future learning than if he had composed a monograph. An example from another, but related, field is illustrative in this context. As is well known, between 1304 and 1311 and at the request of Aimericus de Placentia († 1327), the master general of the Dominican order,[74] Dietrich of Freiberg (c. 1240–c. 1320) composed an impressive monograph on meteorological phenomena that deals primarily with the formation of the primary and secondary rainbow.[75] Unfortunately Dietrich's masterpiece seems to

73 See Oresme, *Questiones de spera*, Qu. 8 (154–173).

74 Flasch K., *Dietrich von Freiberg. Philosophie, Theologie, Naturforschung um 1300* (Frankfurt a.M.: 2007) 634.

75 Freiberg Dietrich of, "Tractatus de iride et de radialibus impressionibus", ed. M.R. Pagnoni-Sturlese – L. Sturlese, in Pagnoni-Sturlese M.R. et al. (eds.), *Dietrich von Freiberg, Opera Omnia*, Tomus IV, *Schriften zur Naturwissenschaft, Briefe* (Hamburg: 1985) 95–268.

have gone completely unnoticed by medieval natural philosophers. Possibly his ideas would have experienced a different fate had he presented them in the much more common framework of a *questiones*-commentary on Aristotle's *Meteorology*.

(2) Although discussions on natural philosophical issues at medieval universities started almost exclusively from Aristotle's works and despite the high esteem medieval natural philosophers felt for Aristotle, whom they often venerably called merely *the* philosopher, Aristotle was never considered sacrosanct. Quite the contrary, particularly the fourteenth and fifteenth centuries were an age of critique, modification and self-conscious further development of essential Aristotelian tenets.[76]

(3) Disputation was considered not only a means of didactics but of scientific inquiry and investigating the truth. It was far more than a mere intellectual tournament in which to show one's skills in dialectics.[77] In principle any opinion was considered worth discussing and tenable at least *gratia exercitii*.

---

76 Cf. Goddu, *Tradition* 89–136; Grant, *Much Ado About Nothing*; Grant E., *The Foundations of Modern Science in the Middle Ages. Their religious, institutional, and intellectual contexts* (Cambridge et al.: 2006); Maier A., *An der Grenze von Scholastik und Naturwissenschaft* (Rome: 1952); eadem, *Metaphysische Hintergründe der spätscholastischen Naturphilosophie* (Rome: 1955); eadem, *Zwischen Philosophie und Mechanik* (Rome: 1958); eadem, *Die Vorläufer Galileis im 14. Jahrhundert* (Rome: 1966); eadem, *Zwei Grundprobleme der scholastischen Naturphilosophie* (Rome: 1968).

77 For the peculiar character and different forms of medieval disputations, their history, structure, procedure and aims see Bazàn B.C., "La quaestio disputata", in *Les genres littéraires dans les sources théologiques et philosophiques médiévales. Définition, critique et exploitation. Actes du Colloque international de Louvain-la-Neuve 25–27 mai 1981* (Louvain-la-Neuve: 1982) 31–49; idem et al., *Les questions disputées et les questions quodlibétiques dans les facultés de théologie, de droit et de médecine* (Turnhout: 1985); Lawn B., *The Rise and Decline of the Scholastic 'Quaestio disputata'. With Special Emphasis on its Use in the Teaching of Medicine and Science* (Leiden et al.: 1993); Schulthess P., "Die Philosophie im lateinischen Mittelalter", in idem et al. (eds.), *Die Philosophie im lateinischen Mittelalter. Ein Handbuch mit einem bio-bibliographischen Repertorium* (Düsseldorf – Zürich: 2000) 15–349, here 147–158; Weijers O., "De la joute dialectique à la dispute scolastique", *Académie des Inscriptions et Belles-Lettres. Comptes rendus des séances de l'année 1999, Paris 1999* (2000) 509–518; eadem, *La 'disputatio' dans les Facultés des arts au moyen âge* (Turnhout: 2002); eadem, "Quelques observations sur les divers emplois du terme *disputatio*", in Meirinhos J.F. (ed.), *Itinéraires de la raison. Études de philosophie médiévale offertes à Maria Cândida Pacheco* (Louvain-la-Neuve: 2005) 35–48; eadem, "La 'disputatio' comme moyen de dialogue entre les universitaires au moyen âge", in Gilli P. (ed.), *Les élites lettrées au Moyen Âge. Modèles et circulation des savoirs en Méditerranée occidentale (XIIe–XVe siècles)* (Montpellier: 2008) 155–169; eadem, *Queritur utrum. Recherches sur la 'disputatio' dans les universités médiévales* (Turnhout: 2009); eadem, "The various kinds of disputation in the faculties of arts, theology and law (c. 1200–1400)", in Gindhart M. – Kundert U.

Medieval scholars were trained to take over the role of both opponent and respondent and this required them to be as familiar with the arguments of their imaginary or real opponent's views as with their own lines of reasoning, and they were expected to raise doubts and objections against their own view. It is evident from works such as Pietro Pomponazzi's (1462–1525) *Quaestio de reactione* that all this was no mere theoretical demand. In this treatise, Pomponazzi heavily criticises the teachings of the Mertonians but simultaneously discusses them in such detail that his work became a major source for their ideas.[78]

Compared to these favourable medieval conditions, the situation in the sixteenth century changed drastically.

The literary form of the *questio disputata* reflected the medieval culture of disputation with its lines of reasoning, its pro and con arguments, its distinctions, doubts, objections, conclusions etc. Also in the cases in which the author of the *questio* or someone else revised or redacted the text, it kept all the forms of a lively and controversial debate. By contrast, monographs, textbooks, handbooks, compendia and syncretistic and eclectic compilations—all the literary forms that became more and more popular in the Renaissance—represent a completely different approach to expressing and communicating natural philosophical contents. These literary forms are no fertile soil for producing an atmosphere that permits an unprejudiced assessment of the pros and cons of competing new and unusual theories since this kind of literature focuses on the account and presentation of basic and commonly accepted views.

The humanists' critique of scholastic logic, theology and natural philosophy contributed extensively to the decline of the scholastic *questio disputata*.[79] To mention only one example: What we nowadays admire as a consistent application of reason to theology, the humanists ridiculed as hair-splitting, a waste of time and tomfoolery. While 'scholastic theologians were fascinated with all sorts of questions that tested God's powers', Erasmus of Rotterdam (c. 1469–1536) regarded such questions as silly and absurd.[80]

(eds.), *Disputatio 1200–1800. Form, Funktion und Wirkung eines Leitmediums universitärer Wissenskultur* (Berlin – New York: 2010) 21–31.

78 Lawn, *The Rise and Decline* 89.

79 For the humanists' critique of scholastic learning and teaching and further literature on this subject see Grant E., *God and Reason in the Middle Ages* (Cambridge et al.: 2001) 293–304; Lawn, *The Rise and Decline* 107–126; Rummel E., *The Humanist-Scholastic Debate in the Renaissance & Reformation* (Cambridge, Mass. – London: 1995); Nauta L., *In defense of common sense. Lorenzo Valla's humanist critique of scholastic philosophy* (Cambridge, Mass. – London: 2009).

80 Grant E., *God and Reason* 297.

81 Erasmus of Rotterdam, *Erasmi Roterodami Encomium moriae i. e. stultitiae laus. Lob der Torheit. Basler Ausgabe von 1515.* Mit den Randzeichnungen von Hans Holbein d. J. in Faksimile mit einer Einführung herausgegeben von Heinrich Alfred Schmid, 2 vols (Basel: 1931), vol. 2 (facsimile), fol. S4 v.

FIGURE 1.1 Illustration by Hans Holbein the Younger (1497–1543) in a copy of the 1515 edition of *The Praise of Folly* (*Moriae encomium* or *Stultitiae laus*) of Humanism's leading figure Erasmus of Rotterdam (c. 1469–1536).[81] Holbein's drawing shows the soul (*Scoti anima*) of the Franciscan theologian, philosopher and logician John Duns Scotus (1265/66–1308), entering the body of Folly. The drawing refers to Erasmus's statement, speaking as Folly: 'I think it more suitable while I [scil. Folly] play the divine and tread among thorns to beg that the soul of Scotus, itself more prickly than a porcupine or hedgehog, shall come from his beloved Sorbonne and dwell in my breast for a season; and very soon it shall go back whither it pleases or to the dogs.'[82] While Duns Scotus, known as 'doctor subtilis', can rightly be considered one of the most eminent medieval thinkers, for Erasmus he is nothing but a fool. Even more drastically, Holbein depicts Duns Scotus's soul purging its bowels of faeces composed of stupid logical works (*cacat stulta logicalia*).

82 Explanations and English translation from Murdoch J.E., *Album of Science. Antiquity and the Middle Ages* (New York: 1984) 187.

It seems that, because scholastic learning and teaching centred mainly on the *questio disputata* as the dominant form of instruction, the ruin or at least retreat of scholastic learning and teaching under the attack from humanism entailed also the ruin and decline of the *questio disputata*. The humanists tended to use the dialogue, the treatise, the oration or the letter to expound their ideas.[83] When they composed *quaestiones naturales*, they applied the simple, non-scholastic, question-and-answer form.[84]

But even if we look at institutions that we would expect to have preserved the former scholastic *questio* tradition, we must state that what had survived was often nothing more than a poor copy of its ingenious medieval predecessor. The widespread commentaries on Aristotle's works by the Coimbra Jesuits at the end of the sixteenth century are a striking example. While they preserve the outer form of the *questiones* literature, a closer look at the question of 'whether the Earth rests in the centre of the cosmos and what the reason is for its immobility' reveals a stereotypical, unmotivated, uninspired and inflexible approach, devoid of any original ideas.[85] It is nothing more than a list of Aristotelian arguments and 'truths' without the slightest attempt to raise doubts about them. Thinking of the way in which Jean Buridan or Nicole Oresme discussed the possibility of a diurnal rotation of the Earth, it is hard to believe that any student—to say nothing of a professor—at the University of Paris in the second half of the fourteenth century presenting a Coimbra-like discourse would have been considered competent to treat natural philosophical topics.

Similarly, one cannot imagine that medieval university scholars would have accepted the way in which Giovanni Maria Tolosani (1470/71–1549) dealt with Copernicus's physical and cosmological doctrines as a rational and adequate form of discussion. In his treatise *Opusculum quartum. De coelo supremo immobili et terra infima stabili, ceterisque coelis et elementis intermediis mobilibus*,[86] Tolosani dispenses with any detailed physical discus-

83 Lawn, *The Rise and Decline* 85.

84 Ibidem 95–96.

85 Collegium Conimbricense, *Commentarii Collegii Conimbricensis, societatis Iesu, in quattuor libros de coelo, meteorologicos & parva naturalia, Aristotelis Stagiritae* (Cologne, Impensis Lazari Zetzneri: 1603) cols. 417–420.

86 Biblioteca nazionale di Firenze, Ms. Conv. soppr. J. 1. 25, ff. 339r–342r. Tolosani's treatise has been edited with a French translation by M.-P. Lerner, in idem, "Aux origines de la

sion of Copernicus's arguments and limits himself to mentioning the standard arguments for geocentrism, evidently completely ignoring the fact that Copernicus's alternative to Aristotle's account of gravity, levity and natural elemental motion constituted a serious threat to his own beloved traditional view. Instead, he simply reproaches Copernicus for not being sufficiently skilled in physics and dialectics and for not having read and concerned himself with Aristotle's and Ptolemy's arguments.[87]

How internal scholastic developments possibly contributed to the decline of the culture of disputation remains a question for further research. As Weijers summarises:

> It is possible that during the 15th century the disputation became a mere technique that lacks flexibility, a kind of automatism of always repeating the same argument scheme. It probably lost the element of research that was so important in the 13th and 14th centuries. Thus it could be criticised and laughed at by the humanists.[88]

4 The Situation at the University of Wittenberg

While the production of new ideas and theories in the sixteenth century took place primarily outside university contexts, the universities were still the appropriate place to discuss them. Particularly the University of Wittenberg, with its focus on astrology and astronomy, would have been the ideal institution to debate Copernicus's physical and cosmological theses. However, as we have seen, the way in which scholars at the University of Wittenberg dealt with Copernicus's new world system in the second half of the sixteenth century is characterised by an almost complete absence of any deeper discussion of Copernicus's cosmological and physical ideas.

From 1517/1518 to 1523 the arts curriculum at the University of Wittenberg experienced a thoroughgoing reform, the protagonists of which were Luther and Melanchthon. Their common objective was to promote those authors, works and disciplines that they considered useful for the propagation of the

polémique anticopernicienne (I). L'Opusculum quartum de Giovanmaria Tolosani [1547–1548]", *Revue des sciences philosophiques et théologiques* 86 (2002) 681–721, here 692–719.

87 Tolosani, *Opusculum quartum* 701–703, 707–711; cf. Goddu, *Tradition* 414–418.

88 Weijers, "Kinds of disputation" 31.

Reformation and the education of its proselytisers. Between 1517 and 1536, as the symbiosis between humanism and Reformation evolved,[89] Luther's ideal of an intellectual reform of the German universities becomes apparent from a treatise of 1520, in which he addressed himself to the German nobility.[90] Deploring that Holy Scripture and Christian faith are much too little taught at the universities, while at the same time the 'pagan master Aristotle rules', he pleads for the theologians to put aside part of Aristotle's philosophical works, on which they had relied for so long, namely the *Physica*, the *Metaphysica*, the *De anima* and the *Ethica*. Hardly anything about nature and the mind, he argues, can be learned from these works. In contrast, Aristotle's *Logica*, *Rhetorica* and *Poetica* should be kept in use in the Faculty of Arts or replaced with short summaries, because these disciplines are useful for preaching. Furthermore, he values history, mathematics and the three languages fundamental to Christian theology, Latin, Greek and Hebrew.

In contrast to Luther's proposal, Aristotle's *Physica* never vanished from the arts curriculum.[91] However, the *scholastic* lectures on Aristotle's *Physics* were abolished. The Scotistic lecture on Aristotle's *Physics* ended in 1519 and the Thomistic one in 1520/21.[92] In 1518 a new chair in Physics was established for the teaching of Aristotle's *Physics* in its new translation from the Greek.[93] Unfortunately there are no printed or manuscript sources that could supply any information on the content of this new lecture.

Concerning cosmology and astronomy the situation was as follows. The newly founded 'Lectio Pliniana' was intended to cover the whole of Pliny's *Naturalis Historia*, but soon it concentrated on the zoological books and later exclusively on book II, which deals with cosmology.[94] The introduction of Pliny's *Naturalis historia* as a textbook at the University of Wittenberg was 'part of the early Protestant attempt to reform the curriculum and to eliminate the authority of Aristotle'.[95] Melanchthon seems to have attached great value to

89 Kathe H., *Die Wittenberger philosophische Fakultät 1502–1817* (Cologne et al.: 2002) 47–48.

90 Scheible H., "Aristoteles und die Wittenberger Universitätsreform. Zum Quellenwert von Lutherbriefen", in Beyer M. – Wartenberg G. (eds.), *Humanismus und Wittenberger Reformation. Festgabe anläßlich des 500. Geburtstages des Praeceptor Germaniae Philipp Melanchthon am 16. Februar 1997* (Leipzig: 1996) 123–144, here 123; Kathe, *Philosophische Fakultät* 48.

91 Scheible, "Wittenberger Universitätsreform" 124.

92 See Kathe, *Philosophische Fakultät* 465.

93 Ibidem; with a list of the professors holding the chair in physics.

94 See ibidem 71.

95 Nauert C.G., "Humanists, Scientists, and Pliny. Changing Approaches to a Classical Author", *The American Historical Review* 84 (1979) 72–85, here 80.

the 'Lectio Pliniana'. In 1531 he mentions that he knows of no other institution where the second book of Pliny's *Naturalis historia* is treated so extensively.[96] The 1545 statutes of the University of Wittenberg, designed by Melanchthon, mention the second book of Pliny's *Naturalis historia* as obligatory in the study of philosophy.[97] However, around the middle of the sixteenth century it became evident that Pliny's work was inadequate as a textbook.[98]

In the course of time Melanchthon realised that the Reformation desperately needed a reliable natural philosophical foundation. From the spectrum of ancient philosophies, Melanchthon regarded Aristotle's as the most suitable for his purposes.[99] Thus Melanchthon had undergone a fundamental change from being a vehement adversary of Aristotelian philosophy in his youth to becoming its most prominent advocate within the Protestant movement. But we have to bear in mind, as Petersen pointed out in 1921, that Melanchthon's youthful passion against Aristotle was primarily motivated by his dislike of the central role Aristotle played in scholastic theology.[100] Having made theology a philosophy was one of the major and constant Protestant reproaches against scholasticism.[101] But evidently Melanchthon must have realised that it was a mistake to resort to drastic measures. Even Luther seems to have put up with Aristotle's new and dominant role in Protestant philosophy.[102]

The university reforms of 1536, prepared and designed by Melanchthon, saw an extensive restoration of Aristotle. Thus, theoretically, Aristotle's natural philosophical works could have served again as a starting point for disputations on cosmological and other natural philosophical issues. However,

---

96 Melanchthon Philipp, "De ordine discendi", in Bretschneider C.G. (ed.), *Philippi Melanthonis opera quae supersunt omnia, Corpus reformatorum*, vol. XI (Halle [Saale]: 1843) cols. 209–214, here col. 213.

97 *Urkundenbuch der Universität Wittenberg. Teil 1 (1502–1611)*, ed. Historische Kommission für die Provinz Sachsen und für Anhalt, bearbeitet von W. Friedensburg (Magdeburg: 1926), 256–257.

98 Nauert, "Humanists, Scientists" 80.

99 See Petersen P., *Geschichte der aristotelischen Philosophie im protestantischen Deutschland* (Leipzig: 1921) 38–60.

100 See Petersen, *Geschichte* 19–38.

101 Cf. Müller G., "Die Aristoteles-Rezeption im deutschen Protestantismus", in Buck A. (ed.), *Die Rezeption der Antike. Zum Problem der Kontinuität zwischen Mittelalter und Renaissance* (Hamburg: 1981) 55–70, here 58. Concerning the relationship between natural philosophy and theology in the Middle Ages Grant states (*God and reason* 289): 'Medieval theologians imported logic and natural philosophy into theology to such an extent that they effectively secularized it. Theology became an analytical subject because it relied heavily on logico-mathematical techniques'; see also Grant E., *Science and Religion, 400 B.C. to A.D. 1550. From Aristotle to Copernicus* (Baltimore: 2004) 220–224.

102 Ferry, "The Guiding Lights" 278–279.

another indispensable prerequisite for the resumption of this tradition was missing, as the disputation system in all of the faculties at the University of Wittenberg had begun to decline at the end of the second decade of the sixteenth century. Wolf suggested the following reasons for this decline: (1) the humanists' critique of the scholastic way of learning and teaching; (2) an enthusiastic polemic against science as a whole and the academic study with its degrees, in favour of an Early Christian simplicity; (3) a general anti-educational trend, partly as a reaction against the erudite arrogance of the humanists.[103] Attempts to revive the disputation system, as expressed in a foundation charter of elector Johann Friedrich from 5 May 1536[104] or by Melanchthon in the statutes for the Arts faculty in 1545,[105] failed.[106] A report by the Elector's visitors from January 1577 clearly states:

> Die circulares und ordinariae disputationes werden vermöge der statuten nicht gehalten, und wirt vorgewant, der herr Philippus[107] und andere nach ihm haben dieselbige zu vorhütunge zanks und haders abgeschaft.[108]

In our opinion there is yet another reason for the decline of the disputation system at the University of Wittenberg, in addition to the factors mentioned by Wolf. In the sixteenth century, university education seems to have focused more and more on the practical needs the future graduates had to face in their profession as mathematicians, astronomers, astrologers,[109] theologians, lawyers or physicians. Thus, in contrast to the Middle Ages, disputations were no longer a means to investigate the truth, but mere rhetorical exercises focusing on providing skills for mastering the practical problems of everyday professional life. The humanists' ideas of the priority of rhetoric went hand in hand with this trend.

In the sixteenth century a new idea with no potential for solving practical problems stood little chance of being noticed by the scientific community. Astronomers had a considerable interest in using Copernicus's planetary

---

103 Wolf E., "Zur wissenschaftsgeschichtlichen Bedeutung der Disputationen an der Wittenberger Universität im 16. Jahrhundert", in *450 Jahre Martin-Luther-Universität Halle-Wittenberg*, vol. 1, *Wittenberg 1502–1817* (Halle [Saale]: 1952) 335–344, here 338–339.

104 *Urkundenbuch der Universität Wittenberg. Teil 1 (1502–1611)* 172–184, here 177–178.

105 Ibidem 266–277, here 270.

106 Wolf, "Bedeutung der Disputationen" 338–340.

107 That is Philipp Melanchthon.

108 *Urkundenbuch der Universität Wittenberg. Teil 1 (1502–1611)* 405.

109 For the prominent role of the University of Wittenberg in the education of astronomers and astrologers see Brosseder, *Im Bann der Sterne*.

theory as a mathematical model that they believed would enable them to compute the celestial motions more accurately than traditional approaches.[110] By contrast, Copernicus's physical and cosmological arguments for his new world system, which at least constitute the prototype of an alternative theory to Aristotle's physical teachings, went unnoticed. But should we really reproach these astronomers with having ignored, that is with not having perceived or concerned themselves with, Copernicus's cosmological and physical theses? It must be admitted that Copernicus's interesting arguments about how one might neutralise Aristotle's proofs for the immobility of the Earth in the centre of the cosmos were of no practical use for an astronomer's or astrologer's daily routine. As Kremer has stated, the authors of 'the annual astrological calendars, practica or prognostications that poured off European printing presses during the sixteenth and seventeenth centuries', 'undoubtedly the leading consumers of astronomical knowledge in early modern Europe, [...] reveal little if any interest in cosmography or causal explanation in astronomy.'[111]

Now, the decisive point is that evidently nobody at the University of Wittenberg felt a moral or intellectual duty to concern themselves with Copernicus's cosmological and physical theses. The purely intellectual interest in an opponent's theory for the sake of discussion and investigating the truth so typical of medieval university disputation culture had died out. If the medieval disputation culture had survived, such a discussion would most probably have taken place. The medieval university disputation culture did not survive due to the rise of humanism and the focus on the more practical needs of university education. However, knowing how Copernicus's cosmology might be defended against Aristotle's natural philosophical tenets was of no practical importance for the professional praxis of a future astronomer or astrologer.

Selective Bibliography

Aristotle, *Aristotelis opera edidit Academia Regia Borussica. Aristoteles graece ex recognitione Immanuelis Bekkeri*, 2 vols. (Berlin: 1831).

Aristotle, *The Complete Works of Aristotle. The Revised Oxford Translation*, ed. J. Barnes, 2 vols. (New Jersey – Oxford: 1991).

Aristotle, *The Works of Aristotle. Translated into English under the editorship of W.D. Ross*. Vol. II. *Physica. De caelo. De generatione et corruptione* (Oxford: 1930).

110 In reality, as Gingerich ("The Role of Erasmus Reinhold" 53–54) has shown, the results yielded by the Prutenic Tables based on Copernicus's planetary theory were not always better, and sometimes even worse, than those derived from the Alfonsine Tables.

111 Kremer, "Copernicus among the Astrologers" 226.

Bazàn B.C., "La quaestio disputata", in *Les genres littéraires dans les sources théologiques et philosophiques médiévales. Définition, critique et exploitation. Actes du Colloque international de Louvain-la-Neuve 25–27 mai 1981* (Louvain-la-Neuve: 1982) 31–49.

Bazàn B.C. – Wippel J. – Fransen G. – Jacquart D., *Les questions disputées et les questions quodlibétiques dans les facultés de théologie, de droit et de médecine* (Turnhout: 1985).

Brahe Tycho, *Tychonis Brahe Dani opera omnia*, ed. I.L.E. Dreyer, 15 vols. (Copenhagen: 1913–1929).

Brosseder C., *Im Bann der Sterne. Caspar Peucer, Philipp Melanchthon und andere Wittenberger Astrologen* (Berlin: 2004).

Collegium Conimbricense, *Commentarii Collegii Conimbricensis, societatis Iesu, in quattuor libros de coelo, meteorologicos & parva naturalia, Aristotelis Stagiritae* (Cologne, Impensis Lazari Zetzneri: 1603).

Copernicus Nicolaus, *De revolutionibus libri sex*, ed. H.M. Nobis – B. Sticker, *Nicolaus Copernicus Gesamtausgabe II* (Hildesheim: 1984).

Copernicus Nicolaus, *Nicholas Copernicus on the Revolutions*, ed. J. Dobrzycki, translation and commentary by E. Rosen (London: 1978).

Digges Thomas, *A Perfit Description of the Caelestiall Orbes according to the most aunciente doctrine of the Pythagoreans, latelye reuiued by Copernicvs and by Geometricall Demonstrations approued* (London, Thomas Marsh: 1576).

Ferry P.T., "The Guiding Lights of the University of Wittenberg and the Emergence of Copernican Astronomy", *Concordia Theological Quarterly* 57 (1993) 265–291.

Freiberg Dietrich of, "Tractatus de iride et de radialibus impressionibus", ed. M.R. Pagnoni-Sturlese – L. Sturlese, in Pagnoni-Sturlese M.R. et al. (eds.), *Dietrich von Freiberg, Opera Omnia, Tomus IV, Schriften zur Naturwissenschaft, Briefe* (Hamburg: 1985) 95–268.

Gingerich O., *An Annotated Census of Copernicus'* De revolutionibus (*Nuremberg, 1543 and Basel, 1566*) (Leiden et al.: 2002).

Goddu A., *Copernicus and the Aristotelian Tradition* (Leiden – Boston: 2010).

Grant E., *God and Reason in the Middle Ages* (Cambridge et al.: 2001).

Grant E., *Science and Religion, 400 B.C. to A.D. 1550. From Aristotle to Copernicus* (Baltimore: 2004).

Grant E., *The Foundations of Modern Science in the Middle Ages. Their religious, institutional, and intellectual contexts* (Cambridge et al.: 2006).

Henderson J.A., *On the Distances between Sun, Moon and Earth According to Ptolemy, Copernicus and Reinhold* (Leiden et al.: 1991).

Hugonnard-Roche H., "Le problem de la rotation de la terre au XIVe siècle", in *Avant, avec, après Copernic: la représentation de l'univers et ses conséquences épistémologiques. XXXIe semaine de synthèse, 1–7 juin 1973* (Paris: 1975) 62–65.

Johnson F.R. – Larkey S.V., "Thomas Digges, the Copernican System, and the Idea of the Infinity of the Universe in 1576", *The Huntington Library Bulletin* No. 5 (April 1934) 69–117.

Kathe H., *Die Wittenberger philosophische Fakultät 1502–1817* (Cologne et al.: 2002).

Kirschner S. – Kühne A., "Zur Copernicus-Rezeption an der Universität Wittenberg in der Frühen Neuzeit", in Dauben J.W. et al. (eds.), *Mathematics Celestial and Terrestrial. Festschrift für Menso Folkerts zum 65. Geburtstag* (Halle [Saale]: 2008) 507–522.

Kleinert A., "'Eine handgreifliche Geschichtslüge'. Wie Martin Luther zum Gegner des copernicanischen Weltsystems gemacht wurde", *Berichte zur Wissenschaftsgeschichte* 26 (2003) 101–111.

Knox D., "Bruno's Doctrine of Gravity, Levity and Natural Circular Motion", *Physis* 38 (2001) 171–209.

Knox D., "Copernicus's Doctrine of Gravity and the Natural Circular Motion of the Elements", *Journal of the Warburg and Courtauld Institutes* 68 (2005) 157–211.

Kremer R.L., "Copernicus among the Astrologers: A Preliminary Study", in Folkerts M. – Kühne A. (eds.), *Astronomy as a Model for the Sciences in Early Modern Times* (Augsburg: 2006) 225–252.

Lawn B.: *The Rise and Decline of the Scholastic 'Quaestio disputata'. With Special Emphasis on its Use in the Teaching of Medicine and Science* (Leiden et al.: 1993).

Lerner M.-P., "Aux origines de la polémique anticopernicienne (I). L'Opusculum quartum de Giovanmaria Tolosani [1547–1548]", *Revue des sciences philosophiques et théologiques* 86 (2002) 681–721.

Lerner M.-P., "Aux origines de la polémique anticopernicienne (II)", *Revue des sciences philosophiques et théologiques* 90 (2006) 409–452.

Maestlin Michael, "M.<agister> Michael Maestlin Goeppingensis, Candido Lectori S.<alutem>", in Kepler Johannes, *Prodromus Dissertationvm Cosmographicarvm, Continens Mysterivm Cosmographicvm* [...]. *Addita est erudita Narratio M. Georgii Ioachimi Rhetici, de Libris Reuolutionum* [...] *D. Nicolai Copernici* (Tübingen, Georg Gruppenbach: 1596) 86–90.

Maricourt Petrus Peregrinus De, *Epistola Petri Peregrini de Maricourt ad Sygerum de Foucaucourt militem de magnete*, in Hellmann G. (ed.), *Rara Magnetica* (Berlin: 1898) 27–40.

Melanchthon P., "De ordine discendi", in Bretschneider C.G. (ed.), *Philippi Melanthonis opera quae supersunt omnia, Corpus reformatorum*, vol. XI (Halle [Saale]: 1843) cols. 209–214.

Methuen C., "Maestlin's Teaching of Copernicus", *Isis* 87 (1996) 230–247.

Nauta L., *In defense of common sense. Lorenzo Valla's humanist critique of scholastic philosophy* (Cambridge, Mass. – London: 2009).

Oresme Nicole, *Le livre du ciel et du monde*, ed. A.D. Menut – A.J. Denomy, translated with an introduction by A.D. Menut (Madison et al.: 1968).

Oresme Nicole, *Questiones de spera*, ed. G. Droppers (Ph.D. Dissertation, University of Wisconsin, 1966).

Palingenius Stellatus Marcellus, *Marcelli Palingenii Stellati Poetae doctissimi Zodiacus vitae hoc est, De hominis vita, studio, ac moribus optime instituendis, libri XII. Cum indice locupletissimo* (London, Thomas Marsh: 1569).

Pliny, *The Natural History of Pliny*, ed. J. Bostock (London: 1855–1857).

Pliny, *C. Plini Secundi naturalis historiae libri XXXVII*, ed. L. Ian, 5 vols. (Leipzig: 1854–1860).

Reinhold Erasmus, "Commentarius in opus revolutionum Copernici", ed. H.M. Nobis et al., *Receptio copernicana. Texte zur Aufnahme der copernicanischen Theorie, Nicolaus Copernicus Gesamtausgabe* VIII/1 (Berlin: 2002) 189–358.

Reinhold Erasmus, *Prvtenicae Tabvlae Coelestivm Motvvm* (Tübingen, Ulrich Morhard: 1551).

Reinhold Erasmus, *Theoricae Novae Planetarvm Georgii Pvrbachii Germani, ab Erasmo Reinholdo Salueldensi pluribus figuris auctae, & illustratae scholijs, quibus studiosi praeparentur, ac inuitentur ad lectionem ipsius Ptolemaei. Recens aeditae & auctae nouis scholijs in Theoria Solis ab ipso autore. Inserta item methodica tractatio de illuminatione Lunae* (Wittenberg, Ioannes Lufft: 1553).

Rheticus Georg Joachim, "De terrae motu et scriptura sancta", ed. H.M. Nobis et al., *Receptio copernicana. Texte zur Aufnahme der copernicanischen Theorie, Nicolaus Copernicus Gesamtausgabe* VIII/1 (Berlin: 2002) 57–73.

Rheticus Georg Joachim, "Narratio prima", ed. H.M. Nobis et al., *Receptio copernicana. Texte zur Aufnahme der copernicanischen Theorie, Nicolaus Copernicus Gesamtausgabe* VIII/1 (Berlin: 2002) 3–48.

Rotterdam Erasmus of, *Erasmi Roterodami Encomium moriae i. e. stultitiae laus. Lob der Torheit. Basler Ausgabe von 1515*, ed. H.A. Schmid, 2 vols. (Basel: 1931).

Scheible H., "Aristoteles und die Wittenberger Universitätsreform. Zum Quellenwert von Lutherbriefen", in Beyer M. – Wartenberg G. (eds.), *Humanismus und Wittenberger Reformation. Festgabe anläßlich des 500. Geburtstages des Praeceptor Germaniae Philipp Melanchthon am 16. Februar 1997* (Leipzig: 1996) 123–144.

Schulthess P., "Die Philosophie im lateinischen Mittelalter", in Schulthess P. – Imbach R., *Die Philosophie im lateinischen Mittelalter. Ein Handbuch mit einem bio-bibliographischen Repertorium* (Düsseldorf – Zürich: 2000) 15–349.

Seidengart J., "Le copernicianisme de Christoph Rothmann dans sa controverse avec Tycho Brahe et son prétendu 'retrait'", in Granada M.A. – Mehl E. (eds.), *Nouveau ciel. Nouvelle terre. La revolution copernicienne dans l'Allemagne de la Réforme (1530–1630)* (Paris: 2009) 179–205.

Souffrin P., "Oresme, Buridan, et le mouvement de rotation diurne de la terre ou des cieux", in Ribémont B. (ed.), *Terres médiévales* (Paris: 1993) 277–303.

Tolosani Giovanni Maria, *Opusculum quartum. De coelo supremo immobili et terra infima stabili, ceterisque coelis et elementis intermediis mobilibus*, edited with a French translation in Lerner M.-P., "Aux origines de la polémique anticopernicienne (I). L'Opusculum quartum de Giovanmaria Tolosani [1547–1548]", *Revue des sciences philosophiques et théologiques* 86 (2002) 681–721.

Urkundenbuch der Universität Wittenberg. Teil 1 (1502–1611), ed. Historische Kommission für die Provinz Sachsen und für Anhalt, bearbeitet von W. Friedensburg (Magdeburg: 1926).

Weijers O., "La 'disputatio' comme moyen de dialogue entre les universitaires au moyen âge", in Gilli P. (ed.), *Les élites lettrées au Moyen Âge. Modèles et circulation des savoirs en Méditerranée occidentale (XIIe–XVe siècles)* (Montpellier: 2008) 155–169.

Weijers O., "Quelques observations sur les divers emplois du terme *disputatio*", in Meirinhos J.F. (ed.), *Itinéraires de la raison. Études de philosophie médiévale offertes à Maria Cândida Pacheco* (Louvain-la-Neuve: 2005) 35–48.

Weijers O., *Queritur utrum. Recherches sur la 'disputatio' dans les universités médiévales* (Turnhout: 2009).

Weijers O., "The various kinds of disputation in the faculties of arts, theology and law (c. 1200–1400)", in Gindhart M. – Kundert U. (eds.), *Disputatio 1200–1800. Form, Funktion und Wirkung eines Leitmediums universitärer Wissenskultur* (Berlin – New York: 2010) 21–31.

Westman R.S., *The Copernican Question. Prognostication, Skepticism, and Celestial Order* (Berkeley, Calif. et al.: 2011).

Westman R.S., "The Melanchthon Circle, Rheticus, and the Wittenberg Interpretation of the Copernican Theory", *Isis* 66 (1975) 164–193.

Westman R.S., "Three responses to the Copernican theory: Johannes Praetorius, Tycho Brahe, and Michael Maestlin", in Westman R.S. (ed.), *The Copernican Achievement* (Berkeley, Calif. et al.: 1975) 285–345.

Wolf E., "Zur wissenschaftsgeschichtlichen Bedeutung der Disputationen an der Wittenberger Universität im 16. Jahrhundert", in *450 Jahre Martin-Luther-Universität Halle-Wittenberg*, vol. 1, *Wittenberg 1502–1817* (Halle [Saale]: 1952) 335–344.

CHAPTER 2

The Silence of the Wolves, Or, Why It Took the Holy Inquisition Seventy-Three Years to Ban Copernicanism[1]

Gereon Wolters

Abstract

I give two main answers to the question why Copernicus's *De revolutionibus* (1543) had so long been officially ignored in 'Rome'. The first is that cosmological issues were of only peripheral importance in such times of life-and-death contestations with Protestant reformers. The second is that it took—in a very contingent way—the personality of Cardinal Bellarmine, driven by an anti-reformationist emphasis on authority, to unite with a 'synergy effect' two separate dogmatic strands: (1) the old teaching of the superiority of theology over all other forms of knowledge, made binding in the Papal Bull Apostolici Regiminis (1513), and, (2) the Decree of the Council of Trent on the Holy Scripture (1546), which stated that only the Church, and not the single believer (as Luther had claimed), had the authority to bindingly interpret the Scripture in matters of faith and morals. The synergy between the two strands had the surprising epistemic result of any biblical passage becoming a 'matter of faith.'

1 The Roman Wolves Howl (1616)

Our story ends on 5 March 1616 with a Decree of the Holy Congregation of the Index:[2]

> This Holy Congregation has also learned about the spreading and acceptance by many of the false Pythagorean doctrine,[[3]] altogether contrary to

---

1 Thanks to Giora Hon (Haifa) for greatly improving the paper.

2 This Congregation had been in charge of prohibiting 'dangerous' books since 1571. This was done in close cooperation with the senior *Holy Office* (*Sanctum Officium* or *Sant'Uffizio*, the short name of the *Suprema Sacra Congregatio Romanae et Universalis Inquisitionis*).

3 This relates to the conception of Philolaos (c. 420 B.C.) of a 'central fire' surrounded by the heavenly bodies, the sun included. Cf. Carrier M., *Nikolaus Kopernicus* (Munich: 2001) 74f.

 | DOI 10.1163/9789004281127_004

> the Holy Scripture, that the earth moves and the sun is motionless, which is also taught by Nicolaus Copernicus's *On the Revolutions of the Heavenly Spheres* [...] Therefore, in order that this opinion may not creep any further [in]to the prejudice of Catholic truth, the Congregation has decided that the books by Nicolaus Copernicus [...] and Diego de Zuñiga [...] be suspended until corrected; but that the book of the Carmelite Father Paolo Antonio Foscarini be completely prohibited and condemned; and that all other books which teach the same be likewise prohibited [...].[4]

The decree does not mention Galileo Galilei (1564–1642) for the simple reason that he had not yet published on the subject, although it was an open secret in Rome and elsewhere since the publication of his *Sidereus Nuncius* (1610) that he was a staunch supporter of Copernicanism. In this little book Galileo reports the discovery of the four satellites of Jupiter. Galileo's Copernicanism became even more visible after the subsequent discoveries of the phases of Venus and Mercury.[5]

Apart from these public developments, so to speak, two denunciatory letters had reached the Holy Office.[6] They had set in motion a procedure part of which was a short report by no less than eleven 'consultants' (consultores) of the Holy Office:

> [...] Propositions to be assessed:
> (1) The sun is the centre of the world and completely devoid of local motion.
> Assessment: All [consultants, G.W.] said that this proposition is foolish and absurd in philosophy,[7] and formally heretical since it explicitly contradicts in many places the sense of the Holy Scripture, according to the

4 The complete text of the decree is in Finocchiaro M.A., *The Galileo Affair. A Documentary History* (Berkeley: 1989) 148–150, quote 149. Finocchiaro gives English translations of important documents in the context of the ecclesiastical actions against Galilei.

5 The existence of satellites of Jupiter supported the idea that the moon was not a planet but rather a satellite of the earth, while the phases of Venus and Mercury could be explained as confirmation of their orbiting around the sun.

6 The denunciations (1615) of the two Florentine Dominicans Niccolò Lorini and Tommaso Caccini are in Finocchiaro, *Galileo Affair* 134–141.

7 'Philosophy' in the usage of the time was understood as explanatory knowledge, as opposed to descriptive ('historical') knowledge. It included what we call 'science.'—Regarding the conceptual history cf. Kambartel F., *Erfahrung und Struktur. Bausteine zu einer Kritik des Empirismus und Formalismus* (Frankfurt a.M.: 1968) ch. 2.

> literal meaning of the words and according to the common interpretation and understanding of the Holy Fathers and the doctors of theology.
> (2) The earth is not at the centre of the world, or motionless, but it moves as a whole and also with diurnal motion.
> Assessment: All said that this proposition receives the same judgment in philosophy and that with regard to theological truth it is at least erroneous in faith.[8]

Given Galileo's reputation as a Copernican, it was no surprise that Cardinal Robert Bellarmine (1542–1621), one of the Cardinal Inquisitors, summoned him to the Holy Office on 26 February 1516, i.e. still prior to the promulgation of the decree. Bellarmine, among other things, had earned a reputation in the trial that brought Giordano Bruno (1548–1600) to the stake,

> in the name of His Holiness the Pope and the whole Congregation of the Holy Office, ordered and enjoined the said Galileo, who was himself still present, to abandon completely the above-mentioned opinion that the sun stands still at the centre of the world and the earth moves, and henceforth not to hold, teach, or defend it in any way whatever, either orally or in writing; otherwise the Holy Office would start proceedings against him. The same Galileo acquiesced in this injunction and promised to obey.[9]

As we know, Galileo did not obey in the long run and was sentenced to prison for life (1633). The Roman wolves, finally, had howled triumphantly. Why had they been quiet for so many years? Why did it take them so long to identify Copernicanism as a severe menace to faith? This brings us to the beginning of our story.

2 Cosmological World Views

2.1 *The Pre-Copernican Situation*

The story begins with Nicolaus Copernicus (1473–1543) who, after studying in Cracow, Bologna, Padua and Ferrara, spent most of his life in Frauenburg

---

8 Finocchiaro, *Galileo Affair* 146. The assessments of 1616 return almost literally in the judgment of the Inquisition against Galileo of 22 June 1633 (cf. Finocchiaro, *Galileo Affair* 288)—'Formally heretical' refers to an opinion of which the person who utters it *knows* that it is heretical, that is, *explicitly* ('formaliter') refuses to acknowledge the 'truth.'

9 Bellarmino's special injunction against Galilei in Finocchiaro, *Galileo Affair* 147f.

(Frombork),[10] Ermland (Varmia), as an ecclesiastical administrator ('Canon'), without having taken holy orders, however. Apart from this, the humanist Copernicus was active as a medical practitioner and amateur astronomer. This means that the person who launched the 'Copernican Revolution' was not at all a professional astronomer at a university or a court. He was also not much of a revolutionary personally; rather, he was anxious to save as much of the old astronomy and physical cosmology as possible. In order to assess his contribution correctly we must first present the main features of the astronomical theory he wanted to improve by introducing one important change, while at the same time saving its conceptual foundations.[11] Secondly, the methodological aspects of Copernicus's move, related to the philosophy of science, will be discussed.

First we must distinguish three basic disciplines that are interwoven almost inextricably in astronomical practice: (1) astronomy,[12] (2) natural philosophy, consisting of a) cosmology, b) (terrestrial) physics, and (3) theology. There is a progressive order of subordination here in the sense that the ultimate explanatory principles of astronomy are to be found in natural philosophy, while those of natural philosophy are to be found in theology.[13] We will later see that these distinctions were of great importance to the ecclesiastical authorities. Astronomy is the discipline that renders data about the positions of heavenly bodies and develops mathematical means to calculate, for example, their future positions or the time that elapses before the same relative positions return (*vulgo* a year), and so on. Basically, astronomy is a kind of applied

---

10 Upon their first occurrence, place names are given in German and Polish; later, only in German. Here I follow Hooykaas R., *G.J. Rheticus' Treatise on Holy Scripture and the Motion of the Earth* (Amsterdam: 1984) 2.

11 There are several good introductions to Copernicus. I have mostly used Carrier, *Kopernikus*, a reliable book also for the non-specialised reader; it focuses on methodological questions. The best English introduction is arguably Gingerich O., *The Eye of Heaven: Ptolemy, Copernicus, Kepler* (New York: 1993). Also very useful is Goddu A., *Copernicus and the Aristotelian Tradition: Education, Reading, and Philosophy in Copernicus's Path to Heliocentrism* (Leiden: 2010); Swerdlow N.M. – Neugebauer O., *Mathematical Astronomy in Copernicus's De revolutionibus*, 2 vols. (New York: 1984) requires advanced astronomical (and mathematical) understanding.

12 Westman R.S., *The Copernican Question: Prognosticism, Skepticism, and Celestial Order* (New York: 2011) 17f. prefers 'heavenly (celestial) practitioners' to 'role designators like *scientists* or *men of science*' or 'astronomers', for that matter. Westman is certainly hermeneutically correct, but for simplicity's and brevity's sake, as well as to facilitate distinctions, I use the above categories.

13 Cf. Barker P., "The Role of Religion in the Lutheran Response to Copernicus," in Osler M. (ed.), *Rethinking the Scientific Revolution* (Cambridge: 2000) 59–88, particularly 71f.

mathematics, more geometry than physical science. Consequently, astronomy says nothing about the 'true causes' of the observed phenomena. Astronomical reasoning from observed effects can at best lead to possible causes.

By and large, Copernicus was a follower of the astronomy of the Greek scholar Ptolemy (second century A.D.) as laid down in the latter's *Almagest*. We will soon see at which point Copernicus disagreed with his authority.

Natural philosophy is a discipline that, other than astronomy, can give us true causes. As long as natural philosophy is occupied with heavenly motions—that is, as long as it is cosmology—it consists in our context of virtually axiomatic ontological assumptions about the structure of the universe and the motions going on in it. I would like to mention four such assumptions, which were widely shared:[14]

(1) the earth occupies the central position of the whole universe;
(2) all heavenly bodies display a uniform, circular motion around the earth, or their motion is composed of uniform and circular parts;
(3) the 'planets'[15] are fixed on moving, completely transparent crystal spheres;
(4) the fixed stars are fixed to an outermost sphere, which annually revolves around the earth. Natural philosophy, understood as physics, is a metaphysical theory of 'sublunar' motions, i.e. motions of terrestrial bodies. It comes into play above all when one wants to test for terrestrial motions the presumed consequences effected by a moving earth, as one would have to within a heliocentric system.[16]

The primary task of astronomy was to account for the observations recorded in tables and calendars. Basically two strategies were used to achieve this: first, to follow the 'instrumentalist' program of 'saving the phenomena.'[17] Sixteenth-century astronomers widely shared the belief that there was no way

14 Cf. Carrier, *Kopernikus* 29f.

15 Note that in the geocentric system the sun and the moon are also 'planets'.

16 Cf. Carrier, *Kopernikus* 39f.

17 The use of the nineteenth-century (P. Duhem) realism/instrumentalism distinction in the philosophy of science has been justly criticised by Barker P. – Goldstein B.R., "Realism and Instrumentalism in Sixteenth Century Astronomy: A Reappraisal", *Perspectives on Science* 6 (1998), 232–258. The authors claim that the present understanding of the distinction does not cover what was at stake in the sixteenth century. I agree with the authors about the anachronism of applying the realism/instrumentalism distinction. I believe, however, that there is a considerable overlap between this and the *demonstratio quia/demonstratio propter quid* distinction: the fact that knowledge of true causes claimed to give an account

to demonstrate *propter quid*, meaning true causes. One had to be content, so the 'instrumentalist' position went, with *quia* demonstrations, in which reasoning goes from observations to only possible causes. Consequently, for the time being cosmological conceptions had to be regarded as 'hypotheses.' This advisable ontological epoché notwithstanding, there existed, second, a large number of realistic cosmological claims from the very beginning: Ptolemy in the *Almagest* generally followed the hypothetical-instrumentalist option, whereas in his *Planetary Hypotheses* he presented his set of nested spheres as a physical realisation of the universe, that is, as cosmology.

Theology comes into play when astronomy and natural philosophy have to be embedded in what one presumes to know about God's plan of creation.

To account for the above-mentioned 'axiomatic' assumptions, particularly the uniform, circular motion of the heavenly objects, Ptolemaic astronomy in its calculations used the following: (1) the deferent-epicycle-system;[18] (2) the eccentric orbits of 'planets,' that is, orbits around the earth where the earth is not, however, at the centre; (3) so-called equants, introduced by Ptolemy himself.

The deferent-epicycle system, in order to account for the irregular motions of the 'planets,' superimposes two uniform and circular motions: first of all, the 'planet' moves in a circular orbit ('deferent') around the earth as the centre of the universe; second, the planet itself, however, does not move along the deferent but rather the centre of a second orbit ('epicycle'), along which the 'planet' is thought to move (Fig. 2.1).[19]

However, this ingenious system did not suffice in the case of Mars, and was not necessary for the Sun and other planets, whose motion could be described equivalently by simply moving the centre of their orbits away from the earth to an 'eccentric' point close to it (Fig. 2.2), thus rendering the system geostatic rather than geocentric. In fact, Ptolemy in the case of the 'planet' Sun preferred the eccentricity hypothesis as 'simpler' because it involved only one motion and not two, as in the deferent-epicycle model.

The motion of the planet Mars required, finally, a third trick: Ptolemy could uphold the uniform motion of the epicycle's centre only by having it move with uniform angular motion around 'a point equal and opposite from the

---

of how the heavens *really* are, while knowledge of *possible* causes lead only to '*hypotheses*' on how the heavens may *possibly* be.

18 The epicycle-deferent system as well as the idea of orbital eccentricity was based on respective considerations of Apollonius in a lost work (cf. Gingerich, *Eye of Heaven* 7–9.).

19 As indicated earlier in realistically interpreted models, in antiquity it is not the 'planets' themselves which move but rather the spheres to which they are fixed.

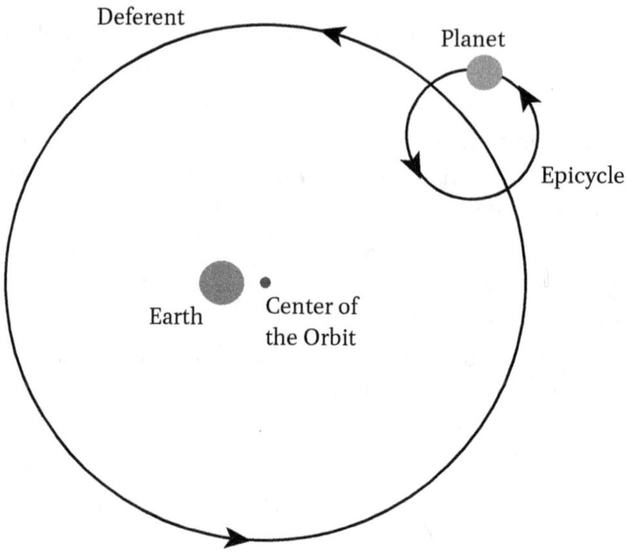

FIGURE 2.1 Siobahn Morgan, *Deferent-epicycle system* (open source lecture course material), taken from: http://www.uni.edu/morgans/astro/course/notes/section1/new3.html.

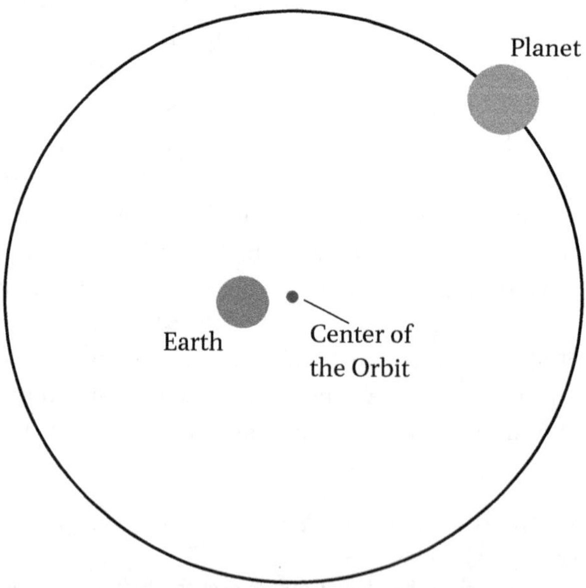

FIGURE 2.2 Siobahn Morgan, *Eccenter model* (open source lecture course material) taken from: http://www.uni.edu/morgans/astro/course/notes/section1/new3.html.

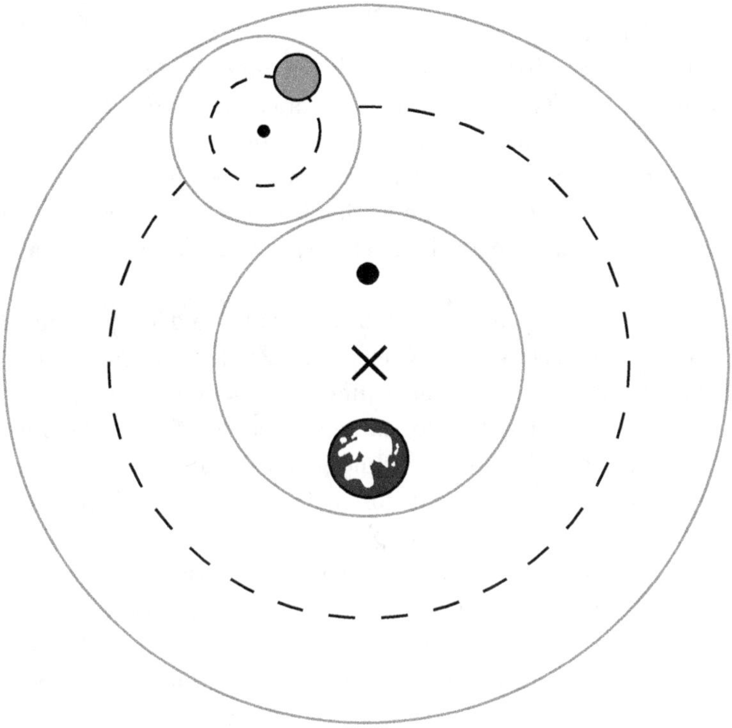

FIGURE 2.3 *The basic elements of Ptolemaic astronomy, showing a planet on an epicycle (smaller dashed circle), a deferent (larger dashed circle), and an equant (larger black dot); "x" marks the eccenter, while the large black dot represents the earth* (taken from Open Source Wikipedia: "Deferent and epicycle" [February 2014]).

earth along the line through the deferent centre.'[20] The equant is shown below in Fig. 2.3 together with the other two ingredients of Ptolemy's attempt to save the phenomena.

This, painted in rather broad strokes, was the situation that Copernicus encountered when he entered the scene. Our question is, what did he change and why?

---

20 Gingerich, *The Eye of Heaven* 10.

2.2 *Copernicus's Innovations*

Copernicus changed, first in his *Commentariolus* (a manuscript sent to a few friends and written prior to 1514)[21] and much later in *De revolutionibus* (1543), the first of the above-mentioned 'axiomatic' assumptions of pre-Copernican thinking—the central position of the earth is now assumed by the sun.[22]

The remaining basic assumptions were retained, except the one concerning the outermost sphere carrying the fixed stars. It no longer revolves but remains at rest.

What did Copernicus gain by placing the sun at rest? I would like to mention three points, each of which shows that Copernicus saw the advantage of his system mainly in its simpler and therefore sounder methodology:

(1) He provides a new explanation for the retrograde motion of the outer planets, i.e. Mars, Jupiter and Saturn.[23] While in the Ptolemaic system one needs separate complicated assumptions for each planet, in the Copernican system the retrograde motions of the outer planets are a direct consequence of the theory, that is, the motion of the earth, when it overtakes an outer planet in its common orbit around the sun.[24] It is this extraordinary explanatory unification, as it is called in contemporary philosophy of science, which is the most important result of heliocentrism in this context and arouses an almost poetic enthusiasm in Copernicus:

> In this arrangement, therefore, we discover a marvellous symmetry of the universe, and an established harmonious linkage between the motion of the spheres and their size, such as can be found in no other way. [...]

---

21 The best English translation of the *Commentariolus*, together with an excellent commentary, is arguably Swerdlow N.M., "The Derivation and First Draft of Copernicus's Planetary Theory: A Translation of the *Commentariolus* with Commentary", *Proceedings of the American Philosophical Society* 117 (1973), 423–512.

22 On the complicated history of the publication of *De revolutionibus* see Gingerich O., *The Book Nobody Read: Chasing the Revolutions of Nicolaus Copernicus* (London: 2004). This book is, incidentally, a fascinating account of Gingerich's sleuthing (even the Russian mafia enters) in order to trace surviving copies of the first (1543) and second (1566) editions of *De revolutionibus*. Gingerich gives an account of his findings, particularly about the most important margins, in idem, *An Annotated Census of Copernicus's De revolutionibus (Nuremberg, 1543 and Basel, 1566)* (Leiden: 2002).

23 Cf. Carrier, *Kopernikus* 89–94.

24 In addition, another old problem, namely the maximum luminosity of the outer planets, is solved: when overtaking a planet, the earth is in its nearest position to that planet. Consequently the planet displays its maximum luminosity.

> All these phenomena [unified under the heliocentric hypothesis, G.W] proceed from the same cause, which is the earth's motion.[25]

(2) He gives a new explanation of the restricted elongation of the inner planets.[26] While the elongation of the outer planets can reach 180°, that of the inner planets does not exceed 28° (Mercury) and 47° (Venus), which is at the same time a consequence of their sequential arrangement in the heliocentric system.[27] While these and other related phenomena require assumptions from one planet to the other in the geocentric system, they follow directly just from the very geometry of Copernican heliocentrism that here again leads to a remarkable explanatory unification.

(3) He asserts that the equant, introduced by Ptolemy in order to save the principle of uniform motion and arguably the most unpopular[28] ingredient of the Ptolemaic theory, is no longer needed in his system and he explicitly rejects it (book V, ch. 25).[29]

Thus far everything looks fine. But when it comes to reconciling this beautiful explanatory unification with the dry figures of astronomical observations, Copernicanism looks no better than the Ptolemaic system in meeting the 'gold standard' (then and now) of theory assessment: empirical adequacy. Rather, Copernicus had to introduce an apparatus of epicycles and eccentrics similarly complex as its Ptolemaic rival, in order to account for the observations. This was mostly due to the fact that he insisted on uniform and circular motion,

25 Copernicus Nicolaus, *On the Revolutions*, trans. E. Rosen, ed. J. Dobrzycki (London: 1978) 22. Nonetheless, Copernicus needs many epicycles in order to determine the exact position of a planet (cf. Swerdlow, "Commentariolus" 510).

26 The elongation of a planet is the angle between the sun and the planet, as viewed from the earth.

27 For B.R. Goldstein in his "Copernicus and the Origin of His Heliocentric System", *Journal of Astronomy* 33 (2002) 219–235, the sequential order of the planets, based on the 'key principle [...] that the periods of planets are longer as their orbs are farther from the centre of motion', that is, the sun (220), was Copernicus's decisive initial motive for proposing the heliocentric hypothesis. This is also true of Westman's *Copernican Question* 139f.

28 The unpopularity of the equant lies in the fact that in a sense it 'cheated' about uniform motion because it 'saved' only uniform angular motion. However, this means that epicycles move at different speeds depending on whether their deferent is closer or farther away from the equant point. In the first case they move more slowly and in the second, faster because the centre of the epicycle must travel along a longer segment on the deferent. A very good illustration is in Gingerich, *The Book* 51–53.

29 One must note, however, that Muslim astronomy had produced 'many models that resolved the problem while maintaining a geocentric framework' (Goldstein, "Copernicus and the Origin" 220).

while planets in actual fact move on elliptical orbits in a non-uniform way, as Kepler would show later.

While the methodological virtues of the explanatory unification and simplicity of Copernicus's system did not make much of an impression on his contemporaries, its instrumental value both for the Gregorian calendar reform and for better astronomic tables was significant and widely respected.[30]

3 Reactions of the Religious Authorities[31]

3.1 *The Preventive Measures of Copernicus & Friends*

In retrospect it is hardly understandable why Copernicus's cosmological innovations could ever collide with religious authority. They were protected by the most venerable teachings of how to read the Bible. In his commentary on the book of Genesis, St. Augustine had already strongly recommended an allegorical, i.e. figurative, understanding of biblical passages that do not relate to faith and morals and are at variance with 'highly certain results of reasoning or with empirical evidence.'[32] By the late Middle Ages this and related approaches had led to the teaching of a 'fourfold sense of Scripture' (*quattuor sensus scripturae*), meaning that a literal reading of the Bible was just one way of understanding its texts; in many cases allegorical, moral or eschatological readings were also possible, and often unavoidable.[33]

Nonetheless, some friends of Copernicus, however, and probably the master himself, saw dangers on the horizon. Most important in this respect is the young Protestant Wittenberg scholar Georg Joachim Rheticus (1514–1574). In 1539 Rheticus went to Frauenburg for almost two and a half years, thus becoming Copernicus's first and only pupil, regarding him as a sort of master and

30 Cf. Carrier, *Kopernikus* 137. However, this did not include an acceptance of the Copernican system.

31 A good presentation of such reactions is Lerner M.-P., "L' « hérésie » héliocentrique: du soupçon à la condamnation", in *Sciences et religions. De Copernic à Galilée (1540–1610)*. Actes du colloque international organisée par l'Ècole française de Rome. En Collaboration Avec L'Ecole Nationale Des Chartes Et L'Istituto Italiano per Gli Studi Filosofici, Avec La Participation De L'Universita Di Napoli Federico II (Rome: 1999) 69–91.

32 *certissima ratione vel experientia* (Aurelius Augustinus, *Über den Wortlaut der Genesis*, vol. I (Paderborn: 1961) 32 (ch. 19, no. 39).

33 Cf. Stuhlmacher P., *Vom Verstehen des Neuen Testaments. Eine Hermeneutik* (Göttingen: 1979) §§ 6,7, which gives an overview of the historical development.

even father.[34] In 1540 Rheticus published a sketch of Copernicus's theory, the *Narratio prima*, which at the same time was the first printed account of Copernicus's heliocentric system.[35] At the same time he was able to convince Copernicus to publish *De revolutionibus*. When in 1541 Rheticus left Frauenburg for Wittenberg where he became the most vocal propagandist of Copernicus's system, the manuscript was in his baggage. Also in his baggage, however, was a defence (*De terrae motu*)[36] against possible theological attacks.

Rheticus's argumentative strategy in *De terrae motus* shows almost all the ingredients one later finds with other defenders of Copernicanism—most prominently Galileo Galilei in his famous letter to the Grand Duchess Christina (1615)—as follows.[37] (1) Presenting oneself as a strict follower of Saint Augustine with respect to the methods of Scripture interpretation.[38] This included admitting the allegorical sense along with the literal sense of the Scripture, and above all emphasising what Hooykaas calls 'accommodation'.[39] This means that the Holy Texts are not scientific textbooks but books of religious and ethical instruction. Where they refer to natural things they must accommodate the understanding of the man in the street (sections 1–13). (2) Quoting passages from the Bible that might support the idea of a moving earth with the sun at rest. This is a clever move, intended to neutralise the strategy of potential opponents, again emphasising accommodation (sections 14–27). (3) Epistemologically and theologically reflecting about sense perception and the Scripture and securing a place for science by stating that, with respect to everything 'beyond the range of the senses (*extra sensum*) [...]

---

34 The reference to Copernicus as his 'master' (*praeceptor*) is found in many places in the *Narratio prima* (see below) and other writings. As to Rheticus's explicit use of the term 'father' with respect to Copernicus cf. the quote in Hooykaas, *Rheticus' Treatise* 150. Rheticus's biological father, a town physician at Feldkirch in what is now Austria, was executed for stealing from his patients when Rheticus was 14 years old. Danielson D., *The First Copernican: Georg Joachim Rheticus and the Rise of the Copernican Revolution* (New York: 2006) provides a good account of the relationship between the two men.

35 Critical edition of the Latin text in Copernicus Nicolaus, *Gesamtausgabe* vol. VIII.1, 5–56.

36 The text that had been printed anonymously only in 1651 was discovered and identified by the Dutch historian of science Reijer Hooykaas and published together with an English translation, notes and an extensive commentary in 1984 (see Bibliography, Hooykaas).

37 English Text in Finocchiaro, *Galileo Affair* 87–118.

38 However, it seems that Copernicus and his associates were not realistic enough to see that Augustine's criteria—'highly certain results of reasoning' or 'empirical evidence'—were still largely missing from the Copernican system.

39 Hooykaas, *Rheticus' Treatise* 33.

we simply accept the Scriptures (*simpliciter Scripturae acquiescimus*)'[40] (sections 28–32). (4) Showing that many geographical passages of the Bible had long ago been proven to be wrong without anyone shouting 'heresy' (sections 33–43).[41] Finally (6), again stressing accommodation as the adequate means of interpreting passages that seem to support the centrality of a non-moving earth. Such passages are better understood allegorically rather than literally (sections 44–64).

For the rest, Rheticus emphasises six times 'that he remains within the bounds of the Catholic Faith'.[42] This 'catholicity' of the Lutheran Rheticus shows, in addition to the tactical aim of possibly protecting his master, that the Reformation process at the time of the publication of *De revolutionibus*, i.e. around 1540, had not yet reached a point of complete rupture from 'Rome'. Many still regarded the Reformation as a movement within the Church. At the same time, however, those forces that envisaged a complete separation were at work on both sides. At the end of the Council of Trent (1563), 24 years after Rheticus's arrival at Frauenburg, it seems inconceivable that a Lutheran from Wittenberg would choose a Catholic Canon as his personal teacher and harmoniously live with him for more than two years.

Copernicus most probably knew about Rheticus's preventive defence (against the theologians) of the conception of a moving earth. From his own pen he added a dedication to *De revolutionibus* for Pope Paul III, asking pontifical support against 'ignorant babblers' (*mataiologoi*) who pronounce judgements on cosmology 'badly distorting some passage of Scripture for their purposes.'[43]

Apart from this, Osiander's preface, too, which presented (against the intention of Copernicus and Rheticus) Copernicus's system as 'hypothetical,' that is, as a mathematical device directed at nothing more than 'saving the phenomena,' would serve as a firewall to keep the Roman wolves at bay, although several persons had noted its contradiction with the text that followed.

Copernicus supplemented these defensive devices—agreement with the exegetical principles of tradition, the dedication to the Pope and Rheticus's

40 Ibidem 52 (80).

41 E.g., the flatness of the earth, the non-existence of antipodes, etc. On the antipodes, cf. Mayaud P.-N., Le Conflit entre l'astronomie nouvelle et l'Écriture Sainte aux XVIe et XVIIe siècles. Un moment de l'histoire des idées autour de l'affaire Galilée, 6 vols. (Paris: 2005) vol. I, 91–107.

42 Hooykaas, *Rheticus' Treatise* 144f.

43 Copernicus, *Revolutions* 5.

outline of how to win a possible controversy—by inserting into *De revolutionibus* a respectful letter dated 1st November 1536 by Nicolaus Schönberg, then a leading cardinal at the court of Pope Clement VII, asking him to have his work, i.e. the still unpublished *De revolutionibus*, copied at Schönberg's expense and sent to Rome. This letter, in turn, was to do with the fact that Copernicanism had already been introduced to the papal court in 1533, most probably only on the basis of the unpublished Commentariolus. Clement had learned in the Vatican gardens, in the presence of two cardinals, a bishop and the pope's physician, of the new theory from his secretary, the Bavarian Johann Albrecht Widmanstetter (1506–1557).[44] By the time of the publication of *De revolutionibus* (1543), however, both Schönberg and Clement VII had died.

These defensive devices were supported by the fact that in those days the Church was almost exclusively concerned with other problems central to Catholic faith and discipline, and with how to contain the Protestant reformation. If one had to describe the catholic counter-reformation in just one word, it would be authority. The counter-reformation was about regaining authority with respect to discipline, faith and morals within the Church, thereby stopping the advance of Protestantism and possibly regaining lost territory.[45]

3.2 *Two Early Whistle-Blowers Die*

Despite all precautionary measures, however, there were vigilant wolves that howled but were not heard: two Dominican Fathers, the Florentine Giovanni Maria Tolosani (1470/71–1549), and his friend, the influential Master of the Sacred Palace,[46] Bartolomeo Spina († 1547), gave early warnings. Together, the men could immediately have made Copernicanism a target of the authorities had they not died before they could take action—perhaps a work of Providence. Tolosani's respective anti-Copernican manuscript (1646/47) rested undisturbed in what is now the Biblioteca Nazionale of Florence. It is under the title *Opusculum quartum*, constituting a portion of a major bundle of apologetic writings, the first part of which contains a letter of approbation dated

44 As a reward for his lecture in the Vatican gardens, Clement VII presented him with a book; on the title page Widmanstetter proudly reports the story of this pontifical present; cf. Prowe, *Nicolaus Coppernicus* (reprint Osnabrück: 1967) I.2 274.

45 This is my reading of Jedin H., "Katholische Reform und Gegenreform", in Jedin H. (ed.) *Reformation: Katholische Reform und Gegenreformation* (Freiburg: 1967) 449–604.

46 The *Master of the Sacred Palace* was a sort of leading theologian in the Vatican (not the *Holy Office*); among other things he was in charge of the prohibition of books.

August 6, 1546 by Bartolomeo Spina.[47] Tolosani's *Opusculum* is dedicated to the same Paul III to whom Copernicus had dedicated *De revolutionibus*. What was it about the latter that had raised the suspicion of the two Dominicans?

It is easily conceivable that with respect to *De revolutionibus* Spina was alarmed for two reasons: (1) the book was dedicated to the Pope with an appeal for protection against theologians ignorant in astronomy (like Spina himself); (2) the book was printed in Nuremberg, 'citadelle du luthéranisme' (Lerner), and its publication coincided with the ban on importing 'Protestant' books.[48] Apart from this, Spina had clear ideas about the methodological ranking of disciplines: in 1525 he had published a book with the telling title *Treatise on the Preeminence of Holy Theology over All Other Sciences, Above All Those of Human Laws*.[49] Tolosani, in turn, was, among other things, an astronomer. He had contributed to the first attempt at a calendar reform in the context of the fifth Lateran Council, where Copernicus had also been asked to contribute.

Tolosani[50] begins (largely following the anti-Copernican roadmap anticipated by Rheticus) by quoting biblical passages about cosmology all unfavourable to Copernicanism. He then gives a short exposition of Book 1 of *De revolutionibus* and criticises Copernicus for his ignorance of philosophy and theology. He continues by representing the true astronomical and physical account as given by Ptolemy and Aristotle. He concludes with an attitude of resignation:

> The Master of the Holy and Apostolic Palace [i.e. his friend Spina] thought of condemning his [i.e. Copernicus's] book. But first sickness and then his death prevented him from implementing this. I, then, thereafter in

---

47 The Latin text was identified and edited by Garin E., "Alle origini della polemica anticopernicana," *Studia Copernicana* 6 (1973) 31–42. On Tolosani I follow Lerner M.-P., "Aux origins de la polémique anticopernicienne (I): L'*Opusculum quartum* de Giovanmaria Tolosani (1547–1548)", *Revue des sciences philosophiques et théologiques* 86, 4 (2002) 681–722, 683–691.

48 Cf. Lerner, "Origines (I)" 685f.

49 *Tractatus de praeeminentia Sacrae Theologiae super alias omnes scientias & praecipue humanarum legum*, cf. Lerner, "Origines (I)" 685. Here, Spina follows the path of the Papal Bull 'Apostolici regiminis' of 1513, which had succinctly stated that 'every assertion that contradicts illuminate faith is completely wrong'; see Denzinger H. – Hünermann P. (eds.), *Enchiridion Symbolorum definitionum et declarationum der rebus fidei et morum*—Kompendium der Glaubensbekenntnisse und kirchlichen Lehrentscheidungen, 40th edn. (Freiburg: 2005) 483 (no. 1441).

50 Cf. Lerner, "Origines (I)" 688–691.

> opusculum undertook to accomplish it in order to protect the truth for the common benefit of the Holy Church.[51]

Lerner's evaluation that the Church observed a 'neutralité bienveillante' can hardly be supported.[52] The 'Church' was in no way "neutral" but rather simply not involved. Tolosani's text and Spina's willingness to condemn Copernicus remained completely unknown. The first formal Roman *Index of Prohibited Books*, issued in 1559 by Pope Paul IV, also does not mention Copernicus or any other Copernican, and neither does the Index of the Council of Trent (1564).[53] In the same vein, the Index Congregation, established in 1571, was not put into action by the sleeping Roman wolves.[54]

In any case, it is interesting to note that in 'Rome' it was not only a lack of interest and competence that hindered dealing with Copernicanism, but also contingent factors such as the premature death of two of its most dogged enemies.

3.3 *An Abject Attempt to Ban Copernicanism*

Apart from Tolosani and Spina, there is only one rather insignificant indication of anti-copernican activities in the Archives of the Holy Office and the Congregation of the Index. In the introduction to the section on Copernicus, in their extremely valuable and meritorious collection of documents on the Church and science, Baldini and Spruit give a fitting evaluation:[55]

---

51 'Cogitaverat magister sacri et apostolici paltii eius improbare librum, sed prius infirmitate, deinde morte praeventus, hoc implere non potuit. Quod ego postea in hoc opusculo perficere curavi pro veritate tuenda in communem ecclesiae sanctae utilitatem.' Lerner, "Origines (I)" 719.

52 Lerner, "Origines (I)" 682.

53 I follow Wolf H., *Index. Der Vatikan und die verbotenen Bücher* (Munich: 2006) 26f., a 'near-entertaining' book written by arguably the greatest expert on the Archives of the Holy Office and the Congregation of the Index.

54 Only in 1613 did Tommaso Caccini (1574–1648), a Dominican preacher of hate in Florence, use Tolosani's text in a sermon in *Santa Maria Novella* to stir up hatred against Copernicanism and Galileo (Cf. Westman, *Copernican Question* 197).

55 Baldini U. – Spruit L. (eds.), *Catholic Church and Modern Science: Documents from the Archives of the Roman Congregations of the Holy Office and the Index*, vol. I (Sixteens Century Documents) in 4 parts (Vatican City: 2009) I.2 1473–1480, quotes on 1473. Joseph Ratzinger, then Prefect of the Congregation for the Doctrine of the Faith, made the archives of the Holy Office and the Congregation of the Index accessible to researchers only in 1998.

> There is frequently a large gap between the importance of an author, at least as it appears today, and the amount of Censorial interventions regarding his works, usually depending upon the intellectual and cultural views of the Censors, upon the intrinsic limitations of available information, and upon a lack of competence in specific fields other than theology and scholastic philosophy. Among the authors and doctrines virtually ignored by the Roman Congregations during the sixteenth century, Copernicus (1473–1543) and heliocentrism represent without doubt the most salient case.

Indeed, in the Archives has survived only an 'extremely short expurgatory censura' of the second edition of *De revolutionibus* (Basel 1566), proposed by a censor in Naples in 1598. It has neither Copernicus's name right, nor the title of the book, and was directed not against Copernicus and *De revolutionibus* but rather the 'heretic' Rheticus and his Narratio prima with which it was bound together:

> Ex operis Nicolai Caprearij volutionibus orbium, Basileae, Henrici etc deleatur[56] 1566. Deleatur narratio per Georgium Ioachinum Raicum [i.e. Rheticum, G.W.] Ereticum.[57]

Given that no action of the Holy Office resulted from this short note from Naples, the editors come, finally, to the fair summary that,

> the Congregations seem to have been virtually unaware of the significance of Copernicanism, and the very few members who were acquainted with Copernicus's central views, as a rule, deemed the latter as utterly absurd, and certainly not a substantial challenge to the traditional interpretation of Biblical cosmology and its intimate link to Aristotelian natural philosophy.[58]

56 The 'deleatur' refers to cancelling the name of Basel and the printer's name.

57 Baldini – Spruit, *Catholic Church* I.2 1481 comment: 'The scribe was probably an ecclesiastic with a fairly limited education and with poor knowledge of Latin, who did not grasp what he was copying.'

58 Baldini – Spruit, Catholic Church I.2 1478. The Naples censura is not mentioned in Mayaud P.-N., La condamnation des livres Coperniciens et sa révocation à la lumière de documents inédits des Congrégations de l'Index et de l'inquisition (Rome: 1997). The first three chapters of the book inform about the structure and activity of the Congregation of the Index.

3.4 *Bellarmine and the Authoritarian Closing of the Ecclesiastical Mind*

The hardening of the ecclesiastical line against Copernicanism has its main roots in the Council of Trent (1545–1563), which aimed at reforming the Catholic Church, thereby giving rise to the counter-reformation by strengthening the principle of authority.[59] The first session (1546) was devoted to the Scripture. This was indeed an issue that deeply divided Catholics and Protestants. To be sure, both sides shared the humanist dedication to philological scholarship that was directed to and favourable of a literal understanding of the texts. That Protestant reformers based faith on the Scripture alone (sola scriptura) corresponded well with this. They understood the Scripture as a literal expression of 'God's word,' accessible to everybody without any interference from theological or other authorities. The Catholic Church, however, reacted to this exegetic anarchy in the Protestant camp: the Scripture together with its authoritative interpretation by the Church and the Fathers of the Church were declared the basis of Catholic faith:

> Furthermore, to control petulant spirits, the Council decrees that, in matters of faith and morals pertaining to the edification of Christian doctrine, no one, relying on his own judgment and distorting the Sacred Scriptures according to his own conceptions, shall dare to interpret them contrary to the sense which Holy Mother Church, to whom it belongs to judge of their true sense and meaning, has held or does hold, or even contrary to the unanimous agreement of the Fathers, even though such interpretations should never at any time be published. Those who do otherwise shall be identified by the ordinaries and punished in accordance with the penalties prescribed by the law.[60]

Note that the ecclesiastical authority with respect to controlling the Scripture is still restricted to 'matters of faith and morals'—quite in the spirit of St. Augustine and the subsequent exegetical development. Although the text of the Council says nothing about methods of interpretation and only interpretative authority, 'some theologians of the late sixteenth century read this passage to assert the primacy of literal interpretation.'[61] It is above all Bellarmine who

---

59 The Council met in three sessions (1545–47, 1551–52 and 1559–63).

60 Translation in Blackwell R.J., *Galileo, Bellarmine, and the Bible. Including a Translation of Foscarini's* Letter on the Motion of the Earth (Notre Dame: 1991) 183. Latin original in Denzinger – Hünermann, *Enchiridion* 498 (no. 1507).

61 This is the observation of Blackwell, *Galileo* 12.

theologically advances the primacy of literal meaning.[62] At the same time, Bellarmine is the man behind the condemnation of Copernicanism. But something prevents him from striking right away: Copernicanism as a cosmological view does not yet relate to faith and morals.

In order to achieve this, the second dogmatic strand was used very effectively. It concerns the old question of 'double truth' that had come up only recently in the context of the timely condemnation of Pietro Pomponazzi's (1462–1525) *De immortalitate animae*. In the Papal Bull *Apostolici regiminis* (1513) in the context of the Fifth Lateran Council, Leo X, with an eye on the immortality of the soul, had arrived at a rather general statement:

> Since truth cannot contradict truth, we define that every statement contrary to the enlightened truth of the faith is totally false.[63]

This seems to be the boldest and most authoritarian epistemological claim the Church has ever made in an officially binding document. What was still needed, however, was to bring 'Trent' and Leo X together. It took the smart and at the same time authoritarian Bellarmine to accomplish this feat. In my view, his stroke of genius was to interpret the 'illuminated faith' of Apostolici regiminis as being directly communicated by the Holy Spirit. In a letter to Foscarini (1615), who together with Copernicus was to end up on the Index a year later, Bellarmine declares:

> Nor can one reply that this [i.e., Copernicanism] is not a matter of faith, because even if it is not a matter of faith because of the subject matter, it is still a matter of faith because of the speaker [i.e. the Holy Spirit through the authors of Scripture]. Thus anyone who would say that Abraham did not have two sons and Jacob twelve would be just as much of a heretic as someone who would say that Christ was not born of a virgin, for the Holy Spirit has said both these things through the mouth of the Prophets and the Apostles.[64]

62 Blackwell, *Galileo* ch. 2.

63 Denzinger – Hünermann, *Enchiridion* 483 (no. 1441).—The Latin text as an appendix is also in Constant E.A., "A Reinterpretation of the Fifth Lateran Council Decree *Apostolici regiminis* (1513)", *Sixteenth Century Journal* 33, 2 (2002) 353–378. The English translation of the quote is on p. 363.

64 Blackwell, *Galileo* 105; see also Finocchiaro, *Galileo Affair* 68.

Blackwell draws the correct conclusion that according to this position,

> every statement in Scripture, once its correct literal meaning has been established, is a de fide truth' and includes 'also historical and empirical matters in the Bible on which there happens to be agreement in the Church or among the Fathers as to the meaning. [...] As a result geocentrism has become a "matter of faith" according to the principle of de dicto truth![65]

These are, indeed, remarkable synergetic dynamics, fuelled by authoritarianism, of two Church teachings that up to that point had not yet come together.

What of St. Augustine's principle to give up the literal sense of the Scripture as soon as it contradicts 'highly certain results of reasoning or empirical evidence'? Bellarmine does not contest this exegetic principle, and rightly responds that Copernicanism was far from being able to claim such an elevated epistemological status. Therefore, it is the word of the Scripture that counts.

Now we also understand why the Church in the condemnation of Copernicanism in 1616 (and later in 1633) could insist on judging this cosmological theory not only as 'formally heretical' and 'erroneous in faith' but also 'foolish and absurd in philosophy.'[66] Because Aristotelian natural philosophy is a kind of systematic presentation of everyday experience, it is to a large degree at least compatible with the respective cosmological ideas of the authors of the Scripture that also largely express everyday experience. Thus a special philosophical (scientific) view of nature is commanded by the Scripture, an idea so bizarre and so contingent on Bellarmine's authoritarian character that it comes as no surprise that it took seventy-three years to develop.[67] Given these contingencies, it is, moreover, easily conceivable that the condemnation of Copernicanism may never have happened.

Selective Bibliography

Augustinus Aurelius, *Über den Wortlaut der Genesis—De genesi ad litteram libri duodecim. Der große Genesiskommentar in zwölf Büchern*, vol. 1., trans C.J. Perl (Paderborn: 1961).—The Latin text is available at: http://www.sant-agostino.it/latino/ genesi_lettera/index2.htm.

65 Blackwell, *Galileo* 105f.

66 'Philosophy' includes 'science,' cf. fn. 7.

67 Bellarmine's intellectual strength—he was at that time seventy-four—might also no longer have matched his institutional power in the *Holy Office*.

Baldini U. – Spruit L. (eds.), *Catholic Church and Modern Science: Documents From the Archives of the Roman Congregations of the Holy Office and the Index*, vol. I (Sixteenth Century Documents) in 4 parts (Vatican City: 2009).

Barker P., "The Role of Religion in the Lutheran Response to Copernicus", in Osler M. (ed.), *Rethinking the Scientific Revolution* (Cambridge: 2000) 59–88.

Barker P. – Goldstein B.R., "Realism and Instrumentalism in Sixteenth Century Astronomy: A Reappraisal", *Perspectives on Science* 6 (1998) 232–258.

Blackwell R.J., *Galileo, Bellarmine, and the Bible. Including a Translation of Foscarini's Letter on the Motion of the Earth* (Notre Dame: 1991).

Carrier M., *Nikolaus Kopernikus* (Munich: 2001).

Constant E.A., "A Reinterpretation of the Fifth Lateran Council Decree Apostolici regiminis (1513)", *Sixteenth Century Journal* 33, 2 (2002) 353–378.

Copernicus Nicolaus, "On the Revolutions", trans. E. Rosen, in Dobrzycki J. (ed.), *Complete Works*, 3. vols. (London: 1978) vol. 2.

Copernicus Nicolaus, "Minor Works", trans. E. Rosen, ed. Czartoryski P. (ed.), *Complete Works*, 3. vols. (London: 1985) vol. 3.

Copernicus Nicolaus, *Gesamtausgabe*, Nobis H.M. – Folkerts M. – Kirschner S. – Kühne A. (eds.): Vol. II: *De revolutionibus* Libri Sex, Nobis H.M. – Sticker B. (eds.) (Hildesheim: 1984); Vol. VI.1: Documenta Copernicana – Briefe: Texte und Übersetzungen, Kühne A. et al. (eds.) (Berlin: 1994); Vol. VIII.1: Receptio Copernicana—Texte zur Aufnahme der Copernicanischen Theorie, Nobis H.M. – Pastori A.M. (eds.) (Berlin: 2002).

Danielson D., *The First Copernican: Georg Joachim Rheticus and the Rise oft he Copernican Revolution* (New York: 2006).

Denzinger H. – Hünermann P. (eds.), *Enchiridion Symbolorum definitionum et declarationum der rebus fidei et morum—Kompendium der Glaubensbekenntnisse und kirchlichen Lehrentscheidungen*, 40th edn. (Freiburg: 2005).

Finocchiaro M.A., *The Galileo Affair: A Documentary History* (Berkeley: 1989).

Garin E., "Alle origini della polemica anticopernicana", *Studia Copernicana* 6 (1973) 31–42.

Gingerich O., *The Eye of Heaven: Ptolemy, Copernicus, Kepler* (New York: 1993).

Gingerich O., *An Annotated Census of Copernicus's De revolutionibus* (Nuremberg, 1543 and Basel, 1566) (Leiden: 2002).

Gingerich O., *The Book Nobody Read: Chasing the Revolutions of Nicolaus Copernicus* (London: 2004).

Goddu A., *Copernicus and the Aristotelian Tradition: Education, Reading, and Philosophy in Copernicus's Path to Heliocentrism* (Leiden: 2010).

Goldstein B.R., "Copernicus and the Origin of His Heliocentric System", *Journal of the History of Astronomy* 33 (2002) 219–235.

Hooykaas R., *G.J. Rheticus' Treatise on Holy Scripture and the Motion of the Earth with translation, annotations, commentary and additional chapters on Ramus-Rheticus and the development of the problem before 1650* (Amsterdam: 1984).

Jedin H., "Katholische Reform und Gegenreform", in idem (ed.), *Reformation: Katholische Reform und Gegenreformation* (Freiburg: 1967) 449–604.

Kambartel F., *Erfahrung und Struktur. Bausteine zu einer Kritik des Empirismus und Formalismus* (Frankfurt a.M.: 1968).

Lerner M.-P., "L''hérésie' héliocentrique: du soupçon à la condamnation", in *Sciences et religions. De Copernic à Galilée (1540–1610)*. Actes du colloque international organisée par l'Ècole française de Rome. En Collaboration Avec L'Ecole Nationale Des Chartes Et L'Istituto Italiano per Gli Studi Filosofici, Avec La Participation De L'Universita Di Napoli Federico II (Rome: 1999) 69–91.

Lerner M.-P., "Aux origins de la polémique anticopernicienne (I): L'Opusculum quartum de Giovanmaria Tolosani (1547–1548)", *Revue des sciences philosophiques et théologiques* 86, 4 (2002) 681–722.

Mayaud P.-N., *La condamnation des livres Coperniciens et sa révocation à la lumière de documents inédits des Congrégations de l'Index et de l'inquisition* (Rome: 1997).

Mayaud P.-N., *Le Conflit entre l'astronomie nouvelle et l'Écriture Sainte aux XVIe et XVIIe siècles. Un moment de l'histoire des idées autour de l'affaire Galilée*, 6 vols. (Paris: 2005).

Prowe L., *Nicolaus Copernicus*, 2 vols. (Berlin: 1883/84; reprint, Osnabrück: 1967) vol. 1 in 2 parts.

Rosen E., *Three Copernican Treatises: The Commentariolus of Copernicus; The Letter against Werner; The Narratio Prima of Rheticus* (New York: 1939, 1971).

Stuhlmacher P., *Vom Verstehen des Neuen Testaments. Eine Hermeneutik* (Göttingen: 1979).

Swerdlow N.M., "The Derivation and First Draft of Copernicus'ss Planetary Theory: A Translation of the Commentariolus with Commentary", *Proceedings of the American Philosophical Society* 117 (1973) 423–512.

Swerdlow N.M. – Neugebauer O., *Mathematical Astronomy in Copernicus's De revolutionibus*, 2 vols. (New York: 1984).

Westman R.S., *The Copernican Question: Prognostication, Skepticism, and Celestial Order* (Berkeley: 2011).

Wolf H., *Der Vatikan und die verbotenen Bücher* (Munich: 2006).

CHAPTER 3

A Natural History of the Heavens: Francis Bacon's Anti-Copernicanism

Dana Jalobeanu[1]

Francis Bacon's anti-Copernicanism was, for a long time, one of the embarrassing stories of the history of science. The major promoter of the new science denied the movement of the earth, promoted a semi-Paracelsian and vitalistic cosmology, and argued against the use of mathematics in cosmology. Meanwhile, as has repeatedly been shown, Francis Bacon was not ignorant when it came to the novelties and discoveries of the New Astronomy. Quite the contrary: he tried to integrate all the new astronomical discoveries into a properly constructed and properly organised natural history of the heavens. This paper proposes a clarification of the seemingly paradoxical situation described above. In the first part of the paper I discuss the major criticisms formulated by Bacon with respect to astronomy in general and Copernican astronomy in particular. I show that they are motivated by more general concerns regarding the relation between disciplines and by a challenging and quite novel view regarding the role of mathematics in physics. In the second part of the paper I discuss Bacon's proposal for a new natural history of the heavens suitable to ground a 'proper' theory of the heavens.

1 Introduction

Francis Bacon's was one of the last anti-Copernicans; he denied the motion of the Earth and constructed an 'uncompromisingly geostatic and geocentric' semi-Paracelsian and vitalistic cosmology;[2] he argued against the use of geometry in cosmology. And he did all these not only at the end of the

---

1 University of Bucharest, dana.jalobeanu@celfis.ro. The research for this paper was financed from the grant PN-II-ID-PCE-2011-3-0719, *From natural history to science*.

2 See Rees G., "Francis Bacon on Verticity and the Bowels of the Earth", *Ambix* 26 (1979) 202–211, 203. For the comprehensive reconstruction of Francis Bacon's semi-Paracelsian cosmology see Rees G. "Francis Bacon's Semi-Paracelsian Cosmology", *Ambix* 22 (1975) 81–101; Rees G., "Francis Bacon's Semi-Paracelsian Cosmology and the *Great Instauration*", *Ambix* 22 (1975)

 | DOI 10.1163/9789004281127_005

sixteenth-century, but also in his most productive years, 1620–1626, apparently unaware of the major progresses of the physical and mathematical astronomy everywhere in Europe. Bacon did not only populate his works with a good number of standard anti-Copernican arguments; his famous doctrine of the idols pictured Copernicus as a major originator of intellectual diseases. He even went so far as treating mathematical astronomy *in toto* as an idol-generating theory. Earlier studies have tried to explain Bacon's attitude by the lack of 'scientific training',[3] by his general distaste of mathematics[4] and by his exclusive interest in matter-theory.[5] Also invoked was Bacon's isolation from contemporary philosophical developments, his failure to keep track with the scientific achievements of the day. Alternatively, Graham Rees has offered a remarkable reconstruction of Bacon's cosmological thinking showing how, stemming the tides of mathematical astronomy, Francis Bacon developed the project of a vitalistic and alchemical cosmology. Eventually, this was an unfinished project, left in manuscript and published too late to impress any of Bacon's late seventeenth-century readers.[6]

In this paper, I propose to take a look at Bacon's views on astronomy, astrology and the constitutions of the heavens from a different perspective, in the larger context of Bacon's own proposal for a reformed 'Living Astronomy'. I claim that what Rees called Bacon's 'semi-Paracelsian cosmology' was just one small part of a much larger project. Little has been done so far to

---

161–73, and Rees G., "Matter Theory: A Unifying Factor in Bacon's Natural Philosophy", *Ambix* 24 (1977) 110–125.

3 Bacon Francis, *The Works of Francis Bacon: Baron of Verulam, Viscount St. Alban, and Lord High Chancellor of England*, ed. J. Spedding – R.L. Leslie – D.D. Health, 14 vols. (London: 1857–1874; reprint, New York: 1968) vol. III, 716; henceforth SHE, followed by volume number and page.

4 This is, for example, the position expressed by Thomas Kuhn in a seminal article. Kuhn claims that 'Those critics who ridicule him [Bacon] for failing to recognize the best science of his day have missed the point. He did not reject Copernicanism because he preferred the Ptolemaic system. Rather, he rejected both because he thought that no system so complex, abstract, and mathematical could contribute to either the understanding or the control of Nature.' Kuhn T.S., "Mathematical versus Experimental Traditions in the Development of Physical Science", in idem, *The Essential Tension. Selected Studies in Scientific Tradition and Change* (Chicago: 1977) 46. In fact, as it will become clear in the next section of this article, Bacon claims rather the opposite: astronomy as a system of calculus is too simple to do justice to the complexity of the physical 'system' which is the universe.

5 Rees, "Semi-Paracelsian Cosmology". See also Gaukroger S., "The Role of Matter Theory in Baconian and Cartesian Cosmologies", *Perspectives on Science* 8 (2000) 201–222; Gaukroger S., *Francis Bacon and the Transformation of Early Modern Natural Philosophy* (Cambridge: 2001).

6 Rees G., "The Fate of Bacon's Cosmology in the Seventeenth Century", *Ambix* 24 (1977) 27–38.

reconstruct this project and even less to read it contextually, against the background theories and debates of the late sixteenth and early seventeenth centuries. Significantly, no one has so far attempted to offer a contextual reconstruction of Bacon's criticism of the received astronomy and of his plans for a novel, reformed and 'Living astronomy'. Although not an astronomer himself, Bacon was deeply interested in the possibilities of astronomical observations and measurements. As a result, I claim, his plans and proposals bear numerous signs of his interest in contemporary debates.

In the first part of the paper I discuss Bacon's critical positions vis-à-vis astronomy in general and Copernican astronomy in particular. I will show that these positions are well informed, plausible and motivated by attempts to propose an alternative 'living astronomy' that would belong to the 'noblest part of physics'.[7] In the second part of the paper I show that Bacon's plans and sketches for a 'proper' natural history of heavens meet some of the criteria he set for such a 'living astronomy'. I discuss several such criteria and I emphasise the similarities they bear with other late sixteenth-century projects to develop a proper observational basis for astronomy. A word of *caveat* is perhaps necessary. Although this paper attempts to bridge a gap between fields of research that were so far set apart, it is not a comprehensive reconstruction of Bacon's sources or a historical identification of Bacon's partners of dialogue. It is merely an attempt to argue for such a 'proper' contextual reading, which will eventually enlarge and deepen our understanding, while also correcting some of the traditional 'idols' so widespread in the field of Baconian scholarship.

2 Prometheus, the 'Stuffed Ox' and the Desiderata of a 'Living Astronomy'

As with many of the received sciences, Bacon is highly critical of received astronomy. 'The philosophy of the heavenly bodies extant hitherto has no soundness', he claims in the *Descriptio globi intellectualis*, a manuscript written probably in 1611 or 1612.[8] It should be thoroughly reformed, not only because it is full of mistakes and false assumptions, but mainly because it poses dangers to the soundness of intellect. This is a common and recurrent warning in Bacon's works. For example, in *De agumentis scientiarum* Bacon claims that

7 SEH IV, 349.

8 Bacon Francis, *Philosophical Studies c.1611–c.1619*, ed. G. Rees – M. Edwards (Oxford: 1996) 117. I will refer to the volumes of the Oxford Francis Bacon edition as OFB, followed by the volume number and page. The *Descriptio globi intellectualis* remained a manuscript and was not published until the mid-seventeenth century.

traditional astronomy is a source of idols;[9] indeed, some may say it is almost an idol itself:

> Certainly astronomy offers to the human intellect a victim like that which Prometheus offered in deceit to Jupiter. Prometheus, in the place of a real ox, brought to the altar the hide of an ox of great size and beauty, stuffed with straw and leaves and twigs. In like manner astronomy presents only the exterior of the heavenly bodies (I mean the number of the stars, their positions, motions and periods), as it were the hide of the heavens; beautiful indeed and skilfully arranged into systems; but the interior (namely the physical reasons) is wanting.[10]

This is an interesting criticism, and one that has not been thoroughly discussed so far. It is often assumed that Bacon simply disliked the fictional character of astronomical constructions and, more generally, the application of mathematics to physics. In fact, what the above paragraph is saying is somewhat different. Bacon claims that what astronomy has to offer is a *simulacrum*: an image of likeness, which, in this particular religious context of the Promethean fable, is both deceiving *and* dangerous. The 'stuffed ox' is not only a fake: it is purposefully deceptive. It offers the exterior appearance of a sacrificial victim while being nothing of the sort. In his interpretation of the Promethean fable, developed in *De sapientia veterum*, Bacon identifies the 'stuffed ox' with hypocritical religion (as opposed to the true religion signified by the 'real' ox).[11] In this case, the subject of the metaphor is not religion, but astronomy. Bacon seems to see all received astronomical theories as simulacra, i.e. deceiving visible images of what a 'real' theory should look like, namely

> a theory [...] which would not merely satisfy the phenomena (of which kind many might with a little ingenuity be contrived), but which would set forth the substance, motion, and influence of the heavenly bodies as they really are.[12]

9 For Bacon, idols are the 'deepest fallacies of the human mind', 'corrupt and ill-ordered predispositions of the mind' (SEH IV, 431). The idols are partly innate, partly acquired and tend to reoccur in any judgment. They 'take possession of the mind, and cannot be wholly removed' (SEH IV, 431), only kept under close surveillance and fought off at every step, with the help of a discipline of the intellect.

10 SEH IV, 347–348.

11 Bacon Francis, "The Wisdom of the ancients", SEH VI, 750–751.

12 Bacon Francis, "De Augmentis Scientiarum", SEH IV, 348.

This theory, Bacon claims, would be the true 'living astronomy', namely the 'noblest part of physics'; 'in distinction from that stuffed ox of Prometheus which was an ox in figure only'.[13]

In order to build up the 'living astronomy', however, the philosopher has to begin with a *pars destruens*. He has to counteract the idolatrous effects of the 'stuffed ox' and to explain the ways in which traditional astronomy has ensnared human understanding. Only after that he can embark into building up a new theory of the heavens.[14] The first, critical part, is that to which I will now turn, analysing Bacon's criticisms of the received astronomical theories and attempting to place them in the context of contemporary debates.

3 'Wings' versus 'Weights' for the Mind: Bacon's Criticisms of Received Astronomy

There are, I think, at least three different reasons why traditional astronomy is, in Bacon's view, a major source of idols. First, astronomy is 'stuffed' with techniques and devices working to introduce 'too much order' in the universe (this belongs to what Bacon calls the 'idols of the tribe').[15] Second, astronomy makes use of ad-hoc assumptions and hypotheses, encouraging the natural tendency of the mind 'to delight in the spacious liberties of generalities'.[16] It encourages the interference of personal idiosyncrasies (or what Bacon calls the 'idols

13 SEH IV, 349.

14 In fact, the two have to go in parallel, since the operation of warning against idols is not an 'eradication' of the idols but at best a programme of disciplining the mind and keeping it in check in the subsequent constructive phase of the programme. See Jalobeanu D., "Idolatry, Natural History and Spiritual Medicine: Francis Bacon and the Neo-Stoic Protestantism of the Late Sixteenth Century", *Perspectives on Science* 21 (2012) 207–226, and Jalobeanu D., "Experimental Philosophers and Doctors of the Mind: The Appropriation of a Philosophical Tradition", in Alexandrescu V. – Theis R. (eds.), *Naturel et surnaturel. Philosophies de la nature et metaphysique aux XVI–XVII siècles* (Hildesheim: 2010) 37–63. For a more general discussion on the relation between natural philosophy and the medicine of the mind see Corneanu S., *Regimens of the mind. Boyle, Locke and the Cultura Animi Tradition*, (Chicago: 2011).

15 The idols of the tribe are '[...] false appearances imposed upon us by the general nature of the mind[...].' See SEH III, 395 and the discussion in Francis Bacon, *Novum organum and associated texts*, ed. G. Rees – M. Wakeley (Oxford: 2000), hereafter OFB XI. The cited line is at OFB XI, 79–81.

16 Bacon Francis, "Advancement of Learning", OFB IV, 88.

of the cave')[17] into the sober research of nature. Also, astronomy is not free from allegiances to an ancient doctrine or another (or what Bacon calls the 'idols of the theatre').[18] It is filled with theory-related but empty concepts, such as 'planetary orbs', which, for Bacon, are perfect examples of the idols of the market.[19] All these ad-hoc assumptions, new concepts and idiosyncratic hypotheses make received astronomical theories really dangerous for any sound mind. The third way in which traditional astronomy can distort the understanding is through its associated 'machinery', i.e. simplified systems of spheres destined to act as heuristic, visual and pedagogical devices. Each of these criticisms deserves a separate discussion, and I will take them in order.

On the more general level, for Bacon, every theory containing assumptions about the simplicity, order and homogeneity of the natural world is suspicious and bears the marks of the idols of the tribe. Astronomers, Bacon claims, are particularly prone to introducing 'more order and equality in things' than there actually is.[20] In this, they are doing little more than instantiating one of the most common idols of the human mind, namely:

> That the spirit of man, being an equal and uniform substance, doth usually suppose and feign in nature a greater equality and uniformity than is in truth. Hence it cometh that the mathematicians cannot satisfy themselves, except to reduce the motions of celestial bodies to perfect circles, rejecting spiral lines, and labouring to be discharged of eccentrics.[21]

---

17 OFB XI, 81; idols of the cave 'belong to the particular individual'; they can be both innate and acquired. According to Bacon, these idols arise usually from an insufficient mental discipline; for the lack of the good skepticism which teaches one to 'distrust whatever ravishes and possesses his intellect'; their immediate causes are either a 'mastering passion,' or certain mental attitude and tendencies, such as an 'excess of composition and division, or from love of studying particular periods, or contemplating large or small-scale objects' (OFB XI 93). For a discussion of the relation between the idols and the passions of the mind see Corneanu, *Regimens of the mind*, Chapter 1.

18 OFB XI, 81–3.

19 OFB XI, 93; the idols of the market 'have slipped into intellect through the alliance of words and names'; as a result 'words turn and bend *their* power back upon the intellect; and that has made philosophy and the sciences sophistical and inactive'. See also OFB XI, 94–5.

20 OFB XI, 83.

21 SEH III, 395. See also SEH IV, 432.

In other words, Bacon ascribes Copernicus's attempts to 'simplify' the astronomical theory by eliminating eccentric circles to an inner tendency of the human mind to 'feign' in nature a 'greater equality and uniformity' than there is. Other astronomical speculations and fictions belong to the same idols, such as for example, 'the fiction that *in the heavens everything moves in perfect circles*'[22] or the assumption that there is 'so much immobility in nature, by laying down that the Sun and the stars are immobile'.[23] Bacon thus classifies the fundamental assumptions of Copernican astronomy as being nothing else than idols of the mind, originating in the 'uniform' and 'simple' nature of the fallen intellect. The result of constructing a theory according to the idols is not only 'false'; it is also 'dangerous': in the name of the same such principles as simplicity and harmony, Copernicus felt entitled to introduce supplementary (and equally idolatrous) fictions and assumptions, such as the presupposition of the 'triple motion' of the Earth:[24]

> [...] or that he wanted the Moon to cling to the Earth as it were on an epicycle; and these and some other things which he assumed are the marks of a man who thinks nothing of making up anything he likes in nature, provided his calculations turn out well.[25]

Bacon seems to believe that astronomy is particularly prone to encourage the introduction, via ad-hoc assumptions and idiosyncratic hypotheses, of some such idols of the cave into astronomical calculation devices. Astronomy is not free of the third kind of idols either: for Bacon, 'planetary orb' is an empty concept, a clear example of an idol of a market, i.e. a word originating in a false theory, by which the intellect is ensnared and deceived.[26]

In conclusion, Bacon finds in the received astronomy, both Ptolemaic and Copernican, all his favourite categories of idols. In discussing them, he proves well informed of the foundations of the Copernican 'reformation' and their metaphysical, epistemological and mathematical presuppositions. In fact, his criticism rehearses some of the contemporary critiques to the 'new' Copernican astronomy: the apparent ad-hoc status of its hypotheses, the problematic relation between physics and mathematics, the arbitrary centre of the universe

22 OFB XI, 83.
23 OFB VI, 121, 125.
24 OFB VI, 121.
25 OFB VI, 123.
26 OFB XI, 95.

(the 'mean' sun) etc.[27] Before discussing Bacon's objections in a wider context, however, I will now turn to the third way in which Bacon seems to believe that idols creep into astronomical calculation, namely with the very introduction of visual aids and devices. Again, Bacon proves to be well informed and well aware of the techniques of mathematical astronomy. He points to the fact that what astronomers are doing, in fact, is to construct 'mathematical demonstrations' and a 'machinery' capable to produce them.[28] For the astronomer,

> all the labour is spent in mathematical observations and demonstrations. Such demonstrations however only show how all theses things may be ingeniously made out and disentangled, not how they may truly subsist in nature; and indicate the apparent motions only, and a system of machinery arbitrarily devised and arranged to produce them.[29]

This passage is in fact a good description of the astronomical practice of the late sixteenth century. As has been shown, sixteenth-century mathematical astronomy worked with *theoricae*, namely models of spheres and teaching devices destined to allow the calculation of relative positions (angular positions) and relative distances of celestial bodies.[30] Their status was that of imperfect tools, destined to 'help' the understanding. Some astronomers even made an explicit connection between the weakness of the human intellect and the needs for *theoricae*,[31] hypotheses of other calculating devices in astronomy.[32] It was precisely because of the weakness of the human intellect

27 Barker P., "Constructing Copernicus", *Perspectives on Science* 10 (2002) 208–227.

28 Westman R., *The Copernican Question: Prognostication, Skepticism, and Celestial Order* (Berkeley: 2011); idem, "The Astronomer's Role in the Sixteenth Century: A Preliminary Study", *History of Science* 18 (1980) 105–147; Barker P. – Goldstein B.R., "Realism and Instrumentalism in Sixteenth-Century Astronomy: A Reappraisal", *Perspectives on Science* 6 (1998) 232–258.

29 SEH IV, 348–349.

30 Barker – Goldstein, "Realism and Instrumentalism" 232–258.

31 Ibidem; for a discussion of the use of mechanical concentric spheres as instruments for the mind see also Recorde Robert, *The Castle of Knowledge* (London, Reginalde Wolfe: 1556) 279ff.

32 In fact, this might have been a standard opening in many books by the mid-sixteenth century. One example is Digges Leonard, *A Prognostication Everlasting* (London, Thomas Marsh: 1576), which opens with a preface to the reader discussing the current debates on the status of astronomy (quotings Guido Bonatus, Schonerus, Melanchthon and Cardano). At the heart of the debate is, whether astronomy is a proper science of the stars or just a mathematical art destined to offer a 'model' and to calculate predictions.

that the astronomer could not understand the real motions of the celestial bodies nor grasp their real causes. Astronomical devices provided imperfect instruments; sometimes they did little else than assisting the intellect in understanding the celestial motions with the aid of mental images of orbs similar to the visual images provided in a *theorica*.[33]

In questioning the status and relevance of such heuristic devices or 'machinery' Bacon is, again, in good company. In fact, the extent to which such 'theoricks' are nothing more than provisional and imperfect devices for our understanding, bearing little resemblance to the real motions of the heavens is one of the important subjects of debate in late sixteenth-century astronomy. There are a couple of important points in this particular discussion. One is, to what extent does the adoption of a model take into consideration current physical theories or physical constraints.[34] Is it a physical or a mere geometrical model? Does geometry 'approximate' a physical reality, and if so, why? The other is, how much is this 'model' tributary to metaphysical considerations such as the value of order, simplicity, the perfect nature of the circle etc. The third important point concerns the role of *theorica* as an imperfect and provisional device, subject to perfectibility.[35]

All these three points, illustrated, in Bacon's case, by his categories of idol-generating forms of hypothetical reasoning, were vividly discussed in the second half of the sixteenth century. For example, in late sixteenth-century England, one of the first Copernicans, Thomas Digges, attempted to provide answers to all the three of them.[36] He described, for example, Copernicus's

---

Digges' take on the debate is on the side of seeing astronomy as one of the mathematical arts, given to us in order to discipline and elevate our mind, with 'good', 'sound' and 'profitable' learning. See Digges, *A Prognostication Everlasting*, fol1B ff. See also Westman, *The Copernican Question* 269ff. and Barker – Goldstein, "Realism and Instrumentalism" 236–237.

33 As in the case of Peucer Kaspar, *Hypotiposes orbium coelestium*, ed. excudebat Theodosius Rihelius (Strasbourg, Theodosius Rihel: 1568), see Barker – Goldstein, "Realism and Instrumentalism" 237.

34 This is what has been called the Averroist challenge to traditional astronomy. According to Barker, for example, one can take the Averroist challenge to be the driving force behind the formation and acceptance of a Copernican theory. See Barker, "Constructing Copernicus" 208–227, and Barker P. – Goldstein B.R., "Theological Foundations of Kepler's Astronomy", *Osiris* 16 (2001) 88–113.

35 See Barker – Goldstein, "Realism and Instrumentalism".

36 Thomas Digges (c.1546–1595), mathematician, astronomer, one of the practical mathematicians developing trigonometry, military fortifications, astronomical instruments and

work as an attempt to provide an 'anatomy' of the heavenly machine,[37] in direct opposition to a more traditional astronomy which provided a mere 'model' of the heavens. This 'anatomy' meant going beyond mere probable hypotheses into looking for physical grounds and causes, observing new phenomena and using mathematics as a tool to provide more 'probable' and 'better' arguments.[38] The 'more probable' and the 'better' arguments were to be assessed with the help of dialectics.[39] Digges' use of dialectics in astronomy went as far as putting together, side by side, the two 'theoricks': the ancient Ptolemaic one and the 'new' one elaborated by Copernicus. In a book published in 1573, *A Prognostication Everlasting* he simply set the Copernican 'theorick':

---

a theory of measurement, was also gentleman, a Member of Parliament and a member of the entourage of Robert Cecil, Lord Burghley and, later, of Robert Dudley, Earl of Leicester. He published a couple of treatises on practical mathematics, some authored by his father, Leonard Digges, also a skilled mathematician. His first book, *Pantometria* (1571) comprised a treatise by Leonard on trigonometry and land-measurement and Thomas's treatise on the five Platonic solids. He wrote then a treatise in Latin, *Alae seu scalae mathematicae* (1573), dedicated to Lord Burghley. His best known book, quite popular and going through a large number of editions was, however, *A prognostication everlasting* (1576), a treatise on astrological predictions. Appended to it one can find the first English (abridged) translation of a substantial chunk of book I of Copernicus's *De revolutionibus*. The translation is prefaced by Digges' famous diagram of a Cosmos where the stars are distributed not on a fixed sphere but randomly, inside of an 'infinite solid'. On Thomas Digges see Pumfrey S., "'Your astronomers and ours differ exceedingly': the controversy over the 'new star' of 1572 in the light of a newly discovered text by Thomas Digges", *British Journal for the History of Science* 44 (2011) 29–60; Westman, *The Copernican Question* 271ff.; Tredwell K., "The Melanchthon Circle's English Epicicle", *Centaurus* 48 (2006) 23–31; Johnson F.R. – Larkey S.V., "Thomas Digges, the Copernican System and the Idea of the Infinity of the Universe in 1576", *The Huntington Library Bulletin* 5 (1934) 69–117.

37 See Westman, *The Copernican Question* 271. Digges claims that Copernicus's attempt to provide a faithful anatomy of the heavens lead to the abandonment of the 'monstrous' Ptolemaic system of eccentric spheres. See *Alae seu mathematicae*, Prefatio.

38 See Westman, *The Copernican Question* 271ff.; Tredwell, "The Melanchthon Circle's" 23–31.

39 For discussions concerning the role and type of demonstration in astronomy see Barker – Goldstein, "Realism and Instrumentalism". For Digges' use of dialectics see Westman, *The Copernican Question*, and Pumfrey S., "'Your astronomers and ours differ exceedingly': the controversy over the 'new star' of 1572 in the light of a newly discovered text by Thomas Digges", *British Journal for the History of Science* 44 (2011) 29–60.

> [...] together with the olde Theorik also publish this, to the ende such noble English minds [...] might not be altogether defrauded of so noble a part of Philosophy.[40]

As has been shown, Digges' publication strategy was far from innocent.[41] To a book on prognostication (practical astronomy and astrology) where calculations are made with the 'old Theorick' he added an appendix translating and rearranging what he took to be the 'essential' chapters of Copernicus's *De revolutionibus* book I dealing with the planetary order and the philosophical answers to Aristotelian counter-claims to the motion of the Earth.[42] It was, in other words, a 'dialectic reasoning' meant to add what Digges sometimes called 'grave reasons Philosophicall' to 'demonstrations Mathematicall'. It was also one of Digges' biggest contributions to the 'new astronomy': the idea that better observations, physical theories and better mathematics can add strong arguments in favour of the new theorick.[43]

> However, I cannot here set a limit to again urging, exhorting and admonishing all students of Celestial Wisdom, with respect to how great and how hoped-for an opportunity has been offered to Earthdwellers of examining whether the Monstrous System of Celestial globes, which was fashioned by the ancients, has been fully corrected and amended by that divine Copernicus of more than human talent, or whether there still remains something else to be further considered. This, I have considered, cannot be done otherwise than through most careful observations, now if this Most Rare Star, now of the rest of the wandering stars and through various changes in their appearances, and all this done in the various regions of this dark and obscure Terrestrial Star, where, wandering as strangers, we lead, in a short space of time, a life harassed by varied fortunes.[44]

---

40 Digges, *A Prognostication Everlasting* M1.

41 Westman, *The Copernican Question* 271ff.; Pumfrey S., "Controversy over the 'new star'".

42 Digges offers an abridged version of chapters 10, 7, 8, 9 (in this order) of Copernicus's *De revolutionibus*. See Digges, *Prognostication* N1–P1 and Westman *The Copernican Question* 271–274. On Digges understanding of Copernicus see also Tredwell, "The Melanchthon Circle's" 23–31, and Pumfrey, "Controversy over the 'new star'" 29–60.

43 Johnson F.R. – Larkey S.V., "Infinity of the Universe" 69–117.

44 Digges Thomas, *Alae seu scalae mathematicae* (London, Thomas Marsh: 1573), trans. in Johnson – Larkey, "Infinity of the Universe", 111.

This long and vivid description of the ways in which observations and new observational and mathematical techniques can play a role in deciding in favour of the 'new' and 'old' theorick comes from a Latin treatise published in 1573 and suggestively entitled *Mathematical Wings or Ladders with which to ascend to the furthest Theatres of the Heavens and, with a new and unheard-of Method, to explore the paths of all the Planets, and then [only then] to find out the Distance, Position and immense Magnitude of that portentous Heavenly Body shining with remarkable brightness in the Northern Part of the World.* The mathematical 'wings' are Digges' techniques of parallax calculation: what he calls his 'surest demonstrations' for detecting, for example, whether a comet is a sublunary or a supralunary phenomenon.[45] The ladders are, most probably, the novel and better instruments that would allow more precise measurements.[46] Interestingly enough, both 'wings' and 'ladders' are seen as instruments fit to lift the mind in a better position for the contemplation of heavenly order and beauty. This provides a particular understanding of the relation between physics (natural philosophy), mathematical techniques and calculation devices. More generally, it raises the question regarding the role of geometry and arithmetic in the description of the universe. Digges sees the astronomer as working with nothing more than a provisional and perfectible model. By developing new mathematical techniques *and* by enlarging his basis of observations, the astronomer can correct and modify this model.

This dynamic understanding of astronomy was only improved in the last decades of the sixteenth century, in the works of Tycho Brahe[47] and Johannes Kepler.[48] Both Tycho and Kepler elaborated extensively on the role of novel observations and the observational basis of astronomy and discussed the

45 Johnson – Larkey, "Infinity of the Universe"; Westman, *The Copernican Question.*

46 Digges was very much a practical mathematician, interested in perfecting instruments for determining distances, angles etc.

47 See for example the following statement: 'Our purpose was to rid this field of mistakes and superstition, and to obtain the best possible agreement with the experience on which it is based. For I think that it will hardly be possible to find in this field a perfectly accurate theory that can come up to mathematical and astronomical truth.' Brahe Tycho, *Tycho Brahe's Description of his Instruments and Scientific Work as given in Astronomiae Instauratae Mechanica*, trans. and ed. H. Raeder – E. Stromgren – B. Stromgren (Copenhagen: 1946) 118.

48 On the role of Kepler's instruments in developing ideas about accuracy and successive approximations see Hon G., "On Kepler's Awareness of the Problem of Experimental Error", *Annals of Science* 44 (1987) 545–591, and Hon G. – Zik Y., "Kepler's *Optical Part of Astronomy* (1604): Introducing the Ecliptic Instrument", *Perspectives on Science* 17 (2009) 307–345.

ways in which observational astronomy can correct the mathematical model. Moreover, Kepler would have probably agreed with Digges that 'mathematical wings' and 'observational ladders' help the soul to 'ascend' to the heavens, to get closer and therefore in a better position to see the true 'anatomy' of the universe.[49]

This special status attributed to mathematics is perhaps not surprising. Moreover, as has been shown, the Platonic metaphor of the 'wings' of the mind echoes a famous passage from one of Melanchthon's prefaces in praise of mathematics:

> Arithmetic and geometry are, therefore, the wings of the human mind [...]. Raised to heaven by their might, you will be able to illuminate with your eyes the natural universe of things, to perceive the distances and measurements of the greatest bodies, to see the fateful conjunction of the stars, in short to perceive the causes of the greatest things which happen in this human existence.[50]

For Melanchthon, and with him, for the whole school of what has been called the 'Wittenberg interpretation'[51] of astronomy, arithmetic and geometry are

49 Barker – Goldstein, "Theological Foundations" 88–113.

50 Melanchthon, *In aritmeticen prefatio quoted*, in Methuen C., "The Role of the Heavens in the Thought of Philip Melanchthon", *Journal of the History of Ideas* 57, 3 (1996) 385–403, here 393–394.

51 As some scholars have emphasised, Copernicus himself was sometimes ambiguous about the physical status of his hypotheses. After Copernicus, what was labelled the 'Wittenberg interpretation' of Copernicus aimed at an astronomy at least constrained by, if not directly determined by, physics. In the Lutheran framework in which Tycho Brahe developed his novel astronomy at the end of the sixteenth century, one of the premises was that physical world operated according to a single providentially designed pattern. In consequence, the principles of astronomy and those of physics were identical or at least consistent. On the cosmological consequences of the Wittenberg interpretation see Moseley A., "Heaven and Earth in the Late-Sixteenth Century: Tycho and Kepler on the Sub- and Supra Lunary", in Granada M.A. – Mehl E. (ed.), *La Révolution copernicienne dans l'Allemagne de la Réforme (1530–1630)* (Paris: 2009) 142–154. On the diffusion of the Wittenberg interpretation in England at the end of the sixteenth century see Westman, *The Copernican Question*, and Tredwell, "The Melanchthon Circle's" 23–31. See also Barker, "Constructing Copernicus" 216; Pérez-Ramos A., "Francis Bacon and Astronomical Inquiry", *British Journal for the History of Science* 23 (1990) 197–205, and Barker – Goldstein, "Realism and Instrumentalism" 232–258.

'wings' of the mind in a very precise sense.[52] Mathematical truths are the few un-scattered seeds of the natural light still left to the human mind after the Fall; they can be developed and grown into 'wings' that would elevate the soul to the contemplation of the law-like order of Creation. There is something 'mathematical' in both the macroscopic universe and in the human soul. Meanwhile, because human mind is fallen, there is a limit to this mathematical development or the 'true' contemplation of the nature of things. Even the heaven itself only displays 'vestiges' of the divinity. Clearly, therefore, the 'wings' of arithmetic and geometry are not enough to lead the soul to the contemplation of Divine truths let alone to salvation. Meanwhile, what they can do is to put together the visible signs, observations and 'experiences' of the visible order of the universe with the potentialities of 'mathematics' and give, through astronomy, law-like regularities and 'prognostications' of future events. It is in this 'marriage' of empirical observations of the heavens and the development of mathematical techniques that Digges follows the Wittenberg interpretation.[53] He even goes, perhaps, one step further, into giving preference not to the 'wings' of the soul but to the 'ladders' of the mind. Digges' treatise *Mathematical Wings and Ladders* is divided into two parts. The first part develops various techniques and proves theorems of spherical geometry destined to improve the parallactic method inherited from Regiomontanus. The second part deals with astronomical observations. It is a 'plea for the adoption of an experimental method in astronomy'.[54] It is also a plea for the use of the 'ladders' of the intellect, namely astronomical instruments destined to help the senses, rightly calibrated by mathematical techniques and by a better understanding of a theory of vision. Two decades before Kepler's *The Optical Part of Astronomy* (1604), Digges emphasised the importance of a better understanding of the optical theory, of a better knowledge of the anatomy of the eye in order to achieve 'better' observations, reduce errors and provide, in this way, a basis of phenomena for improving the hypotheses and astronomical theories. As it has been shown, Digges' writings were influential in the development of astronomy, mechanics and practical mathematics in late sixteenth-century

52 For Melanchthon's promotion of mathematics in his writings see Methuen, "Role of the Heavens", and idem, "The German Reformation and the Mathematization of the World", *Theology and Science* 9 (2011) 35–44.

53 Tredwell, "The Melanchthon Circle's" 23–31.

54 Johnson F.R., "The Influence of Thomas Digges in the Progress of Modern Astronomy in Sixteenth-Century England", *Osiris* 1 (1936) 390–410, 399.

England.[55] His emphasis on building instruments and mathematical techniques for acquiring observational precision can also be found among later practitioners, such as Thomas Harriot and his circle. The first decade of the seventeenth century clearly saw, in England, the development of a cluster of 'experts' interested in both the 'wings' and the 'ladders'. By 1610 it became customary not only for astronomical treatises but also for manuals of navigation and cartography to contain sections on the calibration of instruments, problems of measurement and error. Recent studies have suggested that a proper way to read and interpret Bacon's 'scientific' (i.e. natural-historical) writings would be to place them in an Elizabethan culture of expertise.[56] Like Digges and Harriot, Bacon was interested in the possibility of reforming astronomy with the help of mathematics and better observations. Interestingly enough, he was perfectly willing to emphasise the role of mathematics in educating, disciplining and training the soul.[57] He also emphasised, however, the danger of endowing the Fallen soul with 'wings'. In view of all its deeply rooted idols, what the intellect needs are not 'wings [...] but rather [...] leaden weights to curb all jumping and flying up';[58] meanwhile, he was perfectly willing to work on and improve the 'ladders'. In fact, much of Bacon's proposal for a novel and 'living astronomy', which would replace the 'stuffed ox' of the received mathematical astronomy, was based on the same project of constructing a well regulated and well organised data-base of phenomena, a natural history of the heavens.

55 Johnson, "The Influence of Thomas Digges"; Pumfrey S., "Controversy over the 'new star'"; Westman, *The Copernican Question*.

56 Ash E.H, *Power, Knowledge and Expertise in Elizabethan England* (Baltimore: 2004) 201ff. For a somewhat different perspective see also Harkness D., *The Jewell House: Elizabethan London and the Scientific Revolution* (Yale: 2007). See also Pastorino, C., "Francis Bacon and the Institutions for the Promotion of Knowledge and Innovation", *Journal of Early Modern Studies* 2 (2013) 9–33.

57 It is perhaps interesting to emphasise the way in which Bacon also changes the place of mathematics in his trees of sciences from being a major branch of natural philosophy on a par with physics and metaphysics (in 1605) to being an 'annex' to physics, metaphysics, mechanics and natural magic (in 1623). In the earlier *The Advancement of Learning* (1605) however, mathematics is said to be very able to 'remedy and cure many defects in the wit and faculties intellectual.' (SEH IV, 360).

58 OFB XI 163.

4 'The Marriage of Mathematics with Physics' and the Natural History of Heavens

The common denominator of Bacon's criticism of the 'stuffed ox' of mathematical astronomy is a particular interpretation of mathematics, very similar to what one could read in the famous 'Mathematical Preface' of John Dee:

> No man, therefore, can doute, but toward the atteyining of knowledge incomparable, and Heavenly Wisedome, Mathematicall Speculations, both of Numbers and Magnitutes: are meanes, aydes and guides: readie, certaine and necessary.[59]

Bacon resisted such 'Mathematical Speculations' and argued against the use of geometrical 'fictions' in astronomy. He claimed that by imposing geometrical 'fictions' on the heavens—i.e., circular or elliptic motion, a unique centre of rest of the whole universe—the astronomer creates a coherent, closed and error-proof theory. One can find in Bacon's writings explicit warnings against the incapacity of mathematical astronomy to detect and eliminate its own errors. In fact, this very problematic and idolatrous status of astronomy was used by Bacon as a sad proof of what happens when sciences are 'cut off' from each other. The mathematical astronomer builds a theory based on too few phenomena and observations; the result is an ungrounded and invalid 'anticipation of nature'. How can one decide in favour of either the Copernican or the Ptolemaic hypothesis? They both explain the same phenomena. They might, however, be contrary to the common background knowledge in natural philosophy.[60]

> As for the hypotheses of astronomers, their refutation is more or less useless because no one claims that they are true in themselves and because they can be different and contradict one another just so long as they save and sort out the phenomena equally. If it please you, let it therefore be arranged between astronomy and philosophy, as if it were by a timely and lawful covenant, that astronomy should supply those hypotheses which are most useful for cutting down calculation, philosophy those which come closest to the truth of nature, and that the hypotheses

59 Dee John, *The Mathematicall Preface to the Elements of Geometry of Euclid of Megara*, ed. A. Debus (New York: 1975) aiij.

60 Bacon, *Advancement of Learning*, OFB IV, 92. See also OFB VI, 188–91.

> of astronomy should not be prejudicial to the truth of the matter, while the determinations of philosophy should be such as to be reconcilable with the phenomena of astronomy.[61]

What Bacon seems to propose here is a continuous interplay and check-up between mathematical astronomy and natural philosophy; a direct confrontation of their hypotheses. A somewhat difficult task, since both astronomy and natural philosophy lack exactly the common base of experiments and observations, the 'history' that would allow such a confrontation. This is what Bacon proposed to construct, and never succeeded. However, he left a good number of texts on how to write such a natural history of the heavens. The most developed draft was written in 1611–1612 and published in the mid-seventeenth century under the name *Descriptio globi intellectualis*. In 1620, at the end of *Novum organum* he referred to such a natural history of the heavens, both in his *Preparative to a natural history* and in his list of 130 natural histories to be written (of which two are a 'history of celestial phenomena, or astronomical history' and a 'history of the configurations of the heaven and its parts towards the Earth and its parts, or cosmographical history'). Furthermore, undated manuscripts designed for incorporation into his *Sylva Sylvarum* also discuss elements of a natural history of the heavens. References to it can be found in the published posthumous version of *Sylva* (1626) and in the *New Atlantis*.

The setting of Bacon's natural history of the heavens is quite traditional. Cosmographies and cosmographical natural histories proliferate during the Renaissance. The attempt to build a 'history of celestial phenomena' that precedes the astronomical theory is also a classical gambit. It is how Osiander's (in)famous preface introduces Copernicus's project:

> For it is the duty of an astronomer to compose the history of the celestial motions through careful and expert study. Then he must conceive and devise the causes of these motions or hypotheses about them. Since he cannot in any way attain to the true causes, he will adopt whatever suppositions enable the motions to be computed correctly from the principles of geometry for the future as well as for the past.[62]

---

61 OFB VI, 188–191.

62 Copernicus Nicolaus, *On the Revolutions of the Heavenly Spheres*, trans. E. Rosen (Baltimore: 1979) xix.

Therefore, only after the astronomer has 'composed the history of the celestial motions' will he go on and 'conceive and devise the causes of these motions or hypotheses about them'. Also, in the *Astronomie pars optica* (1604) Kepler identifies three different areas arising from observations: one is 'the mechanical part, concerning instruments fit for observing the celestial motions and the manner of making use [of them]'. Another is 'the historical part', containing, for example, 'the twenty-four books of most meticulous observations left by Tycho, encompassing nearly the past forty years'. The third is the optical part. There is also a fourth part, the physical part of astronomy, but that goes beyond observation into investigating the physical causes, the formal causes and the material causes of the motions of the heavens (as Kepler will later say in the *Astronomia Nova*).[63]

Although Bacon sets up his plans for a natural history of the heavens in a traditional setting, and although he identifies the natural history of the heavens as 'extant' in his tree of disciplines, what he seems to have had in mind in subsequent plans and projects of how a proper natural history of heavens would look like was something rather new, in agreement with his general views on natural history:[64]

> The kind of Natural History I am seeking is one from which natural causation can be understood, on which Philosophy can be based, which is faithful to sense-evidence and proved by works.[65]

Meanwhile, the natural history of the heavens must be 'simple and not imbued with dogmas', made of phenomena 'plain' and 'separated' from theories, phenomena collected through precise observations.[66] In fact, the natural history of the heavens, as all Baconian natural histories, must begin with a collection of 'facts' from books. Bacon's choice of books is, however, intriguing.

---

63 Kepler Johannes, *Optics: Paralipomena to Witelo & Optical Part of Astronomy*, trans. W.H. Donahue (Santa Fe: 2000) 15, and Boner P.J., "Life in the Liquid Fields: Kepler, Tycho and Gilbert on the Nature of the Heavens and Earth", *History of Science* 46 (2008) 275–297, 276.

64 Jalobeanu D., "Francis Bacon's Natural History and the Senecan Natural Histories of Early Modern Europe", *Early Science and Medicine* 17 (2012) 197–229; Jalobeanu D., "The Philosophy of Francis Bacon's Natural History: A Research Program", *Studii de ştiinţă şi cultură* 4 (2010) 18–37.

65 Bacon Francis, "Cogitationes de natural humana", in Farrington B. (ed.), *The Philosophy of Francis Bacon. An Essay on Its Development from 1603 to 1609 with New Translations of Fundamental Texts* (Liverpool: 1964) 42. See also OFB VI, 111.

66 OFB VI, 111.

> [...] the best history of the Heavenly Bodies would be that which could be extracted and elicited from Ptolemy, Copernicus and the more learned writers on astronomy, if you completely stripped the art from experiment and also added the observations of more recent authorities.[67]

The more 'recent authorities' invoked here are very important: they are the observational astronomers who 'by means as it were of the skiffs and boats of optical instruments have begun just recently to do new trade with celestial phenomena'.[68] Therefore, the *Descriptio globi intellectualis* refers to the passage of Mercury before the Sun (observed by Kepler in 1607), to the observations published by Galileo in his *Sidereus nuncius* (1611): Jupiter's satellites, the large number of stars in the Milky Way, the mountains on the Moon. Bacon also mentions the sunspots and seems to be at least partly acquainted with the dispute between Galileo and Scheiner over their nature, although Galileo's letters on sunspots were not published until 1613. In short, the *Descriptio* bears all the marks of a text written by someone who was closely following the novelties and current disputations in astronomy.

The structure of Bacon's proposed natural history also bears the mark of its genre: it was supposed to be dialectical, constructed around questions such as 'Whether the world is a system, or whether the universe is composed of many systems, scattered and disconnected?' or 'Whether the Earth or the Sun is the Centre of this System' etc. In summing up the arguments on each side of the debate, Bacon lists ancient and modern 'discoveries', experiments and facts. He also suggests novel observations and experiments that would perhaps advance the question. In many places, the text bears the mark of Bacon's own hypotheses, such as the denial of any kind of separation between the sublunary and the supralunary regions, the assertion of a common theory of matter etc. It is however worth emphasising that such hypotheses are equally integrated in the dialectal structure of the whole and have to be supported by observational and experimental 'arguments'. It is beyond the purpose of this paper to enterprise a full-length study of this fascinating text. It is however worth emphasising that this is an enterprise for the future. We do not have a contextual reading of this text yet. For the present purpose of this paper, however, it is sufficient to stress that here Bacon proves to be equally informed about the latest discoveries of observational astronomy and about the cosmological theories of Patrizzi, Telesio, Tycho and Gilbert. His discussion on the fiery nature of the heavens, for example, bears interesting resemblances to Tycho Brahe's similar theories.

67 OFB VI, 111.
68 OFB VI, 115.

However this may be, it is important to stress the general outline of Bacon's natural history of the heavens: on a traditional topical and dialectal structure, open-ended questions are addressed to the *nature* and *motions* of heavenly bodies. They are not supposed to be solved. Instead, observational and experimental evidence is gathered on either side. Most of the observational evidence is still rather thin, a fact that Bacon recognises willingly, pointing to ways to improve this observational basis in the future. The resulting, complete natural history of the heavens will be, Bacon claims, the very basis on which one can build a 'living astronomy', i.e. a proper theory that can safely replace the 'stuffed ox' of the traditional astronomy.

This 'new' ideal natural history of the heavens arises here and there in Bacon's writings in connection with the best way to do measurements[69] or in connection with what Bacon takes to be the true importance of mathematics for physics. For example when, in his general directions on how to write natural history, Bacon insists that everything should be 'set down, counted, weighted, measured and defined':

> For we are after works not speculations and, indeed, a good marriage of Physics and Mathematics begets Practice. And for this reason we should investigate in detail and thoroughly record, in the History of Heavenly Bodies, the precise returns and distances of the planets [...]. But where precise proportions are not available to us we must for sure fall back on rough estimates and comparisons, as, for instance (if we happen to distrust the astronomers' calculation of distances) that the Moon stands within the Earth's shadow; that Mercury is above the Moon, and the like. Again, where average proportions are not available, let us set down the extremes [...].[70]

Bacon's plan for a 'proper' natural history of the heavenly bodies, therefore, seems to reproduce the ideals of accuracy expressed by Tycho Brahe's measurements. It also contains hints to his own theory of measurement and 'approximation': when precise observations are not available, the natural historian is required to estimate, approximate, compare and find limiting cases or inflexion points in his collection of phenomena.[71] In addition, Bacon also

69 See for example OFB VI, 115.

70 *Parasceve* OFB XI, 467.

71 More on the quantitative aspects of Bacon's natural histories can be found in Rees G., "Quantitative Reasoning in Francis Bacon's Natural Philosophy", *Nouvelles de la republique des lettres* 1 (1985) 27–48; Rees G., "Mathematics and Francis Bacon's Natural Philosophy",

seems to propose a class of novel observations: observations not constrained by a 'theory' of the heavens or by the accepted 'geometry' of the universe. In the *Descriptio* at least, he seems to believe that he found them in the new discoveries of telescopic astronomy. In other words, he stresses the importance of a thorough study of comets and other such 'rare' astronomical occurrences. A thorough study of comets, he claims in the *Novum Organum*, can perhaps furnish crucial instances that would help falsify one or both of the received theorics: Ptolemaic and Copernican.

5 Conclusion

The purpose of the present paper was to take a fresh look at some of Bacon's writings dealing with astronomy, astrology and the constitution of the heavens in a larger context provided by Bacon's own proposal for a reformed 'living astronomy'. I have tried to show that what Graham Rees has called Bacon's 'semi-Paracelsian cosmology' is just a small part of a larger project comprising a *pars destruens*, namely Bacon's criticism of Copernican and Ptolemaic 'theorics', and a *pars construens* meant to comprise a well organised natural history of the heavens upon which to construct a sound 'theory of the heavens'. Little has been done so far to reconstruct this project and even less to read it contextually, against the background theories and debates of the late sixteenth and early seventeenth centuries. In my paper I have offered the first fruits of a research in progress. My main target has been to show how much we can gain in terms of insight and understanding by reading Bacon's criticism and proposals in the wider context of similar proposals for constructing a new astronomy. Although not an astronomer himself, Bacon was deeply interested in the possibilities of astronomical observations and measurements. He reacted promptly to Galileo's discoveries; he followed the debate around the nature of sunspots, he proved aware of Galileo's theory of tides etc. His proposed natural history of the heavens used such new discoveries in the context of a larger, dialectal structure, topic- and question-oriented, and meant to provide the basis for future astronomical research. In devising such a structure, Bacon also proved

Revue internationale de philosophie 40 (1986) 402–407, and Pastorino C., "Weighing Experience: Experimental Histories and Francis Bacon's Quantitative Program", *Early Science and Medicine* 16 (2011) 542–570. See also Jalobeanu D., *The Hunt of Pan: Francis Bacon's Art of Experimentation and the Invention of Science* (Bucharest: 2014 forthcoming); Jalobeanu D., "The Marriage of Mathematics with Physics: Francis Bacon on Measurement, Mathematics and the construction of Mathematical Physics" (forthcoming).

aware of the importance and role of mathematics and argued for a 'proper' marriage of mathematics and physics. Bacon, in fact, devised a new role for mathematics in physics, sketching a novel way to talk about mathematising nature. He rejected the realist interpretation of the geometrical 'models' of the universe and proposed, instead, a mathematical theory of approximation and measurement, which would endow physics with accuracy, and would be able to 'beget practice.

Selective Bibliography

Ash E.H., *Power, Knowledge and Expertise in Elizabethan England* (Baltimore: 2004).

Bacon Francis, *Sylva Sylvarum or a Natural History in Ten Centuries* (London, J.H.: 1627).

Bacon Francis, *The Oxford Francis Bacon*, vol. IV, *The Advancement of Learning*, ed. M. Kiernan (Oxford: 2000).

Bacon Francis, *The Oxford Francis Bacon*, vol. VI, *Philosophical Studies c.1611–c.1619*, ed. G. Rees – M. Edwards (Oxford: 1996).

Bacon Francis, *The Oxford Francis Bacon*, vol. XI, *The Instauratio magna Part II: Novum organum and Associated Texts*, ed. G. Rees – M. Wakley (Oxford: 2004).

Bacon Francis, *The Oxford Francis Bacon*, vol. XII, *The Instauratio magna, Part III: Historia naturalis et experimentalis: Historia ventorum and Historia vitae et mortis*, ed. G. Rees – M. Wakley (Oxford: 2007).

Bacon Francis, *The Oxford Francis Bacon* vol. XIII, *The Instauratio magna: Last Writings*, ed. G. Rees (Oxford: 2000).

Bacon Francis, *The Works of Francis Bacon: Baron of Verulam, Viscount St. Alban, and Lord High Chancellor of England*, ed. J. Spedding – R.L. Ellis – D.D. Heath, 14 vols. (London: 1857–1874; reprint, New York: 1968).

Bacon Francis, *The Major Works*, ed. B. Vickers (Oxford: 1996).

Barker P., "Constructing Copernicus", *Perspectives on Science* 10 (2002) 208–227.

Barker P. – Goldstein B.R., "Theological Foundations of Kepler's Astronomy", *Osiris* 16 (2001) 88–113.

Barker P. – Goldstein B.R., "Realism and Instrumentalism in Sixteenth-Century Astronomy: A Reappraisal", *Perspectives on Science* 6 (1998) 232–258.

Brahe Tycho, *Tycho Brahe's Description of his Instruments and Scientific Work as Given in Astronomiae Instauratae Mechanica*, trans. and ed. H. Raeder – E. Stromgren – B. Stromgren (Copenhagen: 1946).

Boner P.J., "Life in the Liquid Fields: Kepler, Tycho and Gilbert on the Nature of the Heavens and Earth", *History of Science* 46 (2008) 275–297.

Copernicus Nicolaus, *On the Revolutions of the Heavenly Spheres*, trans. E. Rosen (Baltimore: 1979).

Corneanu, S., *Regimens of the Mind. Boyle, Locke and the Cultura Animi Tradition* (Chicago: 2011).

Digges Leonard, *A Prognostication Everlasting* (London, Thomas Marsh: 1576).

Farrington B., *The Philosophy of Francis Bacon. An Essay on Its Development from 1603 to 1609 with New Translations of Fundamental Texts* (Liverpool: 1964).

Gaukroger S., "The Role of Matter Theory in Baconian and Cartesian Cosmologies", *Perspectives on Science* 8 (2000) 201–222.

Gaukroger S., *Francis Bacon and the Transformation of Early Modern Natural Philosophy* (Cambridge: 2001).

Harkness D., *The Jewell House: Elizabethan London and the Scientific Revolution* (Yale: 2007).

Hon G., "On Kepler's Awareness of the Problem of Experimental Error", *Annals of Science* 44 (1987) 545–591.

Jalobeanu D., "Idolatry, Natural History and Spiritual Medicine: Francis Bacon and the Neo-Stoic Protestantism of the Late Sixteenth Century", *Perspectives on Science* 21 (2012) 207–226.

Jalobeanu D., "Experimental Philosophers and Doctors of the Mind: The Appropriation of a Philosophical Tradition", in Alexandrescu V. – Theis R. (eds.), *Naturel et surnaturel. Philosophies de la nature et metaphysique aux XVI–XVII siecles* (Hildesheim: 2010) 37–63.

Jalobeanu D., "Francis Bacon's Natural History and the Senecan Natural Histories of Early Modern Europe", *Early Science and Medicine* 17 (2012) 197–229.

Jalobeanu D., "The Philosophy of Francis Bacon's Natural History: A Research Program", *Studii de ştiinţă şi cultură* 4 (2010) 18–37.

Johnson F.R. – Larkey S.V., "Thomas Digges, the Copernican System and the Idea of the Infinity of the Universe in 1576", *The Huntington Library Bulletin* 5 (1934) 69–117.

Lerner M.P., *Le monde des spheres*, 2 vols. (Paris: 1996).

Methuen C., "The Role of the Heavens in the Thought of Philip Melanchthon", *Journal of the History of Ideas* 57 (1996) 385–403.

Methuen C., "The German Reformation and the Mathematization of the Created World", *Theology and Science* 9 (2011) 35–44.

Mosley A., *Bearing the Heavens: Tycho Brahe and the Astronomical Community of the Late Sixteenth Century* (Cambridge: 2007).

Mosley A., "Heaven and Earth in the Late Sixteenth century: Tycho and Kepler on the Sub and Supra-lunary", in Granada M.A. – Mehl E. (eds.), *Nouveau Ciel, Nouveau Terre. La revolution copernicienne dans l'Allemagne de la Reforme* (Paris: 2009) 143–155.

Pastorino C., "Weighing Experience: Experimental Histories and Francis Bacon's Quantitative Program", *Early Science and Medicine* 16 (2011) 542–570.

Pastorino C., "Francis Bacon and the Institutions for the Promotion of Knowledge and Innovation", *Journal of Early Modern Studies* 2 (2013) 9–33.

Pérez-Ramos A., "Francis Bacon and Astronomical Inquiry", *British Journal for the History of Science* 23 (1990) 197–205.

Pumfrey S., "'Your astronomers and ours differ exceedingly': the controversy over the 'new star' of 1572 in the light of a newly discovered text by Thomas Digges", *British Journal for the History of Science* 44 (2011) 29–60.

Rees G., "The Fate of Bacon's Cosmology in the Seventeenth Century", *Ambix* 24 (1977) 27–38.

Rees G., "Francis Bacon's Semi-Paracelsian Comsology", *Ambix* 22 (1975) 81–101.

Rees G., "Francis Bacon's Semi-Paracelsian Cosmology and the *Great Instauration*", *Ambix* 22 (1975) 161–73.

Rees G., "Matter Theory: A Unifying Factor in Bacon's Natural Philosophy", *Ambix* 24 (1977) 110–125.

Rees G., "Quantitative Reasoning in Francis Bacon's Natural Philosophy", *Nouvelles de la republique des lettres* 1 (1985) 27–48.

Rees G., "Mathematics and Francis Bacon's Natural Philosophy", *Revue internationale de philosophie* 40 (1986) 402–407.

Tredwell K., "The Melanchton Circle's English Epicicle", *Centaurus* 48 (2006) 23–31.

Westman R.S., *The Copernican Question: Prognostication, Skepticism, and Celestial Order* (Berkeley: 2011).

Westman R.S., "The Astronomer's Role in the Sixteenth Century: A Preliminary Study", *History of Science* 18 (1980) 105–147.

Westman R.S., "The Melanchthon's Circle: Rheticus and the Wittenberg Interpretation of Copernican Theory", *ISIS* 66 (1975) 165–193.

CHAPTER 4

Hume's Copernican Turn

Tamás Demeter

Abstract

In his *Treatise of Human Nature*, David Hume proclaims that 'moral philosophy is in the same condition as natural, with regard to astronomy before the time of Copernicus', and he considers his project to improve moral philosophy so as to reach its post-Copernican phase. In the present paper I intend to explore Copernicus's relevance for Hume's project, the science of man. I shall suggest that Hume's allegiance to Copernicanism means a commitment to searching for principles of human nature underlying various human phenomena, just like Copernicus explored the general principles of explanation for the motions of the planets. Moral philosophy, Hume implies, enters its post-Copernican phase by taking methodological commitments to explanatory reductionism and analogical reasoning. Although his praise for Copernicus is due mainly to methodological considerations, I will also argue that Hume's project has central features that make it similar to Kant's critical project after Kant's Copernican turn. Hume also understands his own project as foundational: a critical work that should be done before immersing ourselves into other cognitive enterprises. Similar to Kant's project, Hume's science of man aims to explore the limits and the conditions of possibility of human knowledge, the main difference being that Hume follows a naturalistic as opposed to a transcendental method. Thus, while a *Copernican turn* means different things in Hume and Kant, its consequences entail important similarities in their philosophical positions.

1 Introduction

David Hume considered his contribution to moral philosophy in his *Treatise of Human Nature* (1739/40) as amounting to a revolution comparable to that achieved by Copernicus in natural philosophy.[1] For Hume, a Copernican

1 The crucial passage in this context is Hume David, *Treatise of Human Nature*, ed. D.F. Norton – M.J. Norton, 2 vols. (Oxford: 2007) 2.1.3.6–7, which I will discuss in detail bellow. S.K. Wertz argues convincingly that Hume understood the term 'revolution' in its modern sense as 'a radical departure or fundamental change'; idem, "Hume and the Historiography of Science", *Journal of the History of Ideas* 54 (1993) 411–436, 416–419.

 | DOI 10.1163/9789004281127_006

revolution in moral philosophy consisted in setting a new aim to the discipline and putting it on a new methodological footing.[2] The need for a revolutionary transformation that breaks up with the continuity of previous moral philosophies is expressed early in Hume's 1734 *Letter to a Physician* in which he complains about the disappointing status of the philosophical tradition his age inherited:

> I found that the moral Philosophy transmitted to us by Antiquity, labour'd under the same Inconvenience that has been found in their natural Philosophy, of being entirely Hypothetical, & depending more upon Invention than Experience. Every one consulted his Fancy in erecting Schemes of Virtue & of Happiness, without regarding human Nature, upon which every moral Conclusion must depend.

No wonder then that in moral and natural philosophy 'there is nothing yet establisht [...] & that they contain little more than endless Disputes, even in the most fundamental Articles.' Overcoming this situation and improving the cognitive standing of philosophy required, in Hume's eyes, breaking new grounds in these disciplines with 'a certain Boldness of Temper [...] which was not enclin'd to submit to any Authority in these Subjects'.[3] Detached from ancient authorities, Hume's new approach as announced in the letter sets the proper aim of moral philosophy to be the study of human nature, which should be conducted with empirical, as opposed to hypothetical, methods.

Renewing moral philosophy by adopting successful models from natural philosophy is a characteristic aspiration of Scottish moral philosophers in the Enlightenment period. The most significant inspiration came from the success of Newton's natural philosophy: Francis Hutcheson attempted to elaborate a 'canon' which contained, as it were, the mathematical principles of moral philosophy and George Turnbull set up an axiomatic framework to analyse moral phenomena.[4] Some commentators have interpreted Adam Smith's achievement in economics and moral philosophy as influenced by Newton,[5] and

2 See e.g. Schliesser E., "Copernican Revolutions Revisited in Adam Smith by way of David Hume", *Revista Empresa y Humanismo* 13 (2010) 213–248.

3 See Hume David, *The Letters of David Hume*, ed. J.Y.T. Greig (Oxford: 1932) vol. 1, 16 and 13.

4 See Hutcheson Francis, *An Inquiry into the Original of Our Ideas of Beauty and Virtue*, ed. W. Leidhold (Indianapolis: 2008) 128–134, and Turnbull George, *The Principles of Moral Philosophy*, ed. A. Broadie (Indianapolis: 2005) 48–50.

5 See Raphael D.D., "Newton and Adam Smith", in Stayer M.S. (ed.), *Newton's Dream* (Montreal: 1988) 36–49, 43–44.

Hume's study of human nature is also frequently interpreted as Newtonian in some respects.[6] Hume's case, however, is rather complicated. While most commentators derive Newton's influence on moral philosophy from the *Principia*, I have argued elsewhere that Hume's science of man is more aptly interpreted in the context of the experimental tradition that flourished in the aftermath of the *Opticks*.[7]

In this paper I intend to lay open the meaning of Hume's references to Copernicus, thereby showing that his understanding of Copernicus's significance is consonant with the experimental methodology. This methodology is at the heart of Hume's reform of moral philosophy that was intended to redeem the shortcomings it had inherited throughout the ages. While exploring Copernicus's significance for Hume I will proceed as follows. First, I briefly explore the place of Copernican ideas in the Scottish Enlightenment and show that Copernicus's meaning in this context was not primarily methodological; instead his reception was focused on the new model of the universe and it was constrained on the field of natural philosophy. It was Hume who placed emphasis on the methodological significance of Copernicanism, and drew his conclusions on the field of moral philosophy. Accordingly, in the next step I turn to exploring the details of Copernicus's methodological significance for Hume focusing on those passages that shed light on the method of Hume's project of a science of man. I will attempt to show that the methodological lessons Hume draws from Copernicus are consonant with the basic tenets of his experimental study of human nature that signify his detachment from the traditional framework in which work in *moral philosophy* had been conceived. Finally I draw attention to the different meanings the phrase *Copernican turn* might have in Hume and Kant. As I will argue, the different ways in which they perceived Copernicus's significance is due, on the one hand, to Hume's fairly

6 See e.g. Noxon J., *Hume's Philosophical Development* (Oxford: 1973); Waxman W., *Hume's Theory of Consciousness* (Cambridge: 1994) 183–184; Capaldi N., *Hume's Place in Moral Philosophy* (Frankfurt a.M.: 1989) 20–21; Force J.E., "Hume's Interest in Newton and Science", *Hume Studies* 13 (1987) 166–216. See also Stroud B., *Hume* (London: 1977) 8; Mounce H.O., *Hume's Naturalism* (London: 1999) 15–18; Dicker G., *Hume's Metaphysics and Epistemology* (London: 1998) 2–4; Pitson A.E., *Hume's Philosophy of the Self* (London: 2002) 6, 14, 152; Beebee H., *Hume on Causation* (London: 2006) 5, 183–185; Hazony Y., "Newtonian Explanatory Reduction and Hume's System of the Sciences" in Biener Z. – Schliesser E. (eds.), *Newton and Empiricism* (Oxford: 2014) 138–170.

7 See e.g. Demeter T., "Hume's Experimental Method", *British Journal for the History of Philosophy* 20 (2012) 577–599; idem, "Enlarging the Bounds of Moral Philosophy: Newton's Method and Hume's Science of Man", in Biener Z. – Schliesser E. (eds.), *Newton and Empiricism* (Oxford: 2014) 171–204.

idiosyncratic interpretation of Copernicus's heritage as mainly methodological and, on the other hand, to Kant's sticking to the common understanding of Copernicus as providing a new model of the universe. Despite this difference, however, Hume's *Treatise* can also be seen as establishing a Copernican turn in philosophy similar to Kant's—albeit one following a naturalistic rather than a transcendental path.

2 Copernicus in Scotland

In the second half of the seventeenth century Copernicus was standard and critically acclaimed material in the curricula of Edinburgh University. The discussion of Copernicus was conducted in the context of abandoning Aristotelian natural philosophy for the sake of Cartesianism just to be rapidly superseded, by the end of the century, by Newtonianism.[8] From the 1660s in the lectures of John Wishart, commentaries on Aristotle's *Physics* had been replaced by a critical discussion of modern developments in natural philosophy, albeit still along the lines of Aristotelian physics. On this basis Wishart challenged the Copernican model of the universe questioning its intelligibility, alleged simplicity, empirical adequacy and its compatibility with the Scripture. Questions of theological compatibility were in the forefront of his natural philosophical interests in general: in his lectures he borrowed some insights from Hobbes and Descartes, but he saw their teaching as threatening either with atheism or with the limitation of God's power—and on this basis he rejected them both.

From the 1680s onwards, Gilbert McMurdo and Alexander Cockburn adopted Cartesian ideas without such reservations, and they spread the mechanical worldview among their students, as was the case with most of their fellow regents in Edinburgh at that time. Until about 1690 Copernicanism prevailed in its Cartesian version in which rotating transparent matter caused the planets to orbit in the same direction, and similar vortices were invoked to explain why objects are falling toward the earth's surface. After the publication of Newton's *Principia*, forces quickly populated the Copernican universe and

8 See Russell J.L., "Copernican System in Great Britain", in Dobrzycki J. (ed.), *The Reception of Copernicus' Heliocentric Theory* (Boston: 1973) 231–235; Shepherd C.M., "Newtonianism in Scottish Universities in the Seventeenth Century", in Campbell R.H. – Skinner A. (eds.), *The Origins and Nature of the Scottish Enlightenment* (Edinburgh: 1982) 65–85; Barfoot M., "Hume and the Culture of Science", in Stewart M.A. (ed.) *Studies in the Philosophy of Scottish Enlightenment* (Oxford: 1990) 151–190; Wilson D.B., *Seeking Nature's Logic: Natural Philosophy in the Scottish Enlightenment* (University Park, Pa.: 2009) 1–32.

gravity replaced vortices in the explanation of planetary motions. This transformation of Cartesian Copernicanism into a Newtonian one took place in Edinburgh fairly rapidly, and by about 1710 the triumph of Newtonianism was eminent at the other Scottish universities as well.

The process of transformation is informatively documented in William Law's astronomy lectures he had given as regent between 1692 and 1704, before he took up the chair in moral philosophy in 1708 that was newly established as a result of the university reform replacing the regent system of Edinburgh University with a professorial structure. Given that he occupied the professorship of moral philosophy until 1728, he might have taught David Hume in that capacity. Law's lectures in the 1690s reflect gradual detachment from, and increasing criticism of, Cartesian vortex theory while approaching Newton's astronomical ideas with an unqualified approval. By 1704 Law's initial criticism of Newton for failing to provide his findings with satisfactory explanations by the standards of mechanical philosophy disappeared from the lectures.[9] In his lectures of 1701, i.e. towards the end of the process of his Newtonian conversion, Law's discussion of Descartes and Newton was situated in a Copernican framework which was contrasted with Ptolemy's and Tycho Brahe's theories, the latter being represented as a middle course between the two models. Ptolemy's system was criticised mainly for its empirical inadequacy in explaining the movement of Mercury and Venus, and also for lacking the epistemic virtue of simplicity: if compared to Copernicus, Ptolemy is too complicated because his model relies on epicycles and eccentrics. Copernicus was also criticised mainly on account of intelligibility because he had ascribed the earth's movement to the influence of the sun, and also because of the rapidity of the planets' motion.

When Hume attended Edinburgh University in the early 1720s, the culture of science was dominated by Newtonianism, and due to the work of David Gregory, John Keill and Colin Maclaurin, the influence of Scottish Newtonians extended well beyond the Scottish borders.[10] Generally speaking, Hume was disappointed with the education he received at the university and he had a very low opinion on the knowledge to be acquired there. One exception, to some extent, seems to be the natural philosophy class, which was taught to him by Robert Steuart.[11] In Steuart's class Hume presumably was required to study

9 See Shepherd, "Newtonianism" 67–71.

10 See Wilson, *Seeking Nature's Logic* 34–59.

11 See Stewart M.A., "Hume's Intellectual Development, 1711–1752", in Frasca-Spada M. – Kail P.J.E. (eds.), *Impressions of Hume* (Oxford: 2005) 25.

Keill's introductions to natural philosophy and astronomy, Gregory's introduction to optics and astronomy along with certain passages from Newton's *Opticks* and *Principia*.[12] Also, he probably made good use of the Physiological Library Steuart had established, which might have provided him with all the relevant literature he needed for an introduction to the problems of contemporary natural philosophy, including those related to various versions of Copernicanism.

An important common feature of these introductory texts, and presumably of the accompanying lectures too, is that they focus on the content and virtues of Copernicus's theory rather than its methodology. The most notable thing about Copernicus's theory is the new model of the universe in comparison with its alternatives and its subsequent interpretations in Descartes's and Newton's natural philosophies. But the novelty of Copernicus's system is not derived from some innovative methodology. In connection with the model of the universe Copernicus offers, his theory is credited with various virtues if compared to that of Ptolemy, but these virtues are not derived from the method Copernicus follows—in fact he is not credited with methodological invention at all. In these texts it is generally acknowledged that the *empirical adequacy* of Copernicus's model surpasses that of the rival systems of Ptolemy or Brahe, i.e. it conforms to the facts better, and so it can save more phenomena than its rivals. One can also discern here the traces of 'a great simplistic myth'[13] that emphasises the *simplicity* of Copernicus's theory if compared to Ptolemy and Brahe. On later scrutiny, however, this myth turned out to be untenable, as has the idea of Copernicus's supreme empirical adequacy,[14] yet in Copernicus's

---

12 See Stewart, "Development" 21–22; Barfoot, "Culture of Science" 152–153. See also Keill John, *An Introduction to Natural Philosophy: or Philosophical Lectures Read in the University of Oxford, Anno Dom. 1700. To which are added, the demonstrations of Monsieur Huygens's theorems, concerning the centrifugal force and circular motion. Translated from the last edition of the Latin*, 4th edn. (London, M. Senex, W. Innys et al.: 1740); Keill John, *An Introduction to the True Astronomy: or Astronomical Lectures, Read in the Astronomical School of the University of Oxford* (London, Bernard Lintot: 1721); Gregory David, *Elements of Catoptrics and Dioptrics* (London, E. Curll, J. Pemberton et al.: 1715); Gregory David, *The Elements of Physical and Geometrical Astronomy*, trans. M. Stone, 2 vols. (London, D. Midwinter: 1726).

13 Lakatos I. – Zahar E., "Why Did Copernicus's Research Programme Supersede Ptolemy's?", in Lakatos I. (ed.), *The Methodology of Scientific Research Programmes* (Cambridge: 1978) 168–192, 173.

14 See Kuhn T.S., *The Copernican Revolution* (Cambridge, Mass.: 1972) 172; Lakatos – Zahar, "Research Programme".

Scottish reception these virtues were unanimously associated with his model of the universe.

Adam Smith, in his essay on the history of astronomy (written in 1751, first published in 1795), finds a further epistemic virtue in Copernicus's model, namely 'a superior degree of coherence, which it bestowed upon celestial appearances'.[15] For Smith, *coherence* is not primarily a logical property of theories; it is used in the context of other terms like 'connection' and 'order' whose establishment is the main task of philosophy as 'the science of the connecting principles of nature'. In his view, philosophy is responsible for 'representing the invisible chains which bind together all these disjointed objects, endeavours to introduce order into this chaos of jarring and discordant appearances'.[16] The success of this enterprise is partly measured by the coherence a theory bestows upon disordered phenomena. Coherence thus understood is a matter of degree and it depends on how successful a theory is in establishing connections among various phenomena, whether it needs *ad hoc* hypotheses for establishing connections and on how much it leaves unexplained etc. Coherence is thus related to *simplicity*: a system is less coherent if it allows for the introduction of phenomena that complicate a system without good reason, or if it introduces phenomena for the sake of explaining other phenomena but leaves the newly introduced phenomena without explanation or independent motivation. If measured by these standards, Tycho Brahe was found less coherent than Copernicus[17] because he had been less successful in finding out 'those hidden chains of events which bind together the seemingly disjointed appearances of nature'.[18]

Although in Smith's evaluation the emphasis falls on the epistemic virtues and cognitive content of Copernicus's theory, his emphasis on coherence in this sense and his understanding of the task of philosophy suggest that he has specific methodological ideals in mind that seem to be consonant with Hume's understanding of Copernicus's importance. For Hume, Copernicus is an early representative of efforts made towards 'true philosophy' that is centrally committed to *explanatory reductionism*, i.e. a method of subsuming the variety of complex phenomena under a limited number of principles or laws whose combination results in an explanation.[19] Both in the *Treatise* and in his *History*

15 Smith Adam, "The History of Astronomy", in idem, *Essays on Philosophical Subjects* (1795), ed. W.P.D. Wightman – J.C. Bryce (Indianapolis: 1982) 33–105, 76.

16 Ibid. 45–46.

17 Ibid. 82.

18 Ibid. 48.

19 See Schliesser, "Copernican Revolutions Revisited".

of England, Copernicus is mentioned in the company of those paving the methodological way to 'true philosophy': Bacon, Kepler, Galileo, Boyle and Newton pointed out the way to, and made 'considerable advances' in 'true philosophy'.[20] It is Hume's emphasis on the methodological relevance of Copernicus that distinguishes his evaluation from those of his Scottish contemporaries.

3 Hume's Copernican Turn

As a letter to Henry Home (Lord Kames) written at the end of 1737 testifies, Hume thought that his forthcoming *Treatise* would communicate new 'philosophical discoveries'. In another letter to Home, which he sent shortly after the publication of the *Treatise*, he would consider these discoveries so profound 'that were they to take place, they would produce almost a total alteration on philosophy', i.e. a 'revolution'.[21] At the bottom, Hume's revolution was *methodological*: as the *Treatise*'s subtitle suggests, it consisted in the introduction of the 'experimental method of reasoning' into moral philosophy which was contrasted with aprioristic, hypothetical methods.

Adherence to this kind of reasoning is one of the most permanent features of Hume's thought: his disappointment with traditional methods is obvious from the above-quoted 'Letter to a Physician', and the alternative method is formulated in and applied throughout the *Treatise*. The supremacy of the experimental method is still emphasised in his 1751 *Enquiry into the Principles of Morals*, which indicates that the fundamentals of Hume's method have not changed much:

> we can only expect success, by following the experimental method, and deducing general maxims from a comparison of particular instances. The other scientifical method, where a general abstract principle is first established, and is afterwards branched out into a variety of inferences and conclusions, may be more perfect in itself, but suits less the imperfection of human nature, and is a common source of illusion and mistake in this as well as in other subjects. Men are now cured of their passion for hypotheses and systems in natural philosophy, and will hearken to no arguments but those which are derived from experience. It is full time

20 Hume, *Treatise* 2.1.3.6–7; Hume David, *The History of England from the Invasion of Julius Caesar to the Revolution in 1688*, foreword by W.B. Todd, 6 vols. (Indianapolis: 1983) vol. 5, 153.

21 Hume, *Letters*, vol. 1, 24 and 26.

> they should attempt a like reformation in all moral disquisitions; and reject every system of ethics, however subtle or ingenious, which is not founded on fact and observation.[22]

The experimental method in moral philosophy that Hume advertises throughout his *oeuvre* is perhaps best explained in part I of chapter VIII of his *Enquiry concerning Human Understanding*. It is here that he gives its detailed description as a kind of analysis and synthesis aiming at the explanatory principles of moral phenomena—a method which is congruous with that of natural philosophy and which can be seen as a refinement of our everyday reasoning underlying navigation in the social world.[23]

Hume's central methodological commitment is reductionist: it consists in finding explanatory principles of human phenomena through comparison and analogies revealed among various particular observations. This process results in more and more general laws or principles of human nature,[24] by the combination of which moral phenomena can be explained. Hume's experimental method of finding causes derives from a study of everyday causal reasoning and consists in its more conscious, reflected and sophisticated application. The empirical study of everyday causal reasoning is thus the source of the normative canon of cause-searching which provides the 'logic' equally characteristic of reasoning in moral and natural philosophy—and of course, to a lesser degree of precision and rigor, of everyday reasoning too.[25]

As Hume sees it, this method is first introduced to natural philosophy as a result of Copernicus's achievement, and in moral philosophy a similar Copernican turn should also take place:

> we find in the course of nature, that tho' the effects be many, the principles, from which they arise, are commonly but few and simple, and that 'tis the sign of an unskilful naturalist to have recourse to a different quality, in order to explain every different operation. How much more must this be true with regard to the human mind, which being so confin'd a

22 Hume David, *An Enquiry concerning the Principles of Morals*, ed. T. Beauchamp (Oxford: 1998) 1.10.

23 I discuss this in detail in Demeter T., "Liberty, Necessity and the Foundations of Hume's 'Science of Man'", *History of the Human Sciences* 25 (2012) 15–31.

24 Hume concisely explains the cognitive benefits of his 'science of man' in the first chapter of his *An Enquiry concerning Human Understanding*, ed. T. Beauchamp (Oxford: 2000).

25 See Serjeantson R., "Hume's General Rules and the 'Chief Business of Philosophers'", in Frasca-Spada M. – Kail P.J.E. (eds.), *Impressions of Hume* (Oxford: 2005) 187–212.

> subject may justly be thought incapable of containing such a monstrous heap of principles, as wou'd be necessary to excite the passions of pride and humility, were each distinct cause adapted to the passion by a distinct set of principles? Here, therefore, moral philosophy is in the same condition as natural, with regard to astronomy before the time of *Copernicus*. The antients, tho' sensible of that maxim, *that nature does nothing in vain*, contriv'd such intricate systems of the heavens, as seem'd inconsistent with true philosophy, and gave place at last to something more simple and natural. To invent without scruple a new principle to every new phaenomenon, instead of adapting it to the old; to overload our hypotheses with a variety of this kind; are certain proofs, that none of these principles is the just one, and that we only desire, by a number of falsehoods, to cover our ignorance of the truth.[26]

This quote suggests that entering into the post-Copernican phase of moral philosophy brings along a set of methodological commitments and epistemic virtues the moral philosopher is expected to keep an eye on. The first and basic one is a preference for *simplicity*, meaning a commitment to not introducing new explanatory principles for every newly found phenomenon. In Hume's hands this preference entails *explanatory reductionism*: given that it prohibits introducing new principles for new phenomena, it encourages a) subsuming various phenomena under a limited number of principles, and b) fortitude with respect to established principles by subsuming new phenomena to them. This latter implication also suggests a way of testing theories: in case *ad hoc* hypotheses are needed in our explanations, this indicates reliably that our principles are false.

What Hume seems to imply here is that Ptolemy's followers relied on a heuristics that allowed, at least implicitly, for the introduction of *ad hoc* explanatory principles: they could accommodate any new fact by increasing the number of epicycles and equants. Thus, invoking new *ad hoc* principles of human nature instead of reducing the variety of phenomena in moral philosophy to a limited set of principles would entail similar consequences: incoherence, increased complexity, and loss of explanatory power which is aptly illustrated by Locke's ridiculing the constant appeals to some very specific faculty responsible exclusively for the activity which is to be explained.[27] In a similar vein, Hume rejects the explanatory strategy that readily introduces independent causes to newly

26 Hume, *Treatise* 2.1.3.6–7.

27 Locke John, *An Essay concerning Human Understanding*, ed. P.H. Nidditch (Oxford: 1975) 2.21.17.

discovered phenomena, and prescribes instead a method whose main methodological rule is explanatory reductionism that also brings along simplicity.

Invoking Copernicus's name in this respect seems to be in perfect order: Osiander's preface to *De Revolutionibus* suggests indeed that simplicity, in contrast with truth, is the main epistemic virtue to be ascribed to the work, and Hume, just like Kant several decades later, was probably unaware of the fact that the preface was not written by Copernicus himself.[28] In the dedicatory letter of *De Revolutionibus*, written to Pope Paul III, Copernicus himself also supports Hume's explanatory reductionism indirectly. Here, he complains about the contradictions that arise in various theories due to the introduction of homocentrics, eccentrics and epicycles, and emphasises the importance of explanatory deduction and, in general, of following stable methodological principles.

In this context it may be surprising to see Hume mentioning approvingly the maxim 'that nature does nothing in vain'—an Aristotelian-Scholastic principle that does not seem to fit Hume's experimental method. First, in the Scholastic tradition principles like this were taken to constitute a self-evident universally valid metaphysical and logical foundation of natural philosophy.[29] Given Hume's epistemological commitments, such standing cannot be granted to the maxim 'that nature does nothing in vain' or any other rule of reasoning; yet, such rules can be approved as 'constant and universal principles' of reasoning, and human nature in general, that are known empirically, i.e. from history and the observation of common life.[30] So the maxim can be seen as distilled from observation and as such it expresses the methodological commitment that the principles of nature and human nature are not complex beyond necessity, not superfluous, and therefore nature follows the simplest path. If viewed from this angle, this maxim establishes explanatory reductionism, i.e. a parsimonious search for a set of principles with which the variety of phenomena can be explained.[31]

28 On Kant see Blumenberg H., *The Genesis of the Copernican World* (Cambridge, Mass.: 1985) 609; on Osiander's preface see Jardine N., *The Birth of History and Philosophy of Science* (Cambridge: 1984) 150.

29 See Dear P., *Discipline and Experience* (Chicago: 1995), esp. 15–25.

30 Hume, *Human Understanding* 8.7.

31 For a useful discussion of the centrality of experimental reductionism in Hume's project see Hazony, "Newtonian Explanatory Reduction".

Secondly, in the Aristotelian tradition the maxim 'that nature does nothing in vain' is clearly a teleological principle,[32] which has a central role to play in finding final causes, and as such it has no place in Hume's non-teleological framework. Hume explicitly denies that there could be any other causes than efficient ones,[33] and thereby he leaves no rational place for a commitment that there are ends in nature toward which efficient causes operate. This denial is also extended to the study of human nature. When responding to Francis Hutcheson's worries, Hume declares:

> I cannot agree to your Sense of *Natural*. 'Tis founded on final Causes; which is a Consideration, that appears to me pretty uncertain & unphilosophical. For pray, what is the End of Man? Is he created for Happiness or for Virtue? For this Life or for the Next? For himself or for his Maker? Your Definition of *Natural* depends upon solving these Questions, which are endless, & quite wide of my Purpose.[34]

As opposed to understanding 'natural' in terms of final causes, in Hume's analysis 'natural' is contrasted with terms like 'miraculous', 'unusual' and 'artificial', and drawing on the contrast with the latter, he characterises natural traits and processes as those belonging to a normally functioning human being in itself, i.e. someone exempt from social influences or pathologies.[35]

In studying the normal functioning of human beings, the maxim that 'nature does nothing in vain' expresses a kind of teleological attention of the 'anatomist of human nature',[36] but it is not of the Aristotelian but of a purely *descriptive* kind. In Hume's case it expresses the anatomist's commitment to functional analysis:[37] Hume abundantly talks about our faculty of reasoning and the faculties of memory and imagination etc., and their various principles—not as independently identified causal sources or postulates of some preconceived hypothesis in the framework of which experience is to be

32 For a discussion see Leunissen M., *Explanation and Teleology in Aristotle's Science of Nature* (Cambridge: 2010) 119–135, esp. 121.

33 Hume, *Treatise* 1.3.14.32.

34 Hume, *Letters*, vol. 1, 33.

35 See Hume, *Treatise* 3.1.2.

36 Hume frequently expresses his aspiration to be the 'anatomist of human nature'. See Hume, *Treatise* 1.4.6.23., 2.1.12.2, 3.3.6.6., Abstract 2; and Hume, *Human Understanding* 1.13–14.

37 See Demeter T., "The Anatomy and Physiology of Mind: Hume's Vitalistic Account" in Horstmanshoff H.F.J. – King H. – Zittel C. (eds.), *Blood, Sweat and Tears: The Changing Concepts of Physiology from Antiquity into Early Modern Europe* (Leiden: 2012) 217–240.

interpreted, but as conclusions of comparative functional analyses: the ingredients of human nature whose identity depends on whether the analysis of relevant observations is correct.[38]

For Hume, inquiry does not begin with hypothetical definitions of faculties, and explanations do not proceed from those definitions. Instead, inquiry begins with observations and reveals their systematic connections, which will result in the principles that describe and identify the characteristic activities of faculties. When Hume claims that 'neither man nor any other being ought ever to be thought possest of any ability, unless it be exerted and put in action'[39] he does not mean that abilities cannot be known at all, but that they cannot be known independently of their exercise. And similarly, we can only know faculties through their effects, i.e. know them functionally and inferentially, without the possibility of independent identification.

This is the context in which Copernicus and the maxim 'nature does nothing in vain' reappear in Philo's monologue in the *Dialogues concerning Natural Religion*:

> *That nature does nothing in vain*, is a maxim established in all the schools, merely from the contemplation of the works of nature, without any religious purpose; and, from a firm conviction of its truth, an anatomist, who had observed a new organ or canal, would never be satisfied, till he had also discovered its use and intention. One great foundation of the Copernican system is the maxim, *that nature acts by the simplest methods, and chooses the most proper means to any end* [...].[40]

This passage contains concisely Hume's commitment to distilling rules of reasoning from observation and to functional analysis, which is portrayed here as a natural and appropriate stance of an anatomist. This is further reinforced in the following paragraph in which Philo praises Galen's aspirations for a functional understanding of the muscles.

---

38 As Hume puts it, 'we must distinguish exactly betwixt the phænomenon itself, and the causes, which I shall assign for it; and must not imagine from any uncertainty in the latter, that the former is also uncertain. The phænomenon may be real, tho' my explication be chimerical. The falsehood of the one is no consequence of that of the other'. Hume, *Treatise* 1.2.5.19.

39 Hume, *Treatise* 2.1.10.4.

40 Hume David, "Dialogues concerning Natural Religion", in idem, *Dialogues concerning Natural Religion and Other Writings Dialogues*, ed. D. Coleman (Cambridge: 2007) 12.2.

While pursuing this understanding of mental faculties, Hume follows a method whose origins, as we have seen above, he traces back to Copernicus, and which he considers to be equally uniform in both natural and moral philosophy, i.e. 'to reduce the principles, productive of natural phænomena, to a greater simplicity, and to resolve the many particular effects into a few general causes, by means of reasonings from analogy, experience, and observation.'[41] Finding analogies between different instances gives the chance of explaining causes and reducing them to 'more general principles'.[42]

In the *Dialogues*, and especially in the sections criticising the design argument, Hume seems to be more critical of analogical reasoning than in his other passages of methodological relevance,[43] nevertheless Philo pronounces 'analogies and resemblances' reliable enough to serve as the 'sole proofs of the Copernican system'.[44] Given that in Hume's epistemology the category of proofs provides the highest level of epistemic certainty available for any piece of empirical knowledge, analogical reasoning is a highly esteemed way of reaching theoretical conclusions in exploring the principles of nature and human nature.[45] It is history, natural and civil as well, that provides the pool of observations from which philosophers, natural and moral as well, relying on analogies can establish explanatory principles:

> Its [i.e. history's] chief use is only to discover the constant and universal principles of human nature, by showing men in all varieties of circumstances and situations, and furnishing us with materials from which we may form our observations and become acquainted with the regular springs of human action and behaviour. These records of wars, intrigues, factions, and revolutions, are so many collections of experiments, by which the politician or moral philosopher fixes the principles of his science, in the same manner as the physician or natural philosopher becomes acquainted with the nature of plants, minerals, and other external objects, by the experiments which he forms concerning them. Nor are the earth, water, and other elements, examined by Aristotle, and

41 Hume, *Human Understanding* 4.12.

42 Hume, *Treatise* Appendix 3.

43 See Demeter, "Hume's Experimental Method".

44 Hume, "Dialogues" 2.26.

45 See e.g. Hume, *Human Understanding* 6n10. Hume's doubts in the *Dialogues* concerning analogical arguments from design primarily do not arise from the weaknesses of analogical reasoning, but mainly from the uniqueness of the world which does not provide a sufficient pool of instances among which analogies can be found.

> Hippocrates, more like to those which at present lie under our observation than the men described by Polybius and Tacitus are to those who now govern the world.[46]

This gives the broad outlines of a methodological ideal that Hume suggests be observed by natural and moral philosophers: phenomena are to be collected from history and observation, and then compared; if analogies and similarities are found, they are to be ascribed to principles that are to be also compared, grouped and resolved into more general ones. Central aspects of this methodological ideal are connected to Copernicus: analogical reasoning, functional understanding, and explanatory reductionism are the main Copernican aspects of Hume's science of man, and on Hume's evaluation, they distinguish his enterprise from most of his predecessors and contemporaries.

This thoroughly naturalistic stance distinguishes Hume's project from many of his Scottish contemporaries. Other contemporary Scottish moral philosophers, like Hutcheson, Turnbull and David Fordyce, share a religious *cum* teleological perspective that promises to deliver knowledge of God and the purpose of human beings, and aims at drawing direct normative consequences concerning our duty.[47] Stephen Gaukroger places Copernicus at the beginning of a long struggle for not letting non-scientific disciplines intervene into scientific matters, for an autonomous scientific enterprise whose 'values and norms are open to no refutation from outside'.[48] Hume certainly contributes to this Copernican struggle with his strict adherence to his 'experimental method' in moral philosophy, and rejection of religious and teleological considerations because, as he writes to Hutcheson in his above quoted letter, these are 'unphilosophical', meaning that they are outside of the scope of Hume's purely descriptive and explanatory aspirations.

---

46 Hume, *Human Understanding* 8.7.

47 See e.g. Hutcheson Francis, *Philosophiae Moralis Institutio Compendiaria with A Short Introduction to Moral Philosophy*, ed. I. Turco (Indianapolis: 2007) 24; Fordyce David, *The Elements of Moral Philosophy*, ed. T.D. Kennedy (Indianapolis: 2003) 6; and Turnbull George, *The Principles* 1:10, 1:459.

48 Gaukroger S., *The Emergence of a Scientific Culture: Science and the Shaping of Modernity 1210–1685* (Oxford: 2006) 19.

4 Hume and Kant

Several commentators suggest that Hume announces a Copernican turn in moral philosophy that is similar in crucial respects to Kant's Copernican turn,[49] which he summarises in the second edition of the *Critique of Pure Reason* as follows:

> Up to now it has been assumed that all our cognition must conform to the objects; but all attempts to find out something about them *a priori* through concepts that would extend our cognition have, on this presupposition, come to nothing. Hence let us once try whether we do not get farther with the problems of metaphysics by assuming that the objects must conform to our cognition, which would agree better with the requested possibility of an *a priori* cognition of them, which is to establish something about objects before they are given to us. This would be just like the first thoughts of Copernicus, who, when he did not make good progress in the explanation of the celestial motions if he assumed that the entire celestial host revolves around the observer, tried to see if he might not have greater success if he made the observer revolve and left the stars at rest.[50]

For Kant the relevance of Copernicus's project consists in supposing the observed motions of the planets not to be real motions but appearances generated by the observer's motion. Analogously, Kant suggests metaphysics should take a similar turn by supposing that the activity or constitution of the observer is responsible for a significant share of what a human being experiences. This is typically the case with those features that we can assign to objects *a priori*; i.e. those features that belong to objects because we apprehend them, but we do not apprehend them that way because the objects in themselves (i.e. independently of human cognitive capacities) are that way. Exploring *a priori* conditions on possible objects of human experience is the main task of Kant's transcendental metaphysics, which thus aims at exploring the limits and prospects of human cognition.

---

49 Price H.H., *Hume's Theory of the External World* (Oxford: 1963) 9; Dauer F.W., "Towards a Copernican Reading of Hume", *Nous* 9 (1975) 269–295; Beck L.W., "A Prussian Hume and a Scottish Kant", in idem, *Essays on Kant and Hume* (New Haven: 1978) 111–112, and more recently Mounce, *Hume's Naturalism* 4.

50 Kant Immanuel, *Critique of Pure Reason*, ed. P. Guyer – A.W. Wood (Cambridge: 1998) B xvi.

Hume's project is sometimes interpreted from this Kantian perspective, and there is indeed a certain fundamental similarity in the two projects, namely their undertaking to explore the foundations and thereby the proper limits of human knowledge.[51] As Hume puts it in his introduction to the *Treatise*:

> 'Tis evident, that all the sciences have a relation, greater or less, to human nature; and that however wide any of them may seem to run from it, they still return back by one passage or another. Even *Mathematics, Natural Philosophy*, and *Natural Religion*, are in some measure dependent on the science of MAN; since they lie under the cognizance of men, and are judg'd of by their powers and faculties. [...] There is no question of importance, whose decision is not compriz'd in the science of man; and there is none, which can be decided with any certainty, before we become acquainted with that science. In pretending therefore to explain the principles of human nature, we in effect propose a compleat system of the sciences, built on a foundation almost entirely new, and the only one upon which they can stand with any security.[52]

Something similar is clearly true about Kant's aspirations as well: exploring the features of human cognition in order to reveal the limits of possible human knowledge. In this respect both Hume and Kant inherit the long-standing philosophical aspiration to explore human nature, but they pursue this project with commitments to different philosophical methods and they urge philosophy to take a turn in different directions. Copernicus thus becomes a symbol in these contexts in two very different guises.

As Blumenberg aptly points out, Copernicus's significance for Kant consists in the *model* of the universe he created. It is the model itself that concerns Kant, and not the criteria of evaluating and creating models, i.e. Copernicus's significance for Kant is not methodological: it is Copernicus's vision that matters to him, and quite consistently with this, Kant does not list Copernicus among the heroes responsible for renewing science.[53] Copernicus's vision of the universe serves as a motivation to Kant's model of the cognitive universe: instead of starting from appearances given in experience, Kant proposes to explore first the *a priori* contribution that the human cognitive subjects make

---

51 For a suggestion to read Hume's *Treatise* as foundational in this sense see Boehm M., "Hume's Foundational Project in the *Treatise*", *European Journal of Philosophy* (2013) doi: 10.1111/ejop.12056.

52 Hume, *Treatise* Introduction 4 and 6.

53 Blumenberg, *Genesis* 600–601.

to the experience available to them. A natural consequence of this Copernican perspective is Kant's *transcendental method*, which is used to explore the *normative* constraints that the mind must conform to if it is to represent things and make judgments as it in fact does.

Kant's method is discontinuous with that of experimental natural philosophy. It is exploited to explore the *a priori* constraints and possibilities of cognition, including any empirical cognition, and this investigation yields knowledge of our cognitive capacities, and not the world of external objects in itself, because 'we can cognize of things *a priori* only what we ourselves have put into them'.[54] Applying the transcendental method can thus provide foundational knowledge, i.e. knowledge about the conditions of possibility of any human knowledge, and it means to 'treat the laws that make possible the concept of a nature in general, even without relation to any determinate object of experience, and thus undetermined with respect to the nature of this or that thing in the sensible world'. Thus the *a priori* method with which this foundational knowledge is pursued aims at revealing what the empirical study of nature presupposes, and therefore it 'must always contain solely principles that are not empirical'.[55] The transcendental method therefore is not a method to be generalised with respect to other fields of inquiry. It belongs exclusively to the *a priori* exploration of the conditions to which any inquiry must necessarily be subordinated.

For Hume, in contrast, there is no methodological divide between the study of nature and the conditions of human cognition. This is due to Hume's commitment to the empirical study of both nature and human nature: natural and moral philosophy (the latter being preoccupied with the study of phenomena belonging to moral beings) are methodologically continuous fields of study. This continuity is based on the fact that the phenomena they study are part of the same causal order. In explanations these phenomena are referred to as equal members of the same causal chain, as Hume's famous example of a prisoner shows: his hopes for freedom are equally frustrated by the physical properties of the bars and the determination of the guard[56]—natural and moral properties concur in making the punishment inevitable. Our reasoning about moral and natural phenomena is thus continuous. Inquiry in both fields of study is based on the idea of a necessary connection that arises from the impression we acquire due to experiencing constant conjunctions between

---

54 Kant, *Critique* B xviii.

55 Kant Immanuel, *The Metaphysical Foundations of Natural Science*, ed. M. Friedman (Cambridge: 2004) 5.

56 Hume, *Human Understanding* 8.19.

phenomena. Our natural causal reasoning is based on this necessity, and this is the foundation of all theoretical causal cognition concerning the moral and the natural world.[57] Among the phenomena studied by moral philosophy from human cognition to morality and sociability one can find constant conjunctions and there are also exceptions to the observed regularities—just like in natural philosophy.

Therefore, there is no special method reserved for studying the conditions of human cognition. If our focus is on any aspect of human nature, human history provides us with the variation of circumstances in which the characteristics of human cognition can be identified and studied so as to establish the principles of its causal contribution in particular situations. The principles of human nature that Hume endeavours to explore belong to the hidden parts of nature that can be explored by a method of qualitative analysis and synthesis.[58] Human nature is a compound entity whose ingredients can be revealed only by the experimental method of *reasoning.* This methodological commitment is also fairly clearly expressed in the Abstract of the *Treatise*:

> 'tis at least worth while to try if the science of *man* will not admit of the same accuracy which several parts of natural philosophy are found susceptible of. There seems to be all reason in the world to imagine that it may be carried to the greatest degree of exactness. If, in examining several phaenomena, we find that they resolve themselves into one common principle, and can trace this principle into another, we shall at last arrive at those few principles, on which all the rest depend. And tho' we can never arrive at the ultimate principles, 'tis a satisfaction to go as far as our faculties will allow us.[59]

The method is thus simple. It consists in collecting relevant phenomena, finding analogies between them, and ascribing those analogies to similar causes, thereby reducing a variety of phenomena to regular principles that inform them. But our knowledge cannot transcend what we can infer on an empirical and analogical basis from the effects themselves—and this diagnosis applies to our knowledge of the conditions of human cognition too. On Hume's account there is no way of acquiring the *a priori* knowledge that Kant after his Copernican turn aims to deliver. On the contrary: the core of Hume's Copernican turn consists in the commitment to the exploration of empirically

57 Ibid. 8.5.

58 See Demeter, "Hume's Experimental Method".

59 Hume, *Treatise* Abstract 1.

accessible principles of human cognition—among other principles of human nature.

Although their projects are similar in aspirations, the methods Hume and Kant follow are different, and this explains the difference between the uses they make of Copernicus. For Hume he becomes a symbol of the *methodological* renewal of natural philosophy, and in that role he provides an inspiration for reforming moral philosophy so as to raise its cognitive value to the level of natural philosophy. For Kant, in contrast, Copernicus's relevance is not at all methodological; rather, he becomes the symbol of a new *perspective*, whose model of the universe is transformed into an inspiring metaphor for a new model of the cognitive universe. Although Copernicus did not aspire to be a revolutionary in the modern sense of the term—instead he wanted to restore something that had been lost—,[60] his example provided inspirations for Hume and Kant for exploring human cognitive capacities in radically novel ways, albeit in rather different directions.[61]

Selective Bibliography

Barfoot M., "Hume and the Culture of Science", in Stewart M.A. (ed.) *Studies in the Philosophy of Scottish Enlightenment* (Oxford: 1990) 151–190.

Beck L.W., *Essays on Kant and Hume* (New Haven: 1978).

Beebee H., *Hume on Causation* (London: 2006).

Blumenberg H., *The Genesis of the Copernican World* (Cambridge, Mass.: 1985).

Boehm M., "Hume's Foundational Project in the *Treatise*", *European Journal of Philosophy* (2013) doi: 10.1111/ejop.12056.

Capaldi N., *Hume's Place in Moral Philosophy* (Frankfurt a.M.: 1989).

Dauer F.W., "Towards a Copernican Reading of Hume", *Nous* 9 (1975) 269–295.

Dear P., *Discipline and Experience* (Chicago: 1995).

Demeter T., "Hume's Experimental Method", *British Journal for the History of Philosophy* 20 (2012) 577–599.

Demeter T., "Enlarging the Bounds of Moral Philosophy: Newton's Method and Hume's Science of Man", in Biener Z. – Schliesser E., *Newton and Empiricism* (Oxford: 2014) 171–204.

Dicker G., *Hume's Metaphysics and Epistemology* (London: 1998).

60 Kuhn T.S., *Copernican Revolution* 135; Dear, *Discipline and Experience* 118–119.

61 My research has been supported by the Hungarian Research Fund (OTKA 79193), and by the European Union and the State of Hungary, co-financed by the European Social Fund in the framework of TÁMOP-4.2.4.A/2-11/1-2012-0001 'National Excellence Program'.

Force J.E., "Hume's Interest in Newton and Science", *Hume Studies* 13 (1987) 166–216.

Fordyce David, *The Elements of Moral Philosophy*, ed. T.D. Kennedy (Indianapolis: 2003).

Gaukroger S., *The Emergence of a Scientific Culture: Science and the Shaping of Modernity 1210–1685* (Oxford: 2006).

Gregory David, *Elements of Catoptrics and Dioptrics* (London, E. Curll, J. Pemberton et al.: 1715).

Gregory David, *The Elements of Physical and Geometrical Astronomy*, trans. M. Stone, 2 vols. (London, D. Midwinter: 1726).

Hume David, *An Enquiry concerning Human Understanding*, ed. T. Beauchamp (Oxford: 2000).

Hume David, *An Enquiry concerning the Principles of Morals*, ed. T. Beauchamp (Oxford: 1998).

Hume David, *Dialogues Concerning Natural Religion and Other Writings Dialogues*, ed. D. Coleman (Cambridge: 2007).

Hume David, *The History of England from the Invasion of Julius Caesar to the Revolution in 1688*, foreword by W.B. Todd, 6 vols. (Indianapolis: 1983).

Hume David, *The Letters of David Hume*, ed. J.Y.T. Greig (Oxford: 1932).

Hume David, *Treatise of Human Nature*, ed. D.F. Norton – M.J. Norton, 2 vols. (Oxford: 2007).

Hutcheson Francis, *An Inquiry into the Original of Our Ideas of Beauty and Virtue*, ed. W. Leidhold (Indianapolis: 2008).

Hume David, *Philosophiae Moralis Institutio Compendiaria with A Short Introduction to Moral Philosophy*, ed. I. Turco (Indianapolis: 2007).

Jardine N., *The Birth of History and Philosophy of Science* (Cambridge: 1984).

Kant Immanuel, *Critique of Pure Reason*, ed. P. Guyer – A.W. Wood (Cambridge: 1998).

Kant Immanuel, *The Metaphysical Foundations of Natural Science*, ed. M. Friedman (Cambridge: 2004).

Keill John, *An Introduction to Natural Philosophy: or Philosophical Lectures Read in the University of Oxford, Anno Dom. 1700. To which are added, the demonstrations of Monsieur Huygens's theorems, concerning the centrifugal force and circular motion. Translated from the last edition of the Latin*, 4th edn. (London, M. Senex, W. Innys et al.; 1740).

Keill John, *An Introduction to the True Astronomy: or Astronomical Lectures, Read in the Astronomical School of the University of Oxford* (London, Bernard Lintot: 1721).

Kuhn T.S., *The Copernican Revolution* (Cambridge, Mass.: 1972).

Lakatos I. – Zahar E., "Why Did Copernicus's Research Programme Supersede Ptolemy's?", in Lakatos I. (ed.), *The Methodology of Scientific Research Programmes* (Cambridge: 1978) 168–192.

Leunissen M., *Explanation and Teleology in Aristotle's Science of Nature* (Cambridge: 2010).
Locke John, *An Essay concerning Human Understanding*, ed. P.H. Nidditch (Oxford: 1975).
Mounce H.O., *Hume's Naturalism* (London: 1999).
Noxon J., *Hume's Philosophical Development* (Oxford: 1973).
Pitson A.E., *Hume's Philosophy of the Self* (London: 2002).
Price H.H., *Hume's Theory of the External World* (Oxford: 1963).
Raphael D.D., "Newton and Adam Smith", in Sweet Stayer M. (ed.), *Newton's Dream* (Montreal: 1988) 36–49.
Russell J.L., "Copernican System in Great Britain", in Dobrzycki J. (ed.), *The Reception of Copernicus' Heliocentric Theory* (Boston: 1973) 189–240.
Schliesser E., "Copernican Revolutions Revisited in Adam Smith by way of David Hume", *Revista Empresa y Humanismo* 13 (2010) 213–248.
Serjeantson R., "Hume's General Rules and the 'Chief Business of Philosophers'", in Frasca-Spada M. – Kail P.J.E. (eds.), *Impressions of Hume* (Oxford: 2005) 187–212.
Shepherd C.M., "Newtonianism in Scottish Universities in the Seventeenth Century", in Campbell R.H. – Skinner A. (eds.), *The Origins and Nature of the Scottish Enlightenment* (Edinburgh: 1982) 65–85.
Smith Adam, "The History of Astronomy", in idem, *Essays on Philosophical Subjects* (*1795*), ed. W.P.D. Wightman – J.C. Bryce (Indianapolis: 1982) 33–105.
Stewart M.A., "Hume's Intellectual Development, 1711–1752", in Frasca-Spada M. – Kail P.J.E. (eds.), *Impressions of Hume* (Oxford: 2005), 11–58.
Stroud B., *Hume* (London: 1977).
Turnbull George, *The Principles of Moral Philosophy*, ed. A. Broadie (Indianapolis: 2005).
Waxman W., *Hume's Theory of Consciousness* (Cambridge: 1994).
Wertz S.K., "Hume and the Historiography of Science", *Journal of the History of Ideas* 54 (1993) 411–436.
Wilson D.B., *Seeking Nature's Logic: Natural Philosophy in the Scottish Enlightenment* (University Park, Pa.: 2009) 1–32.

PART 2

New Astronomy: Textual and Graphic Transformations

CHAPTER 5

Arguing for One's World. Copernicus's Theories and Their Reception in Jean Bodin's *Theatrum*

Jonathan Schüz

By using mathematics to describe the world, the bulk of Copernicus's *De Revolutionibus Orbium Coelestium* relies on what Aristotle calls logical syllogisms. Jean Bodin's rejection of Copernicus's theories in his *Theatrum* does not only rely on his belief that Copernicus is a threat to Christian dogmas, Bodin also opposes Copernicus's approach. Bodin's world is structured around a Neoplatonic "Chain of Being". The differences between these two world images are therefore not only apparent at the level of facts and statements but are also inseparable from the rhetorical strategies the authors rely on, which are logical syllogism in the case of Copernicus, and topical argumentation in Bodin's work. This essay examines how these methodological and rhetorical differences become manifest in the texts.

1

Bodin certainly ranks amongst the most prominent readers of Copernicus in the sixteenth century, even if he does not go along with him, being part of what Dorothy Stimson calls the 'gradual reception' of Copernicus.[1]

Writing about the reception of Copernicus in the Early Modern Period always bears the risk of making rather binary distinctions between 'modern' and 'traditional' descriptions of the universe, and especially in the case of Bodin it would seem his position is clear. I will try to avoid placing too much emphasis on this systematic notion of classification by looking beyond the content of Copernicus and Bodin's theories in order to understand his reception (and rejection) of Copernicus by focussing on the level of language.

By writing *De Revolutionibus*, Copernicus first of all created a text that, although it pretends to simply describe reality, needs to follow a specific set of

1 Stimson D., *The Gradual Acceptance of the Copernican Theory of the Universe* (Gloucester Mass.: 1972).

 | DOI 10.1163/9789004281127_007

rhetorical rules in order to convey its content and to convince the reader.[2] It is therefore impossible to separate the content (the facts and statements in the text) from the strategies and modes of presentation it employs, especially if these strategies of language do not only shape the scientific discourse but also the view of the world in general.

In this context, Bodin's *Theatrum* offers several aspects for comparison. Not only is he a prominent (and often quoted) example of the reception of Copernicus in the late sixteenth century; since he rejects Copernicus's theories outright, he also offers an alternative cosmology and view of the world structured according to rhetorical and topical functions.

2

The differences between Bodin's and Copernicus's texts do not merely consist in their differing views on cosmology. They can be located on the very basic level of language and modes of argumentation, which can be traced as far back as Aristotle: In his *Topikê*, Aristotle distinguishes two possible kinds of syllogisms.[3] On the one hand, he defines logical syllogisms, which are apodictic and rely on first (or primitive) and true sentences that cannot be further deduced. Roughly speaking, these syllogisms prove that, in a given system, something has to be the way it is by necessity. A classic example for a syllogism of this kind is the reasoning: Socrates is human, all humans are mortal, and therefore Socrates is mortal.

These syllogisms convince the audience because they cannot be questioned further and point out a seemingly necessary truth. From the viewpoint of rhetoric, the mathematical proof that Copernicus amasses in his work can be classified as logical syllogisms. However, *De Revolutionibus* also shows the problems of this strategy. For readers who are not well trained in the specific methodology of mathematics, the text's line of argumentation is hard to follow if not completely incomprehensible (even to a modern reader). Even if one is able to follow and understand the text in detail, the syllogisms only cover a short distance and lose their persuasive power as soon as larger contexts are concerned. The astronomical data, the measurements and calculations Copernicus presents were convincing for most members of the learned community of the

2 For a more detailed discussion of these 'performative' aspects of science see Nate R., "Inszenierte Wissenschaft. Rhetorik, Mythos und Pragmatik", in Hempfer K. – Traninger A. (eds.), *Macht Wissen Wahrheit* (Freiburg – Berlin: 2005) 183–200.

3 See Aristotele, *Topics* I, 1, 100a, 25–101a, 4.

time. The more wide-reaching deductions from that, among them the proclamation of a heliocentric universe, were significantly less so.[4]

In contrast to logical syllogisms, Aristotle describes dialectic syllogisms, which differ from them in their logical structure. They do not rely on true but rather on probable sentences rooted in commonly shared beliefs. Aristotle here speaks of *endoxa*, of opinions 'which are accepted by everyone or by the majority or by the wise—i.e. by all, or by the majority, or by the most notable and reputable of them.'[5] Therefore the best way to convince an audience by means of dialectic syllogisms is to demonstrate that the authorities share these beliefs, a strategy that can easily be found in Bodin, as will be shown later. While logical syllogisms point out a necessity but are convincing only for the short-term, dialectical syllogisms are exactly the opposite. Logically speaking, they remain fuzzy, since they cannot be traced back to some definite point, but only to beliefs held by more or less everyone. On the other hand, they can easily convince the audience because they never introduce something substantially new. They only link a given proposition with *endoxa*, with sentences the audience already believes in. Although necessarily blurry from a logical perspective, they nevertheless are by far more convincing.

It is obvious that I am simplifying here on a large scale. The thorough study of Goddu shows that Copernicus cannot simply be reduced to being a logician but has to be seen within the history of science, of dialectic and rhetoric.[6] However, I have chosen Aristotle's very basic distinction because it makes it possible to link rhetoric to the relevant questions of scientific methodology without trying to allocate either Copernicus or Bodin to any special school of dialectic in the sixteenth century. Keeping in mind that a scientific text, like any other text, needs to convince an audience of its propositions, it is appropriate to examine Copernicus and Bodin's texts from a rhetorical viewpoint.

---

4 See for example Kuhn T.S., *Die kopernikanische Revolution* (Braunschweig – Wiesbaden: 1982) 191.

5 Aristotele, *Topics* I,1, 100b, 18f. Translation taken from: Aristotle, *The Complete Works of Aristotle. The Revised Oxford Translation*, ed. J. Barnes (Princeton: 1984) vol. 1, 167.

6 Goddu A., *Copernicus and the Aristotelian Tradition. Education, Reading and Philosophy in Copernicus' Path to Heliocentrism* (Leiden – Boston: 2010), esp. part III 275–386. It can even be argued that the diagrams Copernicus uses to explain geometrical phenomena have to be located within the history of rhetoric and have become topical to a extent in the sixteenth century, cf. Müller K., "Vom Absehen und Hinschauen. Modifikationen des astronomischen und kosmologischen Diagramms an der Wende vom Hoch- zum Spätmittelalter", in Dickhut W. – Manns S. – Winkler N. (eds.), *Muster im Wandel. Zur Dynamik topischer Wissensordnungen in Spätmittelalter und Früher Neuzeit* (Göttingen: 2008) 193–215.

2.1

Copernicus strongly asserts that he has written a mathematical book. On the very title page we encounter the mythological inscription of Plato's academy in Athens, which forbids entry for anyone who does not know geometry. In the *praefatio*, which dedicates the text to Pope Paul III, he braces himself for the attacks of non-mathematicians:

> If there are possible boasters who, ignorant of mathematics, still dare to judge the text on the grounds of some misinterpreted passage from the Bible, I will not waste time taking them seriously, since I think little of their rash judgement. [...] It should not take the studied man wonder that such people laugh about us. Mathematics is written for mathematicians [...].[7]

One might think that this bold statement should serve as a *caveat* to keep anyone who is not an expert from judging the text: Copernicus will not waste time considering comments on his text which are not formulated by studied mathematicians and which are based on rhetorical strategies other than those of mathematics, e.g. on misquotations from the Bible.[8]

Indeed, the reception of *De Revolutionibus* shows that the text was taken seriously as far as mathematical and astronomical data were concerned. Even Bodin, in his *Methodus*, mentions Copernicus in the eighth chapter (*De temporis universi ratione*) as a famous mathematician who, with his calculations of the distance of earth and sun, cleared the ground in a lucid manner for generations to come.[9] Of course, this citation also points out where Copernicus

---

7 'Si fortasse erunt ματαιολόγοι qui, com omnium mathematum ignari sint, tamen de illis iudicium sibi summunt, propter aliquem locum Scripturae, male ad suum propositum detortum, ausi fuerint meum hoc institutum reprehendere ac insectari, illos nihil moror, adeo ut etiam illorum iudicium tamquam temerarium contemnam. [...] Itaque non debet mirum videri studiosis, si qui tales nos etiam ridebunt. Mathemata mathematicis scribuntur [...].' Copernicus Nicolaus, *De revolutionibus orbium coelestium, Libri IV* (Nuremberg, Johannes Petreius: 1543) fol. iv–v.

8 In his edition of the text, Zekl points out that the expression *mathematicus* referred to exact sciences in a broader sense; cf. Copernicus Nicolaus, *Das neue Weltbild*, ed. H.G. Zekl (Hamburg: 1990) 221. However, as it will be shown later on, Bodin takes this expression in his 'Theatrum' literally and offers a differentiation of the exact sciences of mathematics and mechanics; cf. Blair A., *The Theatre of Nature. Jean Bodin and Renaissance Science* (Princeton: 1997) 40.

9 'Sed ex omnibus nihil admirabilius est, quam quod ad sempiternam posteritatis memoriam, Copernicus in libris Revolutionum, deinde Reinholdus, post etiam Stadius, methematicis

seems to fall short. The deductions from these data he presents, the heliocentric universe, are not mentioned at all in this early work by Bodin. Here, he sticks to the figures.

Despite his announcement in the *praefatio*, it quickly becomes obvious that Copernicus tries to follow several rhetorical strategies. Especially in the first book of *De Revolutionibus*, he tries to use a more topical argumentation and backs the mathematical line of reasoning by commonplace arguments whenever possible. In order to convince his audience, he 'had to affirm his continuity with the tradition'.[10] In order to prove this continuity, he quotes *endoxa*, which he normally mentions before beginning his own argumentation.[11]

A very striking and often quoted example can be found in chapter 10 on the order of the heavenly spheres (*De ordine caelestium orbium*). Here, Copernicus presents a diagram of his system *in toto*, starting from the outermost sphere of fixed stars and travelling to the centre, to the position of the sun:

> In the very middle of everything resides the sun. Who would put the lamp in this wonderful sanctuary in another or better place, from which it could illuminate everything? Even more, since it is quite aptly called 'light of the world' by some people, 'mind' or 'guide' by others. Trismegistus [calls it] 'visible god', Sophocles the 'all-seeing light'. Therefore the sun rules, as it resides on the royal throne, the family of stars that is moving around it.[12]

This passage seems to contradict Goddu's statement that Copernicus often subordinates natural-philosophical hypotheses to mathematical and

---

nobiles, perspicuis demonstrationibus docuerunt [...]'. Bodin Jean, *Methodus ad facilem historiarum cognitionem* (Amsterdam, Johannes Ravelstein: 1650; reprint Aalen: 1967) 341. A similarly positive remark on Copernicus can be found later in the same chapter, in my edition on p. 347. Bodin here supports Copernicus's claim, following Melanchthon, that the sun had moved closer to the earth since the time of Ptolemy. In his 'Theatrum', he reverses this position; cf. Blair, *Theatre* 249 and Lindberg D. *The Beginnings of Western Science. The European Tradition in Philosophical, Religious, and Institutional Context, Prehistory to A.D. 1450* (Chicago – London: 2007) 260.

10 Goddu, *Copernicus* 302.

11 See ibidem 302–314.

12 'In medio vero omnium residet Sol. Quis enim in hoc pulcerrimo templo lampadem hunc in alio vel melioro loco poneret, quam unde totum simul possit illuminare? Siquidem non inepte quidam lucernam mundi, alii mentem, alii rectorem vocant. Trismegistus visibilem deum, Sophoclis Electra intuentem omnia. Ita profecto tamquam in solio regali Sol residens circumagentem gubernat astrorum familiam.' Copernicus, *De Revolutionibus* civ.

astronomical hypotheses, even if this is not meant as a general methodological principle.[13] Copernicus here backs his theory of the heliocentric universe by quoting Sophocles as a classical humanist authority, and even mentioning Hermes Trismegistus, following the Neoplatonic ideas of his time.[14] I think it is more appropriate to attribute passages like this to the rhetorical principle of *copia*, which ranks the mass of argumentation above its logical coherence.

The fact that the text changes its rhetorical strategy in several places shows first of all that Copernicus is not writing exclusively for mathematicians (in contrast to what he says in the *praefatio*), but that he tries to link his ideas and results to the contemporary horizon of knowledge whenever possible. If it still was necessary to prove that Osiander's preface doesn't support the main idea of the text,[15] these passages show that the text indeed tries to describe reality, and that it tries to strengthen its argumentation by constantly pointing out where it goes along with what is generally believed about the earth and the universe, thus making it easier for readers to agree.

By using different strategies, the text also shows quite clearly the limited effect of logical syllogisms: they are convincing on a small scale, as far as numerical details are concerned, but fail (at least at first) to persuade readers to change their general views of reality.

Fellmann writes that Copernicus, in attempting to outline his picture of the universe, is supported and strengthened by the mathematical logic of astronomy ('gestärkt durch die mathematische Logik der Astronomie').[16] However, this seems to be true only up to a certain point. Mathematical logic is indeed an important strategy for Copernicus, in fact it is the only way to give proof to his theories. But at least in the sixteenth century, this does not seem to lead to a strong and convincing position but to a weak and defensive argumentation that is still able to persuade the audience on a large scale.

---

13 See Goddu, *Copernicus* 317.

14 Hamel J., *Nicolaus Copernicus. Leben, Werk, Wirkung* (Heidelberg: 1994) 241.

15 In his preface *Ad Lectorem*, Osiander chose what Goddu calls a 'pragmatic view' (Goddu, *Copernicus* 420): The mathematical deductions in 'De Revolutionibus' are to be taken seriously, since they make sense within their system. But for that, it is not necessary that they are true, they don't even have to be likely, but it is enough that they form a calculation which fits to the observations ('Neque enim necesse est, eas hypotheses esse veras, immo ne verisimiles quidem, sed sufficit hoc unum, si calculum observationibus congruentem exhibeant;' Copernicus, *De Revolutionibus* Fol. iv.). It is therefore not necessary to consider the result of Copernicus's theories for the view of the universe.

16 Fellman F., *Scholastik und kosmologische Reform. Studien zu Oresme und Kopernikus* (Münster: 1988) IX.

If one takes a closer look at Copernicus's statement in the dedication to the Pope, this defensive position quickly becomes evident. Although he at first boldly states that he writes exclusively for an audience schooled in mathematics and logic, this position is immediately undermined by the following passage in which he tries to advertise the text for reasons that have at first little to do with the heliocentric universe:

> Mathematics is written for mathematicians, who will, if I'm not mistaken, see this work as an advantage for the ecclesiastical community [...]. It has not been long since under Leo X., during the Lateran Council, the question of improving the calendar was debated. The matter has not been decided upon, only because of the reason that the length of years and months as well as the movements of the sun and the moon had not been measured exactly enough. Since that time I have directed my attention to the accurate observation of these things [...]. What I can offer regarding this matter I leave first of all to the judgement of Your Holiness, and then to the judgement of all other studied mathematicians. And, in order not to seem to promise Your Holiness more of the usefulness of this work than I can keep, I now go on with my proceedings.[17]

Copernicus here advertises the usefulness of his data for the reform of the calendar, which had been planned, but not actually been put into effect by the fifth Lateran Council (1512–1517). Copernicus most likely had been asked to give his expertise. And in fact his data were used in the calendar reform under Pope Gregory XIII in 1582.[18] What is surprising now is the fact that Copernicus seems to locate the advantage of his text for the church not so much in a possible correction of the picture of the universe, which would be the result of his observations and calculations, but more in the raw data he provides to back his argumentation which can be used separately from his deductions,

---

17 'Mathemata mathematicis scribuntur, quibus & hi nostri labores, si me non fallit opinio, uidebuntur etiam Reipub. ecclesiasticae conducere aliquid [...]. Nam non ita multo ante sub Leone x com in Concilio Lateranensi uersabatur quaestio de emendando Calendario Ecclesiastico, [quae] tum indecisa hanc solummodo ob causam mansit, quòd annorum & mensium magnitudines atque Solis et Lunae motus nondum satis dimensi haberentur. Ex quo equidem temporem hos accuratius obseruandis animum intendi [...]. Quid autem ea in re, tuae Sanctitatis praecipue atque omnium aliorum doctorum mathematicorum iudicio relinquo, & ne plura de utilitate operis promittere tuae Sanctitari uidear quam praestare possim, nunc ad institutum transeo.' Copernicus, *De Revolutionibus* fol. iv. v.

18 For a detailed description of Copernicus's role in these reforms see Carrier M., *Nikolaus Kopernikus* (Munich: 2001) 137–143.

possibly even contradicting them: the very fact that Copernicus here mentions the movement of the sun (which, according to him, does not exist at all) shows that his measurements do not necessarily conflict with the traditional, Ptolemaic universe.

It seems quite telling that this turning point in *De Revolutionibus* does not only appear in the later reception of the work, but is also mentioned and planned in the text itself. It becomes obvious that the text is very much aware of the fact that mathematical logic has very little influence whatsoever on reality outside the system; its mathematical and logical mode of argumentation are simply a necessity, but are nonetheless not an entirely convincing rhetorical strategy.

2.2

This dilemma can be seen in Bodin's reception of Copernicus. Although Bodin mentions Copernicus as a serious mathematician in his *Methodus*, he outright refutes his mathematical deductions in his *Theatrum* of 1596. In Book Five, he presents a considerable list of arguments against Copernicus's theories, 'which can easily be refuted by their shallowness':[19]

> The human mind cannot grasp the incredible speed of the heavenly spheres and especially of the tenth sphere which must be ten times greater than the eighth, for in twenty-four hours it must traverse 469, 562, 845 miles, so that earth seems like a dot in the universe. This is the chief argument. Besides this, we get rid entirely of epicycles in representing the motion of the planets [...]. There is one argument which they [i.e. other critics] have omitted but which seems to me more efficacious than any, viz.: rest is nobler than movement, and that celestial and divine things have a stable nature while elementary things have motion, disturbance and unrest; therefore it seems more probable that the latter move rather than the former.[20]

---

19 '[...] sed sua leuitate facile refutari potest [...]', cf. Bodin Jean, *Universae naturae theatrum. In quo rerum omnium effectrices causae, & fines contemplantur, & continuae series quinque libris discutiuntur* (Paris, Jacques Roussin: 1596) 580. For the 'Theatrum' quote the translation of Dorothy Stimson, cf. Stimson, *Gradual Acceptance* 45–47.

20 'Quod mentis humanae captum excederet incredibilis orbium coelestium rapiditas, ac potissimum orbis decimi, quem oportet decies maiorem esse orbe octauo, qui tamen horis 24. consicit milliara 469562845. terra vero quasi punctus as vniuersitatem esse videatur. Hoc argumentum praecipuum est: praeterea nullis omnio Epicyclis egeremus

What is presented here is a complex mixture of different strategies of argumentation. For his chief argument, Bodin relies on scepticism: the human mind's understanding of celestial things is very much limited, as can easily be exemplified by referring to the speed of the heavenly spheres. Therefore, one can deduce, it seems futile to use mathematics in order to explain the universe because it can only present numbers but cannot link them to a more common understanding.

The second argument points out the basic problem of the reception of Copernicus: his theories, as all heliocentric systems in the Early Modern Period, go against most things that were taught about astronomy and which generally were taken for granted, here exemplified by the system of epicycles. As a result, one can add, authorities and *endoxa* are lacking, and for those reasons these theories are unconvincing.

Bodin's strategy of argumentation is perhaps best exemplified by the third argument: because rest is nobler than movement, it seems unlikely that earth and with it the divine creation it carries is in fact moving. Here, Bodin refers to *endoxa* in the formulation of a gnomic sentence that does not seem to be in need of any further proof.

And even if this argument fails to convince, it can be easily supplemented by the *copia* of arguments he further presents to argue against Copernicus in more detail. In order to uphold his theories, Copernicus claims that the earth has three or four different motions, whereas in an Aristotelian universe it is clear that a natural body can only have one motion of its own.[21]

These arguments draw on an educated background and seem to be directed at a learned reader. They are backed by the following argumentation, which lists the well-known statements against the movement of the earth:

> No one in his senses or imbued with the slightest knowledge of physics will ever think that the earth, heavy and unwieldy from its own weight

---

ad planetarum motiones denotandas [...]. Vnum est quod illis omniserunt, quo tamen nullum efficacius mihi videtur, scilicet, quietem motu praestabiliorem esse: coelestia ac diuina naturam habere constantem, elementaria mobilem, perturbatam, & inquietam: ac propterea probabilius videre possit, vt haec prorius quam illa moueantur.' Bodin, *Theatrum* 581.

21 'Copernicus terram triplici motu cieri posuit, vt suam hypothesim tueretur [...]: quibus si adungamus motum grauitatis ad centrum, quattuor motiones naturales vni & eide corpori tribuemus, quo dato omnia Physicorum fundamenta ruere necesse est: omnes enim in eo conspirant vnius corporis naturalis vnum esse proprium motum [...].' Bodin, *Theatrum* 581.

> and mass, staggers up and down around its own centre and that of the sun; for at the slightest jar of the earth, we would see cities and fortresses, towns and mountains thrown down. [...] For if the earth were to be moved, neither an arrow shot straight up, nor a stone dropped from the top of a tower would fall perpendicularly but either ahead or behind. With this argument Ptolemy refuted Eudoxus.[22]

The last argument that Bodin presents is taken from the unquestionable authority of the Bible: God himself, the Lord of Wisdom, said that the earth in truth stays ('stat') forever.[23]

Bodin presents his arguments in an order that supports his rhetorical strategy: starting with his chief argument that no human mind can in fact comprehend the movements of celestial bodies, he undermines the mathematical argumentation of Copernicus. And even if one holds onto the argument regarding the limits of the human mind, Copernicus's theories collide with most of what is thought about the universe; in this manner the second and third argument can be read.

This argumentation is followed by a passage referring to *experientia*:[24] everyone knows that on a moving earth nothing could remain fixed. The conclusion aligns this argumentation, based so far more or less on human knowledge, with the divine truth: If mankind, learned or not, goes along with what is written in the Bible, the statement agreed upon must in fact be true.

With this last quotation from the Bible, Bodin makes possible his sceptical approach to the matter: to point out that the human mind is fallible is at first a weak argumentation since Bodin himself refers to common sense most of the time. By showing that divine truth agrees with him on that point, he is able to transgress the limitations of human reason and to reach absolute truth.

---

22 'Terram vero grauitate sua ac pondere brutam ac solidam, sursum deorsum, circa suum centrum, & circa Solem simul ac titubare motu rapitari nemo sanus aut levuissime Physicis disciplinis imbutus vmquam existimabit, cum leuissima terrae concussione vrbes, arces, oppida, montes subrui videamus [...]. Nam si terra moueretur nec sagitta rectu motu sursum emissa, nec lapis e sublimi turre demissus ad perpendiculum caderet, sed prorsum aut retrorsum. Hoc etiam argumento Ptolemaeus Eudoxus refellit.' Bodin, *Theatrum* 582.

23 The biblical sources Bodin supplements in the margin are Ecclesiastes 1 and Psalm 19.

24 In the 'Theatrum', Bodin frequently refers to *experientia* to prove his point. The expression refers, as Blair puts it, 'to an undifferentiated, medieval mix of hearsay, bookish learning, and actual experience.' Blair, *Theatre* 96. See also Greengrass M., "The experiential world of Jean Bodin", in Lloyd H. (ed.), *The reception of Bodin* (Leiden – Boston: 2013) 67–96.

2.3

Although the limitation of the human mind that Bodin refers to had itself become part of a topical argumentation in the sixteenth century, I still want to take a closer look at it, since the *Theatrum* is in fact not the only text in which he makes use of this 'chief argument'.

In the last book of the *Démonomanie*, the *Refutation des opinions de Jean Wier*, a very similar passage can be found, yet in a completely different context. As the title suggests, Bodin in this part of the text directs all his knowledge and argumentative fervour towards a complete refutation of the theories and the person of Johann Weyer, one of the most prominent opponents of witch-hunts in the sixteenth century.[25]

In the 1560s and 1570s, Weyer argued in two publications, *De Praestigiis Daemonorum* and *De Lamiis*, against the common juridical practice of witch trials. The basic argument, which Bodin opposes strongly, is that the deeds witches confess to most of the time are not real but mere melancholic dreams. One example of those dreams is the transportation of witches to other places, which must happen at an impossible speed. According to Bodin, Weyer here proves to be a bad mathematician as well as a bad physicist:

> Puis apres il dict que le transport d'icelles aux assemblees est impossible par nature, & en si peu de temps. [...] Et neantmoins Wier monstre bien qu'il est aussi mauuais Mathematicien, comme Phisicien.[26]

In the argumentation that follows this statement, Bodin argues that there is little sense in continuing to refer to the human mind with regard to natural law, since the universe is full of objects moving at an incomprehensible speed. Here, the eighth sphere is his prime witness: 'Car on voit l'huitiesme ciel auec tous les astres faire son tour en xxiiii heures, lequel tour a plus de cent trente & trois millions de lieuës à deux mil pas la lieuë Geometrique.'[27] In a later calculation, Bodin makes this distance even bigger, adding that regarding the ninth and tenth spheres, these numbers slip completely out of proportion:

> [...] qui est pour tout le circuit du ciel huictiesme, deux cens quarante & cinq millions sept cens nonante & nonante & vn mil quatre cens quarante

25 For a detailed discussion of this topic see Scholz W.G., *Defining Dominion. The Discourses of Magic and Witchcraft in Early Modern France and Germany* (Ann Arbor: 1995) 65–87.

26 Bodin Jean, *De la démonomanie des sorciers* (Paris, Jacques du Puys: 1581) 243v.

27 Bodin, *Démonomanie* 244r.

> lieuës, qui se font en vingt & quatre heures. Le neuf & dixiesme ciel sont encores plus grands.

As in the 'Theatrum', he emphasises the incomprehensibility of these numbers by comparing them to the earth, which, according to Ptolemy and despite its own vast diameter, now seems little more than a negligible dot in the universe: 'Car il est tresbien demonstré par Ptolomee en son Almageste, que toute la terre qui a onze mil cens soixante lieuës de tour, n'est rien que vn point insensible [...].[28]

Bodin's main point is to show that these numbers are overwhelming, despite being mathematically correct. They can be proved but understood, in the sense of being linked to a common horizon of experience.

Bodin includes a discussion of scepticism, under the title of Pyrrhonism, in the preface of the *Démonomanie*. As Popkin points out, this passage is a very early example of the reception of Pyrrhonism after the publication of the *Hypotyposis*' of Sextus Empiricus in 1562.[29] Since even in the sixteenth century it is difficult, if not completely impossible, to prove the existence of witches, Bodin has to find a middle course between a crude mode of empiricism (which he here attributes to Aristotle) and the total scepticism of Pyrrho. The way out of this dilemma is what he calls the common-sense theory of Theophrastus:

> Il fault [!] donc s'arrester à l'opinion de Theophraste, qui a recours au sens commun, qui est moyen entre les sens & l'intellect, & rapporter à la raison comme à la pierre de touche ce qu'on aura veu, ouy, gousté senty. Et d'autant plus qu'il y a des choses si hautes, & si difficiles à comprendre qu'il n'y a que peu d'hommes qui en soyent capables: en ce cas il faut croire chacun en sa science.[30]

The common sense that Bodin refers to is an anatomically exactly located part of human understanding that mediates between direct sense experience and

28 Bodin, *Démonomanie* 244v.

29 Popkin R., *The History of Scepticism. From Savonarola to Bayle* (Oxford: 2003) 77.

30 Bodin, *Démonomanie* 12r. For a more detailed discussion of scepticism in the 'Démonomanie' see Schüz J., *Johann Fischarts Dämonomanie. Übertragungs- und Argumentationsstrategien im dämonologischen Diskurs des späten 16. Jahrhunderts* (Diss. Berlin: 2011) 23–38 [URL: http://www.diss.fu-berlin.de/diss/receive/FUDISS_thesis_000000021283]; as well as Krause V., "Listening to witches: Bodin's use of confession in De la démonomanie des sorciers", in Lloyd H. (ed.), *The reception of Bodin* (Leiden – Boston: 2013), 97–115.

the intellect, using human reason as a touchstone. However, there are things that go beyond individual reason (like witches or the universe). It is then necessary to believe scientists in their area of expertise.

This statement opens the field for an argumentation that relies on *endoxa* and authorities as the main source of probability and persuasiveness. This strategy can be found in the 'Démonomanie' as well as in the 'Theatrum', as I have already shown.

3

The question remains as to why Bodin does not seem to believe Copernicus even when operating in his area of expertise, the field of mathematics. As he describes Copernicus as a decent mathematician in the *Methodus*, this rejection seems even more questionable. The reason, however, is as simple as it seems to be paradoxical at first: 'The disputation of the celestial motions is the purview of the physicist, for nothing is more appropriate to the physicist than motion and nothing more foreign to mathematicians than motion.'[31] This distinction becomes clear in a later passage:

> Mathematicians by thinking separate numbers for their own sake, as well as points, lines and surfaces from bodies, and they stay as far away as possible from motion: but *physici* follow motion and the senses, although the senses are very often wrong.[32]

Bodin here acknowledges that the study of mathematics is more certain since it operates from an abstract point independent of the fallibility of human senses and flaws in observation or measurement. The study of physics always runs the risk of employing false data; however, it remains close to the world of objects and is able to make statements about things mathematics cannot reach: 'The *Theatrum* thus offers only descriptive and casual physics and astronomy with no use of mathematics.'[33]

---

31 '[...] coelestium motuum disputatio Physici propria sit, nihil enim Physici tam proprium est quam motus, nihil à Mathematici motu magis alienum' Bodin, *Theatrum* 3. I here quote the translation of Ann Blair, cf. Blair, *Theatre* 31.

32 'Mathematici numeros per sese, ita puncta, lineas, superficies à corporibus cogitatione segregant, & ab omni motu quàm longissimè recedunt: Physici verò motum ac sensum sequuntur, sensus verò saepissimè fallunt;' Bodin, *Theatrum* 550; cf. Blair, *Theatre* 41.

33 Blair, *Theatre* 41.

In this passage, the underlying difference between Bodin and Copernicus becomes obvious: Copernicus states explicitly that he started pondering about a moving earth because the calculations he had found so far were unsatisfactory: 'Nothing else had made me think about a different movement of the spheres of the earth than the understanding of the fact that the mathematicians themselves do not agree on their examinations.'[34] Copernicus's aim, as stated here, is to correct those mathematical deficiencies in order to create a more exact and thus coherent science of the study of celestial motions. And as a result of that, he had to formulate his theory of a heliocentric universe because that seemed to be the only way the observations and calculations would actually converge.

Bodin does not try to match figures. The subtitle of his *Theatrum* promises a discussion of the phenomena of nature, focussing on the *topoi* of effect and cause.[35] In order to achieve this aim, he needs to rely on his version of physics: 'Mathematics cannot provide casual understanding of real phenomena, which is the most valuable type of knowledge.'[36] Instead of using logical deductions and syllogisms, he implies a system of ten *hypostases*, which, in a hierarchical order of complexity and nobility, contain the whole creation, starting from ash as matter without form, and ending with God.[37] Bodin thus presents the world structured according to rhetorical *topoi*, the common humanist way to organise knowledge before the introduction of alphabetically organised encyclopaedia.[38] Bodin's system mirrors the *Great Chain of Being*[39] and makes it possible to locate the exact place of each phenomenon of nature.[40] By tracing this chain forwards or backwards, it is possible to display the coherence and order of the elements' creation, always keeping God as the final vanishing

34 '[...] me nihil aliud mouisse ad cogitandum de alia ratione subducendorum motuum sphaerarum mundi, quàm quod intellexi, Mathematicos sibi ipsis non constare in illis perquirendis.' Copernicus, *De Revolutionibus* Fol. iii r.

35 In the Frankfurt edition of 1596: 'Universae naturae theatrum. In quo rerum omnium effectrices causae, & fines contemplantur, & continuae series quinque libris discutiuntur.'

36 Blair, *Theatre* 41.

37 Cf. Blair, *Theatre* 32f.

38 See for example Schmidt-Biggemann W., *Apokalypse und Philologie. Wissensgeschichten und Weltentwürfe der Frühen Neuzeit* (Göttingen: 2007) 244.

39 For a discussion of this expression see Lovejoy A., *The Great Chain of Being* (Cambridge – London: 1976) 59 [first published in 1933] and Hintikka J., "Gaps in the Great Chain of Being: an Exercise in the Methodology of the History of Ideas", in Knuuttila S. (ed.), *Reforging the Great Chain of Being. Studies in the History of Modal Theories* (Dordrecht: 1981) 1–17.

40 'in ordine, ac proprie sedibus', Bodin, Theatrum 1.

point of creation in view: 'This chain descends from above, transmitting causation to each below level in a strictly hierarchical fashion.'[41]

Whereas Copernicus' study of the heavens tries to observe and describe the universe, Bodin's version seems more to be a mode of contemplation, always keeping the final hierarchy of creation in mind and ordering natural phenomena accordingly:

> These [observations of nature] bring us not only the sweetest pleasure, but also such a great desire for the Creator that, despite ourselves, stupefied and dumbfounded, we become seized of love for him.[42]

Physics, seen that way, 'forces the impious to acknowledge God.'[43]

4

Starting from Bodin's definition of the methodology and the aims of physics, it becomes obvious that Copernicus misses the point in a number of ways by stating that his text, being mathematical, is written solely for mathematicians. Perhaps first of all, he strays from his field of expertise by making statements about the structure of the universe based on mathematical proof. This implies, secondly, that he inverts the order of the sciences by ranking mathematics above physics and thus locating mathematical hypotheses above natural-philosophical hypotheses: To take astronomy as a mathematical science does 'not only detract from the honour due to natural science, as if withholding a roof from a house, but also wreak havoc on the mathematical arts entirely.'[44] Finally, Copernicus places his emphasis on a rather irrelevant field in trying to make astronomy more exact, pointing out that the descriptions of the universe have so far been deficient.

The physics Bodin implies in his study of the heavens has, as he himself puts it, little need for more exact data, even more since in Copernicus's case these data seem to lead to deductions that collide with almost everything which so far had been held true. By methodising the necessary fuzziness of human

---

41 Blair, *Theatre* 127.

42 'Haec non modo voluptatem nobis suavißimam afferunt, ut obstupefacti atque attoniti in illius amorem inuiti rapiamur', Bodin, *Theatrum* Sig. 4v; cf. Blair, *Theatre* 26.

43 Blair, *Theatre* 41.

44 '[...] non solùm decus scientiae naturalis, quasi tectum ab aedibus dextrarerunt: verumetiam Mathematicas artes omnino conturbarunt', Bodin, *Theatrum* 550; cf. Blair, *Theatre* 41.

observation and implying a topical systematisation of creation, Bodin's physics mirrors rhetorical strategies.

Used in concrete argumentation, the topical ordering systems, optimised to contain as many facts as possible and keeping them at one's disposal, become arsenals for arguments.[45] The final aim of a topical argumentation, however, can never be absolute truth (otherwise, an argumentation would be unnecessary). A topical argumentation generates *probabilitas*, which Schmidt-Biggemann translates as 'Zustimmungseignung', the fact that it is possible to agree with what is said.[46]

Because Bodin's physics result in a contemplation of nature by tracing every phenomenon along the 'Chain of Being' towards God as the final cause, little more than this *probabilitas* can be expected, since human reason falls short facing eternal and divine truth. What is more, Bodin's topical structure offers only very limited possibilities for Copernicus's project, the introduction of a new theory. The very core of topical argumentation, the reliance on *endoxa*, works by embedding every single phenomenon in a pre-stabilised matrix of common sense. The use of topics is a way to systematise what is already known, e.g. to structure the use of pretexts,[47] and to link one's point to this more or less traditional universe of facts and statements. Roughly speaking, there cannot be anything significantly new in topical structures.[48]

It becomes evident that Copernicus's programme and methodology offer little (if any) foothold for a topically structured project like the *Theatrum*. Bodin's rejection of Copernicus's worldview therefore cannot simply be reduced to the fact that he has a more traditional or conservative image of the universe. Moreover, Bodin and Copernicus use different languages which, at least in the sixteenth century, cannot be reconciled.

45 Cf. Schmidt-Biggemann, *Apokalypse und Philologie* 239.

46 Ibid. 229.

47 See Neuber W., "Topik und Intertextualität. Begriffshierarchie und raministische Wissenschaft in Theodor Zwingers METHODUS APODEMICA", in Kühlmann W. – Neuber W. (eds.), Intertextualität in der Frühen Neuzeit. Studien zu ihren theoretischen und praktischen Perspektiven (Frankfurt a.M.: 1994) 253–278, 254.

48 Bornscheuer uses the expression 'Habitualität' for this fact, cf. Bornscheuer L., *Topik. Zur Struktur der gesellschaftlichen Einbildungskraft* (Frankfurt a.M.: 1976) 96f.

Selective Bibliography

Aristotle, *The Complete Works of Aristotle. The Revised Oxford Translation*, ed. J. Barnes (Princeton: 1984–1985).

Blair A., *The Theatre of Nature. Jean Bodin and Renaissance Science* (Princeton: 1997).

Bornscheuer L., *Topik. Zur Struktur der gesellschaftlichen Einbildungskraft* (Frankfurt a.M.: 1976).

Bodin Jean, *De la démonomanie des sorciers* (Paris, Jacques du Puys: 1581).

Bodin Jean, *Methodus ad facilem historiarum cognitionem* (Amsterdam, Johannes Ravelstein: 1650; reprint Aalen: 1967).

Bodin Jean, *Universae naturae theatrum. In quo rerum omnium effectrices causae, & fines contemplantur, & continuae series quinque libris discutiuntur* (Paris, Jacques Roussin: 1596).

Copernicus Nicolaus, *De revolutionibus orbium coelestium, Libri IV* (Nuremberg, Johannes Petreius: 1543).

Fellmann F., *Scholastik und kosmologische Reform. Studien zu Oresme und Kopernikus* (Münster: 1988).

Goddu A., *Copernicus and the Aristotelian Tradition. Education, Reading and Philosophy in Copernicus' Path to Heliocentrism* (Leiden – Boston: 2010).

Greengrass M., "The experiential world of Jean Bodin", in Lloyd H. (ed.), *The reception of Bodin* (Leiden – Boston: 2013) 67–96.

Krause V., "Listening to witches: Bodin's use of confession in De la démonomanie des sorciers", in Lloyd H. (ed.), *The reception of Bodin* (Leiden – Boston: 2013), 97–115.

Lovejoy A., *The Great Chain of Being* (Cambridge – London: 1976).

Müller K., "Vom Absehen und Hinschauen. Modifikationen des astronomischen und kosmologischen Diagramms an der Wende vom Hoch- zum Spätmittelalter", in Dickhut W. – Manns S. – Winkler N. (eds.), *Muster im Wandel. Zur Dynamik topischer Wissensordnungen in Spätmittelalter und Früher Neuzeit, Berliner Mittelalter- und Frühneuzeitforschung 5* (Göttingen: 2008).

Nate R., "Inszenierte Wissenschaft. Rhetorik, Mythos und Pragmatik", in Hempfer K. – Traninger A. (eds.), *Macht Wissen Wahrheit* (Freiburg – Berlin: 2005) 183–200.

Neuber W., "Topik und Intertextualität. Begriffshierarchie und raministische Wissenschaft in Theodor Zwingers METHODUS APODEMICA", in Kühlmann W. – Neuber W. (eds.), *Intertextualität in der Frühen Neuzeit. Studien zu ihren theoretischen und praktischen Perspektiven* (Frankfurt a.M.: 1994) 253–278.

Popkin R., *The History of Scepticism. From Savonarola to Bayle* (Oxford: 2003).

Schmidt-Biggemann W., *Apokalypse und Philologie. Wissensgeschichten und Weltentwürfe der Frühen Neuzeit* (Göttingen: 2007).

Schüz J., *Johann Fischarts Dämonomanie. Übertragungs- und Argumentationsstrategien im dämonologischen Diskurs des späten 16. Jahrhunderts* (Diss. Berlin: 2011) [URL: http://www.diss.fu-berlin.de/diss/receive/FUDISS_thesis_000000021283].

Stimson D., *The Gradual Acceptance of the Copernican Theory of the Universe* (Gloucester Mass.: 1972).

CHAPTER 6

Writing after Copernicus: Epistemology and Poetics in Giordano Bruno's *Ash Wednesday Supper*[1]

Steffen Schneider

Until now, Giordano Bruno's reception of Copernicus in the *Ash Wednesday Supper* has been discussed mainly with regard to his cosmology. As I shall argue, however, Bruno gives at least as much emphasis to the epistemological implications of Copernicus's theory as to the latter's astronomical and cosmological affirmations. Thus, Bruno also discusses the psychological causes that inhibit the full acceptance of the Copernican system and in this context exposes an entire theory of knowledge. This paper analyses Bruno's epistemological reflections by taking into consideration his works on artificial memory. This interpretation will eventually present a new understanding of the literary structure of the *Ash Wednesday Supper* which is interpreted as a textual device aiming to purify the reader's soul from the fantastic ideas of old philosophy and at the same time to implant the 'new phantasm' of Bruno's thought.

1 The 'Copernican Gap'

Giordano Bruno's reception of Copernicus, especially in the *Ash Wednesday Supper*,[2] has produced an impressive number of articles, yet scholars are still

---

1 I am grateful to Paul Strohmaier for reading and carefully correcting this paper.

2 The major references to Copernicus in Bruno's œuvre can be found in the *Ash Wednesday Supper* and in *De immenso et innumerabilibus*. For an outline of the research on this topic see Maspero R., "Scienza e copernicanismo in Bruno: Principali orientamenti della critica dal 1950 ad oggi", *Rivista di storia della filosofia* 1 (1989) 141–162; Maspero disregards, however, such an important work as Blumenberg H., *Die Genesis der kopernikanischen Welt* (Frankfurt a.M.: 1989), and the important paper of Westman R.S., "Magical Reform and Astronomical Reform: The Yates Thesis Reconsidered", in idem – McGuire J.E. (eds.), *Hermeticism and the Scientific Revolution*. Papers read at a Clark Library Seminar, March 9, 1974 (Los Angeles: 1977) 1–91. Among more recent publications are to be considered Gatti H., *Giordano Bruno and Renaissance Science* (Ithaca – London: 1999), esp. 43–98; Tessicini D., *I dintorni dell'infinito: Giordano Bruno e l'astronomia del Cinquecento* (Pisa: 2007); Bönker-Vallon A., "Einleitung", in Giordano Bruno, *De l'infinito, universo et mondi/Über das Unendliche, das Universum und die*

 | DOI 10.1163/9789004281127_008

far from agreement on the topic. For some, like Miguel A. Granada, Bruno was the first genuine disciple of Copernicus—genuine in the sense that he accepted the idea of the earth orbiting the sun as cosmological reality and not just as a mathematical hypothesis.[3] According to others, Bruno used Copernicus's theory for his own philosophical purposes, modifying it in a way, however, that precludes calling him a Copernican.[4] For the majority of scholars, the question of Bruno's Copernicanism concerns his proper or improper understanding of the new astronomy and, beyond that, his role in the history of science. In this paper, I shall instead elaborate on a different, yet important aspect of the *Ash Wednesday Supper* which is also closely connected to Copernicus, but has not yet attracted the due attention of scholars: In the five dialogues of this work, Bruno unfolds the epistemological implications and consequences of Copernicus's theory and his own.[5] His point of departure is what I would like to call the 'Copernican gap': the fact that the concept of an orbiting earth may be scientifically correct, but nevertheless runs counter to everyday sense experience as well as to the convictions and beliefs of most

---

Welten, ed. A. Bönker-Vallon (Hamburg: 2007) xiii–cxxviii, esp. xxxi–xxxviii; Omodeo P.D., "Giordano Bruno and Nicolaus Copernicus. The Motions of the Earth in the *Ash Wednesday Supper*", *Nuncius. Journal of the History of Science* 24 (2009) 35–59.

3 See Granada M.A., *Giordano Bruno. Universo infinito, unión con Dios, perfección del hombre* (Barcelona: 2002) 93. Most contemporaries of Copernicus accepted the view of Osiander, who in his unauthorised preface to Copernicus's *De revolutionibus* held that the heliocentric model was intended only as a hypothesis which aimed to facilitate calculation. See [Osiandri] "Ad Lectorem de Hypothesibis Huius Operis", in Copernicus Nicolaus, *Das neue Weltbild. Drei Texte—Commentariolus, Brief gegen Werner, De revolutionibus I*, ed., trans. and ann. H.G. Zekl (Hamburg: 1990) 60.

4 See e.g. McMullin E., "Bruno and Copernicus", *Isis* 78 (1987) 55–74, here 64, who argues that 'to call Bruno a "Copernican" requires one to empty the label of all content save the assertion that the earth and planets move around the sun.' The famous hermetic interpretation given by Yates F.A., *Giordano Bruno and the Hermetic Tradition* (London: 1964) 235–256, is now of merely historical interest.

5 As far as I can see, only H. Blumenberg ("Einleitung", in Giordano Bruno, *Das Aschermittwochsmahl*, 1981, 11f.) and A. Bönker-Vallon ("Einleitung", in Giordano Bruno, *De l'infinito* xxxvsq.) were aware of the epistemological dimension when talking about the metaphor of the inner eyes in the *Ash Wednesday Supper*. As I am arguing in this paper, this important observation has to be integrated into an account of Bruno's theory of knowledge on the one side and into the literary and rhetorical structure of the *Ash Wednesday Supper* on the other side. For a general reflection on Bruno's epistemology see Blum P.R., "Giordano Bruno's Changing of Default Positions", in Hufnagel H. – Eusterschulte A. (eds.), *Turning Traditions Upside Down. Rethinking Giordano Bruno's Enlightenment* (Budapest – New York: 2013), 13–18.

of Copernicus's and Bruno's contemporaries. At a time when the physical knowledge necessary to sustain Copernicus's astronomy was not yet available and when optical instruments were not yet capable to turn into sensual evidence what Copernicus taught, it was impossible to vindicate the correctness of his theory. How then could the rupture with daily sensual evidence and the traditional system of beliefs be justified? Copernicus never developed further the epistemological fundaments of his thinking. His 'Preface' to *De revolutionibus*, dedicated to Pope Paul III, justifies his decision to go beyond sense experience by quoting two humanistic stock arguments:

> Therefore, having obtained the opportunity from these sources,[6] I too began to consider the mobility of the earth. And even though the idea seemed absurd, nevertheless I knew that others before me had been granted the freedom to imagine any circles whatever for the purpose of explaining the heavenly phenomena. Hence I thought that I too would be readily permitted to ascertain whether explanations sounder than those of my predecessors could be found for the revolution of the celestial spheres on the assumption of some motion of the earth.[7]

The first argument is the reference to ancient authorities who held similar ideas. The second is the notion of the dignity of Man as it was invented in Renaissance humanism. The latter is implied when Copernicus deduces his claim for scientific certainty from the notion that God created the 'world machine' for Man's sake.[8] For the humanists, Man's dignity was founded, to a great extent, in the autonomy of human reason. The gift of reason ennobled Man by enabling him to emancipate himself from created nature.[9] As God's

6 Copernicus refers to ancient authorities mentioned by Cicero and Plutarch, who sustained the thesis of the earth orbiting around the sun. See Copernicus Nicholas, *On the Revolutions*, ed. J. Dobrzcycki, transl. E. Rosen (Warsaw – Krakow: 1978) 4f.

7 Ibidem 5. ('Inde igitur occasionem nactus coepi et ego de terrae mobilitate cogitare. Et quamuis absurda opinio videbatur, tamen quia sciebam alijs ante me hanc concessam libertatem, vt quoslibet fingerent circulos ad demonstrandum phaenomena astrorum, existimaui mihi quoque facile permitti, vt experirer an posito terrae aliquo motu firmiores demonstrationes quam illorum essent, inueniri in reuolutione orbium caelestium possent.' Ibidem 74.)

8 Ibidem 4.

9 For Ficino, reason is the proper faculty of Man's soul, while he is only participating in the intellect. See for example *Theologia Platonica* XIII,2,18. See also Giovanni Pico della Mirandola, *De hominis dignitate*, who equally emphasises the gift of reason.

vicegerent on earth,[10] he was free to design his own world by the power of reason. When Copernicus stresses his freedom to choose a new explanatory model of the world, he applies these humanistic arguments to the quadrivial discipline of astronomy. By stressing the priority of reason over sense perception, he likewise gave an important impulse to scientific methodology. As Immanuel Kant put it, Copernicus taught the epistemological priority of methodological decisions over sensual evidence:

> [...] Kopernikus [versuchte], nachdem es mit der Erklärung der Himmelsbewegungen nicht gut fort wollte, wenn er annahm, das ganze Sternenheer drehe sich um den Zuschauer [...], ob es nicht besser gelingen möchte, wenn er den Zuschauer sich drehen, und dagegen die Sterne in Ruhe ließ.[11]

Kant, however, was not the first philosopher to discover the epistemological and methodological implications of Copernicus's work. So did Giordano Bruno who unfolded these implications in his *Ash Wednesday Supper*—a work which to a great extent is dedicated to the problems of cognition. Bruno's thoughts on epistemology had already been developed in his earlier work, *De umbris idearum* where he made no explicit reference to Copernicus. Here, he laid down the outline of his theory of cognition, which may be summed up in these two propositions: First, Bruno was convinced that no immediate intellectual insight into the ideas of things existed.[12] Second, he concluded from this that intellect, reason and imagination had the potential to approximate true knowledge through the generation of images (this word is to be understood in a broad sense).[13] As ultimate truth remains concealed from Man, Bruno considers the human intellect as an inventive, productive power which projects its images onto the world. These images, however, need to be verified by experience and argument. Though they are never absolutely true, they nonetheless inform about reality. Human thinking is thus situated in between truth and

10 Ficino calls Man the God of all living creatures ('deus animalium') in his *Theologia Platonica* XIII,3,1.

11 Kant Immanuel, *Kritik der reinen Vernunft* (Frankfurt a.M.: 1956) vol. I, 25 (B XVI).

12 He uses the metaphor of *umbra*—'shadow'—to describe the validity of human cognition. See *De umbris idearum*, in Bruno Giordano, *Opere mnemotecniche*, vol. I, ed. M. Matteoli – R. Sturlese – N. Tirinnanzi, (Milan: 2004) 1–380. According to Bruno, we see shadows of the truth, not the truth itself, but we do see shadows and not nothing. See Leinkauf T., "Einleitung", in Bruno Giordano, *De la causa, principio et uno/Über die Ursache, das Prinzip und das Eine*, ed. T. Leinkauf (Hamburg: 2007) xiii–xv; Spruit L., *Il problema della conoscenza in Giordano Bruno* (Naples: 1988).

13 For references see paragraph IV below.

error. Through constant effort, Man may approximate true knowledge, but losing his zeal, he may just as well also fall back into dullness and illusion.[14] When Bruno, beginning with the *Ash Wednesday Supper*, dedicated himself to questions of astronomy and cosmology, this epistemological context remained an important reference. For him, Copernicus was paramount, because he liberated human reason from the dictates of the senses. And yet, the famous astronomer himself was not interested in exposing a theory of knowledge concordant with his astronomical theory. In the *Ash Wednesday Supper*, however, Bruno investigates extensively into the possibility of knowledge under the condition of the 'Copernican gap'.

2 The Literary Structure of the 'Ash Wednesday Supper'

The *Ash Wednesday Supper* is organised as a polyphonic text: the cosmological and the epistemological issues are developed all at once. Each of them dominates the other in an alternating change, sometimes both are equally important. They accompany each other throughout the text. This polyphony creates the considerable complexity of this text, which is often disregarded by researchers interested only in its scientific 'content'. Such disregard is highly problematic since the literary structure of the *Ash Wednesday Supper* cannot be considered as merely ornamental—it is part of the scientific meaning in this work and essential for its proper understanding.[15]

The *Ash Wednesday Supper* tells the story of a supper, which took place in the house of the English nobleman, Sir Fulke Greville in London. The motive for this supper was Greville's wish to hear the Nolan explain 'his Copernicus and other paradoxes of the new philosophy'.[16] The Nolan takes his name from Bruno's city of birth, Nola, near Naples, and therefore is often confused with

14 This struggle for knowledge, this constant effort is the hallmark of the 'heroic' individual, to whom Bruno dedicated his work *De gl'heroici furori*.

15 The literary form of this dialogue is analyzed in Hufnagel H., *Ein Stück von jeder Wissenschaft. Gattungshybridisierung, Argumentation und Erkenntnis in Giordano Brunos italienischen Dialogen* (Stuttgart: 2010); Clucas S., "Galileo, Bruno and the Rhetoric of Dialogue in Seventeenth-Century Natural Philosophy", in *History of Science* 46 (2008), 405–429. See also my *Kosmos—Seele—Text. Formen der Partizipation und ihre literarische Vermittlung. Marsilio Ficino, Pierre de Ronsard, Giordano Bruno* (Heidelberg: 2012), 287–336, 356–365.

16 Bruno Giordano, *The Ash Wednesday Supper*, ed. and transl. S.L. Jaki (The Hague – Paris: 1975) 55. ('il suo Copernico et altri paradossi di sua nova filosofia.') The Italian text is quoted after Bruno Giordano, *Opere italiane*, ed. N. Ordine, 2 vols. (Torino: 2002) here I, 447.

the author Giordano Bruno. Although the Nolan does serve as a mouthpiece of Bruno's philosophy, he should be treated as a literary character whose creation has an important function within the text's intention. Therefore, I shall subsequently refer to the Nolan as a fictional character and to Bruno as a historical person and the author of the text. The Nolan accepts Greville's proposal on condition that he engage two worthy disputants. Greville agrees—and this gives Bruno an excellent opportunity for a harsh and witty polemic against his academic opponents. The invited disputants are two doctors—Nundinio and Torquato—, who are utterly ignorant and incompetent representatives of obsolete knowledge. The *Ash Wednesday Supper*, however, does not represent the dispute in an immediate fashion. Instead, an account of this dispute is given by Theophil to a group of listeners. Thus, the work consists of a mimetic framing dialogue and an internal debate, mediated by a narrator.[17] The central figure in the framing dialogue is Theophil, a follower of the Nolan. He mediates between his own audience and the events at the supper. The Nolan's dispute ended in disaster as he was unable to convince Nundinio and Torquato and, due to their ignorant impoliteness, missed the great occasion to successfully expose his philosophy. Now, it is Theophil's job to put things straight: he explains to his listeners why the Nolan was right and the doctors were wrong. Apart from this, the external dialogue also instantiates the ideal of a successful discussion. It therefore presents a contrast to the narrated dialogue. Especially important among the listeners is the character of Smith, an intelligent, open-minded layman interested in philosophy and science. By formulating his reservations regarding the Nolan's thoughts, he gives Theophil the opportunity to explain it in further detail. In this way, the reader becomes a first-hand witness of Smith's conversion to the Nolan's philosophy. The remaining audience of Theophil's consists of two comical characters: Frullo, the fool, and Prudenzio, the pedant. Their comments and conflicts create comic relief from the gravity of the debate. This textual architecture of the *Ash Wednesday Supper* enables Bruno to develop his cosmological thought and, at the same time, to reflect the epistemological implications of Copernican and Nolan philosophies.

---

17 Mimetic in the sense of Plato, who in *Politeia* III, 392d–393d distinguishes between mimetic and diegetic representations: the mimetic genre reproduces the voice of the figure, while the diegetic genre is narrative: here we hear the voice of the narrator.

3 The Blindness of the Senses and the Eyes of Reason

In the *Ash Wednesday Supper*, epistemological issues are developed mostly in the first dialogue. At the outset, Theophil presents the Nolan's view on traditional astronomy. He then reports the latter's appraisal of Copernicus as a precursor of the Nolan. In a third step he deepens the discussion by considering the role of the senses, of imagination and of institutions such as the Church and the universities in the formation of knowledge.

The critique of astronomy sets in when the Nolan, on the occasion of his invitation, underlines that his thought does not depend on that of Copernicus nor on anyone else: 'the Nolan replied that he does not see either with the eyes of Copernicus, nor with those of Ptolemy, but with his own as far as judgement and conclusions are concerned.'[18] With the distinction between 'eyes' and 'judgement and conclusions' the Nolan refers to the Copernican gap. The exact wording of the sentence manifests a semantic shift in the meaning of the word 'eyes': in the first clause ('he does not see either with the eyes of Copernicus, nor with those of Ptolemy'), it refers to the physical eyes and, metonymically, to the observations made by the two great astronomers. In the second clause, in which the Nolan speaks of 'his own [eyes] as far as judgement and conclusions are concerned', the word signifies, metaphorically, the operation of the rational faculties. Here, a play with the terms of 'blindness' and 'sight' begins that structures the whole text, the sensual eyes being from now on deemed 'blind' because the Nolan considers the senses inapt to correctly assess the nature of appearances. The intellectual faculties are, in turn, called the 'eyes of reason',[19] as they procure the only possibility of secure judgement.

This distinction forms the basis of Theophil's definition of the work of astronomy in contrast to natural philosophy. This becomes evident when he compares the traditional astronomers who collected important observations on the movement of the stars, to 'interpreters who translate words from one language to another.'[20] According to him, astronomers transcribe celestial phenomena into mathematical symbols, but they do not penetrate the real meaning of these by means of their intellect. Yet there are other translators—the natural philosophers and among them, especially, the Nolan—who use these observations as a text in need of more profound elucidation: 'then there

18 Bruno, *Supper* 55. ('Al che rispose il Nolano, che lui non vedea per gli occhi di Copernico, né di Ptolomeo; ma per i proprii quanto al giudizio e la determinazione', *Op.it.* I, 447.)

19 Ibidem 56. ('occhi della raggione', *Op.it.* I, 448.)

20 Ibidem 55. ('interpreti che traducono da uno idioma a l'altro le paroli', *Op.it.* I, 447.)

are others who fathom the meanings and not the words themselves.' Theophil uses another telling comparison to clarify what the Nolan means, likening the astronomers to 'the Theban Manto who saw but did not understand', while the Nolan is compared to 'Tiresias, the blind but divine interpreter'.[21] Tiresias needs information from Manto for his prediction of the future but reaches far beyond any sensual experience in his judgement. From this comparison, Theophil concludes that a true philosopher should be grateful to the astronomers for informing him; but he should trust only his own reason and intellect in judging and explaining the world's structure and the real causes of phenomena: 'we should open our eyes at what they [i.e., the astronomers] observed and saw, and should not give our consent to what they conceived, meant, and set forth.'[22]

The passage just discussed reflects to a certain extent the medieval differentiation between astronomy, understood as a purely hypothetical description of appearances, and of natural philosophy authorised to define and explain the structure of reality.[23] Bruno, however, is no longer satisfied with this distinction. Instead, he aims to integrate astronomy and cosmology into a new vision of the cosmos by leaving aside any religious or philosophical restrictions. This is the point where Copernicus comes into play. In the third dialogue, the Nolan defends the realistic interpretation of Copernicus's theory by polemicising against Osiander's preface to *De revolutionibus*, while Nundinio supports Osiander. The Nolan acknowledges Copernicus as an important precursor of himself, since the astronomer talked about the real structure of the world and did no longer accept the traditional, hypothetical nature of astronomy. Nonetheless, when Smith wants to know about Theophil's opinion on Copernicus, his answer is ambiguous. On the one hand, Theophil extols Copernicus's great gifts and merits. He was 'possessed of a grave, elaborate, careful and mature mind'. He was 'not inferior, except by succession of place and time, to any astronomer who had been before him'—here Theophil alludes to the ancient precursors of heliocentrism—and his 'natural judgement was far superior to Ptolemy, Hipparchus, Eudoxus and all the others who walked in the footsteps of these'.[24] Above all, Theophil underlines Copernicus's effort

21 Ibidem ('la tebana Manto, che vedeva ma non intendeva', *Op.it.* I, 448.)

22 Ibidem 56. ('doviamo aprir gli occhi a quello ch'hanno osservato e visto: e non porgere il consentimento a quel ch'hanno conceputo, inteso e determinato', *Op.it.* I, 448.)

23 On the medieval understanding of astronomy see Carrier M., *Nikolaus Kopernikus* (Munich: 2001) 49–54.

24 Bruno, *Supper* 56f. ('Lui [i.e. Copernico] avea un grave, elaborato, sollecito e maturo ingegno; uomo che non è inferiore a nessuno astronomo che sii stato avanti lui, se non

to overcome the old erroneous axioms of classical astronomy and cosmology: he was 'a man who liberated himself from some false presuppositions of the common and commonly accepted philosophy, or perhaps I should say, blindness.' These efforts were, however, not yet sufficient to complete this liberation:

> But for all that he did not move too much beyond them; being more intent on the study of mathematics than of nature, he was not able to go deep enough and penetrate beyond the point of removing from the way the stumps of inconvenient and vain principles, so as to resolve completely the difficult objections, and to free both himself and others from so many vain investigations, and to set attention firmly on things constant and certain.[25]

Copernicus did not abolish the celestial spheres. Instead, he left the epicycles in their place. As to the form of the planets, he continued to adhere to the classical geometrical ideal of the perfect spheroid and did not believe in the actual infinity of the universe—all these beliefs were then discredited by the Nolan. Finally, Copernicus's heliocentric cosmos was not erected upon the solid ground of philosophy, while the Nolan deduces his radical cosmological ideas from one principle: the notion of an infinite divinity which brings each of its infinite possibilities into being, thus creating an infinity of worlds in an infinite space and time. Bruno's cosmological thought completely absorbed Copernicus's astronomy by eliminating all elements that did not fit his own theory.

In the next step, Theophil returns to his distinction between the physical eyes and the intellectual power of Man and discusses in more detail the causes for the insufficiency of the senses. In the field of astronomy, sense perception is obstructed by the nature of the air. It is impossible to penetrate the sky with human eyes, since the atmosphere precludes any clear sight of celestial phenomena. This limitation does not only concern sense perception but

per luogo di successione e tempo; uomo che quanto al giudizio naturale è stato molto superiore a Tolomeo, Ipparco, Eudoxo, e tutti gli altri ch'han caminato appo i vestigii di questi; al che è dovenuto per essersi liberato da alcuni presuppositi falsi de la comone e volgar filosofia, non voglio dir cecità.' *Op.it.* I, 448f.)

25 Ibidem 57. ('Ma però non se n'è molto allontanato: per che lui più studioso de la matematica che de la natura, non ha possuto profondar e penetrar sin tanto che potesse a fatto toglier via le radici de inconvenienti e vani principii onde perfettamente scioglesse tutte le contrarie difficultà, e venesse a liberar e sé et altri da tante vane inquisizioni, e fermar la contemplazione ne le cose costante e certe.' *Op.it.* I, 449.)

also imagination. According to Theophil, it is the restricted vision that leads human imagination to construct false images—'chimeras'—of the structure of the cosmos:

> The Nolan [...] set free Man's rational soul and cognition which was retained in the narrow prison of the turbulent air, from where as if through some holes it could hardly contemplate the most distant stars; its wings were cut lest it should fly and open the veil of these clouds, to see what is really there, and liberate itself of the chimeras of those who, though originating from the mud and caves of the earth, filled, as if they were Mercuries and Appollinos coming from heaven, the whole earth through many a swindle with endless folly, beastliness and vice, as if with as much virtue, piety and discipline, crushing that light that turned the souls of our ancestors divine and heroic, by approving and confirming the cloudy darkness of the sophists and jackasses.[26]

The chimeras refer to the 'vain' and 'fantastic' constructions mentioned above: the spheres, the spheroids, the fixed stars,[27] geocentrism and the limits of the cosmos. While the range of the eyes is limited, the rational soul ('animo') and its cognition have, in principle, infinite possibilities to approximate the truth. If there is any limitation, it only exists in the sense of a false belief of our rational power in the productions of *phantasia*. Theophil does not conceive *phantasia* as a merely individual capacity. For him, there are agents manipulating *phantasia* and keeping mankind ignorant. They call 'virtue, piety and discipline' what must be considered as 'endless folly, beastliness and vice'. Who are these false Mercuries and Appollinos born 'from the mud and caves of the earth', but pretending to be messengers from heaven? They are the representatives of Christian religion, whom Bruno likes to call 'donkeys' (in the translation: 'jackasses'), and the 'sophists' who are Aristotelian scientists just as Nundinio

26 Bruno, *Supper* 60, translation modified. ('Il Nolano [...] ha disciolto l'animo umano e la cognizione che era rinchiusa ne l'artissimo carcere de l'aria turbulento, onde a pena come per certi buchi avea facultà de remirar le lontanissime stelle, e gli erano mozze l'ali, a fin che non volasse ad aprir il velame di queste nuvole e veder quello che veramente là su si ritrovasse, e liberarse da le chimere di quei che essendo usciti dal fango e caverne de la terra, quasi Mercuri et Appollini discesi dal cielo, con moltiforme impostura han ripieno il mondo tutto d'infinite pazzie, bestialità e vizii, come di tante vertù, divinità e discipline: smorzando quel lume che rendea divini et eroichi gli animi di nostri antichi padri, approvando e confirmando le tenebre caliginose de sofisti et asini.' *Op.it.* I, 453.)

27 The illusion of the fixed stars is destroyed in the fifth dialogue of the *Supper*, see 145–150 (*Op.it.* I, 544–548).

and Torquato.[28] The Nolan, who wants to free the rational powers of the soul, has to fight against the power of the Church and the old university. He has to dissipate the shadows of false imagination and to establish the intellect as the leading faculty in science and philosophy. He teaches not to trust in the sensual eyes but to rely on the inner eyes, the eyes of reason.

This passage makes evident that for Bruno astronomy, philosophy and the problem of power are closely connected. Philosophically speaking, he gives clear priority to rational and intellective thinking over sensual experience and imagination. The real structure of the universe cannot be perceived and it is not immediately evident from the poor data available from remote stars. Only intellectual speculation may destroy the illusions of the imagination and reveal the truth.[29]

4 The Double Nature of 'phantasia' and Imagination

For Theophil and the Nolan, the 'endless folly, beastliness and vice' all originate from misguided human *phantasia* and imagination.[30] Consequently, Bruno employs, throughout the *Ash Wednesday Supper*, the semantic field of *phantasia* to refer to the Aristotelian and Ptolemaic theories that he considers 'chimeras'. Thus, he speaks of the 'fictitious walls of the first, eighth, ninth, tenth spheres, and whatever else might have been attached to these'[31] and alludes to Plato's Allegory of the Cave when he writes that 'our mind is no longer imprisoned in the fetters of the imaginary movables and movers, eight, nine and ten'.[32] In the 'Prefatory Epistle' he calls Torquato and Nundinio 'two fantastic goblins,

28 'Asini' and 'sophisti' are very prominent in Bruno's works. Here is not the place to proceed further into the problem of Bruno's relation to the Christian religion. See "Die Kabbala des Pegasus. Zugänge zu einem Text Giordano Brunos", *Zeitschrift für Ästhetik und allgemeine Kunstwissenschaft* 52, 1 (2007), thematic issue ed. S. Schneider – P. Theisohn.

29 Bruno often alludes to this order of the faculties of the soul in the *Supper*, but he fully explains it in *De l'infinito, universo et mondi* (*On the infinite universe and worlds*), see in: *Op.it.* II, 35–36.

30 Certain currents of early modern thought were hostile to imagination. See for example Gianfranceso Pico della Mirandola's treatise *De imaginatione* (1501) and the good survey of this tradition given by Wels V., "Zur Vorgeschichte des Begriffs der 'kreativen Phantasie'", *Zeitschrift für Ästhetik und allgemeine Kunstwissenschaft* 50 (2005) 199–226.

31 Bruno, *Supper* 61. ('[...] fantastiche muraglia de le prime, ottave, none, decime, et altre che vi s'avesser potute aggiongere sfere', *Op.it.* I, 454.)

32 Ibidem ('Non è più impriggionata la nostra raggione co i ceppi de fantastici mobili e motori otto, nove e diece.' *Op.it.* I, 454.)

two dreams, two shadows and to four-day fevers',[33] whereby both characters come to personify the false use of imagination.[34] Although criticising the fantastic and imaginative faculties, he nonetheless acknowledges their great power—but what does this power consist in? What role does it play in the process of cognition and how does it relate to sense experience?

Bruno denies the intellect's capacity to contemplate truth as such. Rather, he conceives cognition as an active, image-producing process.[35] This theory is exposed in Bruno's writings on artificial memory, where by elaborating an original theory of cognition he goes far beyond the pragmatic end of improving natural memory.[36] In his analysis of *phantasia*, Bruno agrees for once with Aristotle who, in his work *On the Soul* asks whether human cognition depends on images constructed by imagination.[37] Bruno affirms this view and goes on:

> Therefore, not without particular form or figures, which are conceived by the external senses from sensible objects and which are located and digested in the inner senses do we know how we can achieve an operation suitable to our nature.[38]

---

33 Bruno, *Supper* 44 ('due fantastiche befane, doi sogni, due ombre e due febbri quartane', *Op.it.* I, 434).

34 There are more similar passages in the *Supper* and also in other works of Bruno. For more records see Ciliberto M., *Lessico di Giordano Bruno*, 2 vols. (Rom: 1979), notably "fantasia" (I, 429f.) and "imaginazione" (ibidem 550f.).

35 Cf. Ciliberto M.: '"[...] per speculum et in aenigmate [...]"', in Giordano Bruno, *Opere mnemotecniche* II (Milan: 2009) ix–xlvi.

36 For the role of imagination in Bruno's mnemonic writings see Rosa G. De, *Il concetto di* immaginazione *nel pensiero di Giordano Bruno* (Naples: 1997) 99–136; Matteoli M., "Immaginazione, conoscenza e filosofia: l'arte della memoria di Giordano Bruno", in Bassi S. (ed.), *Bruno nel XXI secolo. Interpretazioni e ricerche—Atti delle Giornate di studio (Pisa, 15–16 ottobre 2009)* (Florence: 2012), 17–38. The classical account of Bruno's *ars memoriae* is to be found in Yates F.A., *Giordano Bruno* 211–324, also idem, *The Art of Memory* (London: 1966) 314–319. Yates's interpretation has been effectively criticised and corrected by Sturlese R., "Per un'interpretazione del *De umbris idearum* di Giordano Bruno", *Annali della Scuola Normale Superiore di Pisa*, third series, 22 (1992) 943–968.

37 Cf. Borsche T., "Denken in Bildern. *Phantasia* in der Erkenntnislehre des Giordano Bruno", in Hirdt W. (ed.), *Giordano Bruno. Tragik eines Unzeitgemäßen* (Tübingen: 1993) 93–106, 94.

38 Bruno Giordano, *On the Composition of Images, Signs and Ideas*, transl. C. Doria, ed. and ann. D. Higgins (New York: 1991) 18. ('Ideoque non sine formis quibusdam seu figuris, quae per sensus externos ab obiectis sensibilibus concipiuntur et in interioribus collocantur atque digeruntur, operationem aliquam naturae nostrae convenientem perficere posse cognoscimus.' *De imaginum compositione, Op. mnemotecniche* II, 508f.)

Once *phantasia* has formed images out of sense impressions, imagination assesses their content. In a third step, reason uses these images as information.[39] Bruno calls these images also 'specie intelligibili' or, in Latin, 'species intelligibiles'.[40] Even the highest operations of the soul refer to these images, freely combining and analysing them.

Another important aspect of Bruno's theory of imagination is the active, productive role conferred to it. According to him, human memory is not a store of images. Rather it is formed by a person's outlook on the world. Bruno does not conceive of thinking as an introspective act, i.e. as an analysis of images. Instead, our thought creates and forms images and only through them looks into the external world:

> Consider that we do not need to contemplate the images in us but things themselves through the images in us; if an image presents itself to the soul, we direct our soul's attention not so much onto the image but through it.[41]

This notion of cognition explains why the way in which we imagine determines our mode of thought. The 'chimeras' of Aristotelian cosmography and of Ptolemaic astronomy are based on images that originate from bad habits of our imagination and our reason. What has to be done, then, is purging the human mind from those images and habits.

5 The Role of Emotions

Closely connected to the problem of *phantasia* is the role of emotions. On the one hand, emotions are partially responsible for the formation of false fantasies. On the other hand, they are also useful, since only images charged with emotional energy are powerful enough to excite our cognition and to change our habitual modes of thought. In a passage from *Sigillus sigillorum*, we find

39 Cf. e.g. "Sigillus sigillorum", in *Op. mnemotecniche* II, 208f.

40 Cf. Spruit L., *Species intelligibilis. From Perception to Knowledge*, II: *Renaissance Controversies, Later Scholasticism, and the Elimination of the Intelligible Species in Modern Philosophy* (Leiden – New York – Cologne: 1995) 203–213.

41 'Memento igitur, non ea quae sunt in nobis, sed res ipsas per ea quae sunt in nobis, esse inspiciendas; quamvis enim animae praesens adsit imago, non tanquam ipsam, sed tamquam per ipsam aspicientes intendamus animo.' *Sigillus*, *Op. mnemotecniche* II, 202.

good information about how *phantasia* is linked to emotions and how dangerous it may become if not kept in check by reason:

> [...] let us finally be careful that our rational soul may not be seduced by the images constructed upon sense impression and, fixed on these, be deprived of the joy of intellectual life and, drunk from the wine of physical affects and vulgar authorities [...] be urged to stagger around in the house of assuming ignorance and to sleep and to wander around there, irritated by dreams coming from a disturbed *phantasia* and after the loss of its native wings of intelligence, so that it would never find an adequately formed image—as if it had looked at the face of Proteus. [...] Let us stop, let us stop to admire the images generated by the senses, which are only the shadows of things, and let us be careful to withdraw and to listen to the words uttered by the active intellect which lives in our rational soul.[42]

This passage does not only demonstrate that *phantasia* needs to be bridled. It also shows that emotions operate on it and may even dominate it. Through *phantasia*, physical emotions are linked to the higher faculties of the soul. While this interconnection seems problematic in the quoted passages, there are other passages in which Bruno stresses the importance of emotions for cognition. Thus, in the same work he writes:

> If *phantasia* does not knock energetically with sensually perceivable images, the cognitive faculty will not open the door; and if the cognitive faculty, which is watching the door, does not open it, then the indignant mother of the muses [i.e. the memory] will not receive them.[43]

42 '[...] caveamus ne animus a sensibilibus speciebus illectus, ita sui in ipsis fixionem faciat, ut intelligibilis vitae privetur delitiis vinoque affectuum corporeorum et vulgaris authoritatis [...] ebrius, perpetuo in praesumptuoso ignorantiae domicilio titubando pernoctet ibidemque, turbatae phantasiae velut insomniis exagitatus, amissis connatis alis intelligentiae, proruat et Prothei contemplatus vultum, nunquam concinne formatam, in qua conquiescat, speciem inveniat. Desinamus igitur, desinamus eas, quae veluti rerum sunt umbrae, a sensibus obiectas species admirari, et illud in nobis caveamus, ut ad auscultandum domestici in animo intellectus agentis affatum regrediamur.' *Sigillus*, *Op. mnemotecniche* II, 206.

43 Cf. *Sigillus*, *Op. mnemotechniche* II, 196: 'Ni igitur vivacius phantasia sensibilibus pulsaverit speciebus, cogitatio non aperiet; ostiaria quoque cogitatione non aperiente, easdem indignans Musarum mater non recipiet.'

He explains this as follows:

> Such things excite [the memory], which—in the company of reasoning, thinking and powerful *phantasia*—arouse emotions; by these we are stimulated to esteem, despise, love, hate, to be sad or happy, to admire or to weigh with our senses the impression of the form of the object we want to remember; at the same time, we receive the images of esteem and despise, of love and hate, of sadness and happiness, of admiration and of weighing. The stronger and more intense emotions are, the stronger and more intense impressions are.[44]

The idea that strong emotions lend power to these images comes from the tradition of ancient rhetoric.[45] Bruno, however, attributes to them a fundamental role for reasoning. He stresses the necessity of rational supervision of these images and, at the same time, of their connection with emotions. Furthermore, he acknowledges that emotions are crucial to cognition, because only emotional images 'knock at the door of cognition'. Without emotionally charged images, cognition would be devoid of any content.

6 A Typology of Ignorance

The cloudy sky, the 'turbulent' atmosphere and the absence of optical instruments allow only for weak and uncertain sense impressions. Only the intellect can find out the truth about what remains hidden behind the clouds. Yet, the mighty obstacles of 'vulgar authorities', habitual ways of thinking and unchecked fantasies all impede the dissemination of Copernicus's astronomy and of the Nolan's philosophy. The particular characteristic of the *Ash Wednesday Supper* consists in the fact that the conditions of thinking and the obstacles of dissemination are continuously reflected throughout the work. Theophil, for example, distinguishes different types of ignorance: While some people are by nature unable to understand complex issues,

44 Ibidem: 'Exitent ergo [sc. Memoriam] quae comitante discursu, cogitatione, fortique phantasia movent affectum, quibusque zelantes, contemnentes, amantes, odientes, maerentes, gaudentes, admirantes et ad sensu[r]um trutinam referentes, cum zeli, contemptus, amoris, odii, maeroris, gaudii, admirationis et scrutinii speciebus [cum] memorandae rei forma afficimur. Porro fortiores atque vehementiores fortius consequentia quadam atque vehementius imprimunt.'

45 See for example Rhet. Her. III, 37, where this theory of images is exposed.

others are sufficiently gifted to comprehend but fail for accidental reasons.[46] This class of ignorance is further divided into three subclasses: 1. The first class subsumes obdurate and malicious people, who fancy themselves wise and erudite. Nundinio and Torquato are clearly of this kind. 2. There are fearful people, who are afraid of truth, such as religious fanatics, who are portrayed with bitter sarcasm in the *Cabala*. 3. Finally, there are those who are in possession of both, proper reason and clear senses, and strive to gain knowledge, like Theophil and Smith.[47]

While people of the second class are not relevant for this work, the first class is of particular importance for the dispute in Greville's house, and the third class for the framing dialogue. And while the Nolan's attempt to convince Nundinio and Torquato fails completely—both remain stuck in their malicious ignorance—, Theophil succeeds in converting Smith. This asymmetry between the Nolan's failure and Theophil's success may seem quite odd within a text which tries to convince Bruno's readers of the truth of his philosophy and further contributes to the seemingly incoherent, even illogical structure of the work.[48] This structure makes perfect sense, however, since the characters of Nundinio and Torquato are analysed as embodied pre-Copernican and pre-Nolan chimeras, which have to be exorcised from the reader's soul and replaced by the argument presented by Theophil and Smith. Bruno applies two strategies of purging his reader: one consists in rational argumentation (Smith and Theophil), the other in the construction of emotionally powerful images of 'vulgar authorities', which have to be deposed. Rational argument accompanies Theophil's narration of the dispute: while the doctors interrupt the Nolan,

46 Bruno, *Supper* 63–64: 'We can be teachers only of such ignorant men, and we can be the light only of such blind persons that are called handicapped not because of the inability of natural impotence, or by the absence of ability and self mastery, but solely because of inadvertence and inconsideration; which occurs through mere acts, and not through disposition.' ('Di que' dumque indotti possiam esser maestri, e di quei ciechi illuminatori, che non per inabilità di naturale impotenza, o per privazion d'ingegno e disciplina, ma sol per non avvertire e non considerare, son chiamati orbi: il che avviene per la privazion de l'atto solo, e non de la facultà ancora.' *Op.it.* I, 457.)

47 See Bruno, *Supper* 64.

48 G. Aquilecchia doubted the usefulness of the artistic form of the *Supper* because its 'exuberance continuously risks to blur the logic of the discourse which pretends to be philosophical'. Aquilecchia G., "Introduzione a '*La cena delle ceneri*'", in idem, *Schede bruniane (1950–1991)* (Rome: 1993) 65–96. H. Hufnagel finds involuntary structural ironies in this work, which undermine or 'deconstruct' its intention to enthrone the Nolan as an authority. Hufnagel, *Ein Stück von jeder Wissenschaft* 98–102.

preventing him from explaining his thought, Smith and Theophil reconstruct the proper argument, correct the errors of the doctors and fill in the missing elements of the Nolan's discourse.

Before accepting Theophil's explanations, Smith unfolds a veritable theory of custom and tradition according to which every notion of good and bad, true and false is relative and based on habits and education:

> How can I, who knows nothing, learn to know the difference between dignity and indignity, poverty and richness, between those who think they are wise and those who are esteemed as such? I see well that we all are born ignorant; we believe readily that we are ignorant; as we grow, we are imbued with the discipline and customs of our home, and to no less a degree do we hear criticism of the laws, the rites, the faith, the style of our adversaries and of those opposed to us, than is the case with these things concerning us and our affairs. To no less an extent are planted in us, perforce, the roots of a zeal for our own things, than is the case with those many other and diverse people about their own affairs.[49]

Smith's doubt is perfectly in line with Bruno's theory of *phantasia* and cognition. For him, the habitual modes of imagination and reason are shaped by the customs of a community. Prejudices and habits are imbibed from infancy on and continue to dominate the individual's thoughts throughout his whole life. If this is true, how may a philosopher ever hope to convince his adversaries?[50]

Theophil answers by inviting Smith to be patient and to listen to the Nolan's arguments, as this is the only way of experiencing the power of his philosophy. Though it is difficult, if not impossible, to theoretically define the criteria

49 Bruno, *Supper* 69. ('Come io che non so nulla potrò conoscere la differenza de dignità et indignità, de la povertà e ricchezza, di que' che si stimano e son stimati savi? Vedo bene che tutti nascemo ignoranti, credemo facilmente d'essere ignoranti, crescemo e siamo allevati co la disciplina et consuetudine di nostra casa: e non meno noi udiamo biasimare le leggi, gli riti, le fede e gli costumi de nostri adversarii et alieni da noi, che quelli de noi e di cose nostre. Non meno in noi si piantano per forza di certa naturale nutritura le radici del zelo di cose nostre, che in quelli altri molti e diversi de le sue.' Bruno, *Op.it.* I, 464).

50 Smith asks: 'Or tell me, by what means can the ears of a listener be attuned to you rather than to someone else when his mind is perhaps less inclined to attending to your propositions than to those of thousands of others?' Bruno, *Supper* 70. ('Or dimmi: con quale arte ti conciliarai queste orecchie più tosto tu ch'un altro, essendo che ne l'animo di quello è forse meno inclinazione ad attendere le tue proposizioni, che quelle di mill'altri diverse?' *Op.it.* I, 464).

of legitimate authority, in a given case it can be very easy to discern which of two opposing theories is superior. With this answer, the first dialogue is left open-ended. The remaining text has to reveal whether the Nolan's authority deserves more trust than that of Nundinio and Torquato.

7 Purification from and by Fantastic Images

One striking feature of Bruno's writing is his emotional style. Addressing his readers in his 'Prefatory Epistle', the author enumerates a large scale of emotions and habits which may be caused by his work. The reader may become 'heroic and dejected', 'happy and sad', 'saturnine and jovial', 'facile and ponderous, cringing and liberal, apish and dignified, a sophist with Aristotle, a philosopher with Pythagoras, smiling with Democritus, crying with Heraclitus'.[51] At first glance, it may seem odd that a work of philosophy and science tries to provoke such a multitude of strong emotions and habits, especially to readers who consider philosophy the discipline of thinking guided by reason. As has been shown above, however, for Bruno, thinking, imagining and feeling are closely related. Emotions contribute to the performative goal of the *Ash Wednesday Supper* which does not only consist in convincing through the use of sound argumentation but also in purging the reader's soul from the dangerous and deceptive simulacra of false authority and chimerical astronomy and cosmology. The emotional and powerful style, the Nolan's aggressive way of disputing, the mocking of his adversaries—all these literary and rhetorical devices serve as means to shock and provoke, offend and inspire admiration and, as a result, to reform the reader's imagination and his rational faculties.

The main targets of these literary devices are the doctors. For Theophil and the Nolan, they embody the erroneousness and malignity of traditional scientific and religious culture. They instantiate the persistent power of misguided and misleading imagination. This is why they are called 'two fantastic goblins, two dreams, two shadows and two four-day fevers' in the 'Prefatory Epistle'.[52] Furthermore, they are represented as incarnation of false authority, as their prestige is grounded in the way they dress:

---

51 Bruno, *Supper* 44. ('eroico, dismesso'; 'gaio, triste'; 'saturnino, gioviale'; 'leggiero, ponderoso'; 'canino, liberale; simico, consulare; sofista con Aristotele, filosofo con Pitagora; ridente con Democrito, piangente con Eraclito.' *Op.it.* I, 432).

52 Ibidem 44. (*Op.it.* I, 434).

[Smith] Speak they Latin well?
The[ophil] Yes.
Smi. Honest men?
The. Yes.
Smi. Of good reputation?
The. Yes.
Smi. Learned?
The. Competently enough.
Smi. Well-bred, courteous, polite?
The. In a rather mediocre degree.
Smi. Doctors?
The. Yes, my master; yes my father; yes my lady; yes, by all means; I believe, from Oxford.
Smi. Qualified?
The. Why not? Men of distinction, of long robes, dressed in velvet, one of them with two golden chains shining round his neck, and the other, with that precious hand (which contained twelve rings on two fingers), seemed (by God) to be a very rich jeweller, who almost carved out your eyes and heart as he gesticulated.
Smi. Did they show a knowledge of Greek?
The. And of beer too, *eziamdio*.[53]

The vain splendour unfolded by the doctors represents an analogy to the illusions of astronomy: in both cases, outward appearances deceive reason and seduce the imagination to construct an illusionary equivalence of the appearance and the real character of the two Oxford professors.

The mocking of these only seemingly learned characters is one strategy of stirring up the reader's emotions: one may feel a mixture of indignation and of hilarity when confronted with these doctors. Another important device

53 Ibidem 51–52. ('[SMITHO] – Parlavan ben latino? TEOFILO – Sì. SMITHO – Galant'uomini? TEOFILO – Sì. SMITHO – Di buona riputazione? TEOFILO – Sì. SMITHO – Dotti? TEOFILO – Assai competentemente. SMITHO – Ben creati, cortesi, civili? TEOFILO – Troppo mediocremente. SMITHO – Dottori? TEOFILO – Messer sì, padre sì, madonnasì, madesì: credo da Oxonia. SMITHO – Qualificati? TEOFILO – Come non? uomini da scelta, di robba lunga, vestiti di velluto: un de quali avea due catene d'oro lucente al collo; e l'altro (per Dio) con quella preziosa mano (che contenea dodeci anella in due dita) sembrava uno ricchissimo gioielliero che ti cavava gli occhii et il core, quando la vagheggiava. SMITHO – Mostravano saper di greco? TEOFILO – E di birra eziamdio.' *Op.it.* I, 441–442).

of Bruno's is the abrupt and elliptic manner in which the Nolan presents his theses.[54] He utters the most provocative and far-reaching assumptions without developing them step-by-step.[55] Just because the Nolan does not systematically expose his philosophy, his sayings acquire a high degree of concision and rhetorical efficacy. In the end, they may not convince entirely, but at least they impress and remain memorable. A fine example of this way of reasoning can be found in the third dialogue, in which the Nolan affirms the infinity of the universe against Nundinio who begs the question when saying that the earth could not move, as it were in the centre of the world and because a centre would always be at rest. The Nolan's notion of the actual infinity of the world is perfectly *à propos* insofar as it annihilates the idea of a centre. This affirmation, however, comes as a surprise and is presented without any further argument:

> But we who set our sight not on fantastic shadows but at the very things, we who envision an aerial, ethereal, spiritual, liquid body capable of motion and of rest, though infinite and immense (this we must affirm at least because we do not perceive any limit either sensibly or rationally), and we know for certain that being the effect and product of an infinite cause and infinite principle, it must be infinitely infinite according to its bodily capacity and according to its mode of being.[56]

Bruno's notion of infinity is derived from his metaphysics and ontology as exposed in *De la causa* and *De l'infinito*. In these works, the existence of an infinite number of worlds is philosophically expounded, whereas in the *Ash Wednesday Supper* Bruno gives only a few hints to this background. The reader

---

54 Michel P.-H., *La cosmologie de Giordano Bruno* (Paris: 1962) 47, calls this form of argumentation aphoristic.

55 At one point in the fourth dialogue, the Nolan laments that the discussion turns around details but that there is no opportunity to come to the really important issues: 'if I were to take time to answer this request, we would be here through the whole night without having any disputation and without ever laying the foundations of our claims against the generally accepted philosophy.' Bruno, *Supper* 133. ('Se io metterò tempo per rispondere a questa dimanda, noi staremo qua tutta la notte senza disputare e senza ponere giamai gli fondamenti delle nostre pretensioni contra la comone filosofia.' Bruno, *Op.it.* I, 531).

56 Bruno, *Supper* 112. ('Ma noi che guardamo non a le ombre fantastiche, ma a le cose medesme; noi che veggiamo un corpo aereo, etereo, spirituale, liquido, capace loco di moto e di quiete, sino immenso et infinito (il che dovamo affermare al meno perché non veggiamo fine alcuno sensibilmente, né razionalmente), e sappiamo certo che essendo effetto e principiato da una causa infinita e principio infinito, deve secondo la capacità sua corporale e modo suo essere infinitamente infinito.' Bruno, *Op.it.* I, 508.)

may only speculate about the notion of this 'infinite cause and infinite principle'. The intended effect of this elliptic method is shown in Nundinio's reaction to the Nolan's assertions: 'he remains foolish and astonished like the one who suddenly sees a new phantasm'.[57] That Bruno's new philosophy of infinity is called 'a new phantasm' fits perfectly with the theory of purgation: by representing the Nolan as someone who speaks in frenzy, he becomes the striking image and the incarnation of his philosophy while the doctors, his opponents, embody the obsolete cosmology and astronomy. While the old phantasms are to be abolished by way of mocking and polemicising, Theophil is building a new phantasm, the phantasm of the Nolan as the liberator from the old chimeras.

The Nolan is thus not simply the *alter ego* of Giordano Bruno. By creating the frenetic Nolan and his counterpart Theophil who argues rationally, responds to objections and explains the enigmatic sentences of the Nolan, Bruno invents an image following the same method he recommended in his writings on artificial memory. He charges the Nolan's character and language with emotional power, which again is a prerequisite for the correction of habitual ways of thinking. Another aspect of this 'new phantasm' becomes evident at the end of the work, when the Nolan is involved in the fruitless discussion with Torquato. While Nundinio at least tried to provide arguments, Torquato applies pure rhetorical brutality. In his report of the Nolan's confrontation with Torquato, Theophil highlights all the attributes and gestures of power summoned by the latter:

> This [i.e. Torquato], with an emphatic look with which the *divum Pater* is described in the *Metamorphoses* as sitting in the middle of the counsel of the Gods and fulminating that most severe sentence against the profane Lycaon, after having viewed his golden necklace, [...] and after having looked closely at the chest of the Nolan, where some buttons were possibly missing, straightened himself, pulled his hands from the table, shook his back a little, rasped his voice, adjusted the velvet biretta on his head, twisted his moustache, straightened his perfumed face, curved his eyebrows, widened his nostrils, placed himself in readiness with a backward glance, put his left hand on his right flank, joined the three first fingers of his right hand so as to start his skirmish, and while tracing with his right

57 Ibidem ('rimase stupido et attonito: come quello a cui di repente appare nuovo fantasma', Bruno, *Op.it.* I, 508).

hand through the air began speaking in this way: *Tune ille philosophorum protoplastes?* [Are you the protoplast of all philosophers?][58]

The vanity of Torquato, his authority and wealth contrast with the Nolan's social irrelevance and his utter lack of recognition in the English context. It is precisely this social and academic insignificance, however, his status as a stranger in every sense of the word that contribute to the revelation of his uncorrupted personality. Like so many reformers, the Nolan is proud of being merely humble and poor, without influence and authority. Thus, Bruno invents the myth of the Nolan as an authentic truth-seeker who stands in striking contrast to the obsolete and decadent authorities of his time.

Towards the end, the debate degenerates more and more, turning into an aggressive skirmish between both opponents. Torquato's incompetence is shown extensively, and Theophil, Smith and Frulla call him names like 'jackass' ('asino') and 'big sheep' ('gran pecoraccia').[59] On the level of the framing dialogue, the superiority of the Nolan is evident, but it is not clear within the narrated dispute: though Theophil suggests that everybody was aware of Torquato's ignorance,[60] he also tells us that some people asked the Nolan to apologise to Torquato. Those were not able to recognise the doctor's ineptitude.[61] At the

58 Ibidem 131–132. ('Costui [i.e. Torquato] con un emfatico aspetto, col quale il *divum pater* vien descritto nella *Metamorfose* seder in mezzo del concilio de gli dèi, per fulminar quella severissima sentenza contra il profano Licaone, dopo aver contemplato la sua aurea collana [...] et appresso remirato al petto del Nolano, dove più tosto arrebbe possuto mancar qualche bottone; dopo essersi rizzato, ritirate le braccia da la mensa, scrollatosi un poco il dorso, sbruffato co la bocca alquanto, acconciatasi la beretta di velluto in testa, intorcigliatosi il mustaccio, posto in arnese il profumato volto, inarcate le ciglia, spalancate le narici, messosi in punto con un riguardo di rovescio, poggiatasi al sinistro fianco la sinistra mano; per donar principio alla sua scrima, appuntò le tre prime dita della destra insieme, e cominciò a trar di mandritti, in questo modo parlando: "*Tune ille philosophorum protoplastes* [...] ?"' Bruno, *Op.it.* I, 528–529).

59 Bruno, *Supper* 134–135.

60 Ibidem 134: 'Now none of those around was so ignorant as not to show with look and gesture that he realised that he was a big sheep *aurati ordinis*'. ('Or qua nessuno di circostanti fu tanto ignorante, che col viso e gesti non mostrasse aver capito che costui era una gran pecoraccia *aurati ordinis*.' *Op.it.* I, 531.)

61 'As all rose from table, there were some who in their own language accused the Nolan of being impatient, instead of having before their eyes the barbarous and boorish discourtesy of Torquato and of their own.' Bruno, *Supper* 137. ('Alzati tutti di tavola, vi furono di quelli che in lor linguaggio accusavano il Nolano per impaziente, in vece che doveano aver più tosto avanti gli occhi la barbara e salvatica discortesia del Torquato e propria.' *Op.it.* I, 535.)

end, when the Nolan thinks to have convicted Torquato of a severe error in his reading of Copernicus (while, in fact, Torquato was perfectly right and the Nolan wrong),[62] the present Englishmen exclude him from the debate by switching from Latin to English, a language the Nolan does not speak.[63] The Nolan evidently failed—not intellectually, because there is no doubt that the text supports his philosophy; but he failed according to his own wish to abolish ignorance by teaching his philosophy. This failure, however, raises the sympathies of Theophil's audience and, at least in intent, those of the reader. While hardly anybody would want to identify with Nundinio or Torquato, the character of the Nolan is touching and appealing. The end of the *Ash Wednesday Supper*, thus, contributes perfectly to the effect of purgation, which in turn is based on the epistemological reflections of the work.

Selective Bibliography

Aquilecchia G., *Schede Bruniane (1950–1991)* (Rome: 1993).

Blumenberg H., *Die Genesis der kopernikanischen Welt* (Frankfurt a.M.: 1989).

Borsche T., "Denken in Bildern. *Phantasia* in der Erkenntnislehre des Giordano Bruno", in Hirdt W. (ed.), *Giordano Bruno. Tragik eines Unzeitgemäßen* (Tübingen: 1993) 93–106.

Bruno Giordano, *The Ash Wednesday Supper*, ed. and transl. S.L. Jaki (The Hague – Paris 1975).

Bruno Giordano, *On the Composition of Images, Signs and Ideas*, transl. by C. Doria, ed. and ann. D. Higgins (New York: 1991).

Bruno Giordano, *Opere italiane*, ed. N. Ordine, 2 vols. (Torino: 2002).

Bruno Giordano, *Opere mnemotecniche*, 2 vols., ed. M. Matteoli – R. Sturlese – N. Tirinnanzi (Milan 2004, 2009).

Bruno Giordano, *De la causa, principio et uno/Über die Ursache, das Prinzip und das Eine*, ed. T. Leinkauf (Hamburg: 2007).

Bruno Giordano, *De l'infinito, universo et mondi/Über das Unendliche, das Universum und die Welten*, ed. A. Bönker-Vallon (Hamburg: 2007).

Copernicus Nicholas, *On the Revolutions*, ed. J. Dobrzcycki, transl. E. Rosen, (Warsaw – Krakow: 1978).

---

62 The issue of this debate is the position of the earth in the Copernican system. Torquato puts the earth correctly in the centre of the epicycle of the moon, while the Nolan puts it on the same epicycle as the moon. Bruno, *Supper* 139–142 (*Op.it.* I, 539–542).

63 Bruno, *Supper* 142 (*Op.it.* I, 542).

Copernicus Nicholas, *Das neue Weltbild. Drei Texte—Commentariolus, Brief gegen Werner, De revolutionibus I*, ed., transl., ann. H.G. Zekl (Hamburg: 1990).

Gatti H., *Giordano Bruno and Renaissance Science* (Ithaca – London: 1999).

Granada M., *Giordano Bruno. Universo infinido, unión con Dios, perfección del hombre* (Barcelona: 2002).

Kant Immanuel, *Kritik der reinen Vernunft* (Frankfurt a.M.: 1956) vol. 1, 25.

Maspero R., "Scienza e copernicanismo in Bruno: Principali orientamenti della critica dal 1950 ad oggi", *Rivista di storia della filosofia* 1 (1989) 141–162.

Michel P.-H., *La cosmologie de Giordano Bruno* (Paris: 1962).

Rosa G. De, *Il concetto di immaginazione nel pensiero di Giordano Bruno* (Naples: 1997).

Spruit L., *Il problema della conoscenza in Giordano Bruno* (Naples: 1988).

Sturlese R., "Per un'interpretazione del De umbris idearum di Giordano Bruno", *Annali della Scuola Normale Superiore di Pisa*, third series, 22 (1992) 943–968.

Tessicini D., *I dintorni dell'infinito: Giordano Bruno e l'astronomia del Cinquecento* (Pisa: 2007).

Westman R.S., "Magical Reform and Astronomical Reform: The Yates Thesis Reconsidered", in idem – McGuire J.E. (eds.), *Hermeticism and the Scientific Revolution*. Papers read at a Clark Library Seminar, March 9, 1974 (Los Angeles: 1977) 1–91.

Yates F.A., *Giordano Bruno and the Hermetic Tradition* (London: 1964).

CHAPTER 7

Die Erde als Mond. Kopernikanische Wenden in Raumreiseerzählungen des 17. Jahrhunderts (Kepler, Godwin, Cyrano de Bergerac)

Thomas Rahn

Kopernikanische Wenden? Im Experimentalraum der frühneuzeitlichen Raumreiseerzählungen[1] wird auf ganz verschiedene Weisen probiert, die Kopernikanische Wende ‚durchzuspielen' und die neue Astronomie durch *Inversionen* zu bestätigen. Im ersten Teil des Aufsatzes werden fiktive Blicke auf die Erde in Keplers *Somnium*, Godwins *The Man in the Moone* und Cyrano de Bergeracs *L'Autre Monde* verglichen,[2] die auf die Evidenz kopernikanischer

1 Vgl. Nicolson M.H., „Cosmic Voyages", *English Literary History* 7 (1940) 83–107; Nicolson M.H., *Voyages to the Moon* (New York: 1948); Guthke K.S., *Der Mythos der Neuzeit. Das Thema der Mehrheit der Welten in der Literatur- und Geistesgeschichte von der kopernikanischen Wende bis zur Science Fiction* (Bern – München 1983); Pleithner R., „Zwei Mondreisen des 17. Jahrhunderts: Voyages imaginaires oder Reiseutopien?", in Pleithner R. (ed.), *Reisen des Barock. Selbst- und Fremderfahrung und ihre Darstellung* (Bonn: 1991) 75–87; Campbell M.B., „Impossible Voyages: Seventeenth-Century Space Travel and the Impulse of Ethnology", *Literature and History*, 3rd series 6, 2 (1997) 1–17; Pioffet M.-C., „Godwin et Cyrano: Deux conceptions du voyage", *Dalhousie French Studies* 39/40 (1997) 45–57; Weber A.S., „Changes in Celestial Journey Literature: 1400–1650", *Culture and Cosmos: A Journal of the History of Astrology and Cultural Astronomy* 1 (1997) 34–50; Campbell M.B., *Wonder & Science. Imagining Worlds in Early Modern Europe* (Ithaca, N.Y.: 1999) 111–218; Guthke K.S., „Nightmare and Utopia: Extraterrestrial Worlds from Galileo to Goethe", *Early Science and Medicine* 8 (2003) 173–195; Cressy D., „Early Modern Space Travel and the English Man in the Moon", *American Historical Review* 111 (2006) 961–982; Aït-Touati F., *Fictions of the Cosmos: Science and Literature in the Seventeenth Century* (Chicago – London: 2011); Siebenpfeiffer H., „(Imagining) First Contact—Literary Encounters of the Extraterrestrial Other in 17th and 18th Century Novels", in Blackmore S. – Haekel R. (eds.), *Discovering the Human. Life Science and the Arts in the Eighteenth and Early Nineteenth Centuries* (Göttingen: 2013) 139–155.

2 Drei Texte, die wiederholt als frühneuzeitliche Inauguraltexte des Science Fiction-Genres diskutiert wurden; vgl. Christianson G.E., „Kepler's *Somnium*: Science Fiction and the Renaissance Scientist", *Science Fiction-Studies* 3 (1976) 76–90; Menzel, D.H., „Keplers Place in Science Fiction", *Vistas in Astronomy* 18 (1975) 895–904; Hervier J., „Cyrano de Bergerac et le Voyage spatial: de la fantaisie á la science-fiction", in Blakian A. – Wilhelm J.J. (eds.), *Proceedings of the Tenth Congress of the International Comparative Literature Association*

 | DOI 10.1163/9789004281127_009

Thesen (Erdrotation, Planetenargument, Heliozentrismus) zielen. Der zweite Teil sucht in Keplers *Somnium* auf der Dispositionsebene des Buches nach Perspektivenwechseln und Fiktionalitätsmarkierungen, die als Transformation methodologischer Implikationen des Textes verstehbar sind. Im dritten Teil wird gezeigt, wie in der Zoologie und der Kosmologie der Mondbewohner in Cyrano de Bergeracs *L'Autre Monde* irdische Anthropozentrismen invertiert werden und in einer Prozeßepisode, die auf den Inquisitionsprozeß gegen Galilei anspielt, in eine Geltungskonkurrenz geraten.

1 Evidenz und *evidentia*: Blicke auf die Erde bei Kepler, Godwin und Cyrano de Bergerac

Die Raumreiseerzählungen der Frühen Neuzeit ermöglichen Beobachtungsszenarien, mit denen der großartigen Pseudo-Evidenz des Geozentrismus—dem Kreisen des Himmels um die Erde—durch *evidentia*, d.h. durch ein deskriptives und metaphorisches ‚vor Augen Stellen' von Phänomenen, die sich nur im All betrachten lassen, begegnet werden kann.[3] Die fiktionalen Blicke auf die Erde, die hier untersucht werden,

> sollen Sichtbarkeit ersetzen, selbst Sichtbarkeiten werden. Wer Evidenz produziert, verfolgt die Absicht, bei seinen Adressaten einen Blickwechsel hervorzurufen, ein anderes Sehen und ein alternatives Verständnis zu implementieren.[4]

(New York: 1985) vol. 1: *General Problems of Literary History*, 436–442; Bozetto R., „Kepler's ‚Somnium'; or, Science Fiction's Missing Link", *Science Fiction-Studies* 17 (1990) 370–382; Poole W., „Kepler's *Somnium* and Francis Godwin's *The Man in the Moone*: Birth of Science-Fiction 1593–1638", in Houston C. (ed.), *New Worlds Reflected. Travel and Utopia in the Early Modern Period* (Surrey – Burlington 2010) 57–69.

3 Der Begriff der *evidentia* wird, mit Rücksicht auf die Entstehungszeit der untersuchten Texte, im Folgenden ganz bewußt in dem Sinne des 17. Jahrhunderts verwendet, wie er etwa bei Gerhard Johannes Vossius in dessen einflußreichem Rhetoriklehrbuch *Commentariorum Rhetoricorum, sive Oratoriorum Institutionum libri sex* (1630) begegnet, nämlich sowohl als Bezeichnung für deskriptive/narrative als auch für metaphorische Verfahren. Vgl. Campe R., „Vor Augen Stellen. Über den Rahmen rhetorischer Bildgebung", in Neumann G. (ed.), *Poststrukturalismus. Herausforderung an die Literaturwissenschaft*, Germanistische Symposien. Berichtsbände 18 (Stuttgart – Weimar: 1997) 208–225, hier 213.

4 Rößler H., *Die Kunst des Augenscheins. Praktiken der Evidenz im 17. Jahrhundert*, Wissenschaftsgeschichte 2 (Berlin – Münster – Zürich: 2012) 15.

Die ‚gedankenexperimentellen' Evidenzproduktionen der Raumreiseerzählungen (Evidenz nicht verstanden als simulierte ‚Selbstevidenz', sondern als Zusammenspiel von Imagination und Theorie)[5] versuchen mit einer faktisch unmöglichen Umwendung des Blicks eine Kehre der Weltwahrnehmung zu erreichen.

Die erste literarische (An-)Wendung der ‚Kopernikanischen Wende' in einer Raumreiseerzählung ist eine ‚Keplersche Wende'. Kepler, der das letztlich konservative Forschungsprogramm des Kopernikus durch eine Physikalisierung der Astronomie revolutionierte,[6] simuliert in seinem 1634 publizierten *Somnium, seu opus posthumum de astronomia lunari*[7] die astronomische Wahrnehmung fiktiver Mondbewohner, um durch diesen Perspektivenwechsel ‚einen Beweis zu führen für die Bewegung der Erde, oder eher: die Einwände zu entkräften, die aus dem allgemeinen Widerspruch des Menschengeschlechts gewonnen sind'.[8] Die Idee für dieses Gedankenexperiment geht bis auf die (zurückgewiesene) Dissertation des jungen Kepler zurück; das *Somnium* verfaßte Kepler im Jahr 1609—also aus der vorteleskopischen Wissenssituation der Astronomie heraus—und ergänzte es in den Jahren 1622 bis 1630 um einen umfangreichen Apparat von 223 Endnoten, einen mondgeographischen Anhang, dessen

5 Vgl. zum begrifflichen Konzept von Evidenz in der Frühen Neuzeit Zittel C., „Trügerische Evidenzen. Bild-Lektüren in wissenschaftlichen Texten der Frühen Neuzeit", in Dettmar U. – Lotz S. – Migge B. – Pieper I. (eds.), *Grenzbereiche des Lesens* (Frankfurt: 2005) (Text online: http://publikationen.stub.uni-frankfurt.de/frontdoor/index/docId/4758); ders, *Theatrum Philosophicum. Descartes und die Rolle ästhetischer Formen in der Wissenschaft*, Wissenskultur und gesellschaftlicher Wandel 22 (Berlin: 2009) 248–279. Vgl. ferner zum Verhältnis von Rhetorik und Evidenz im Kontext der frühneuzeitlichen Astrononomie und ihrer Beobachtungstechnik Witthaus J.-H., *Fernrohr und Rhetorik. Strategien der Evidenz von Fontenelle bis La Bruyère*, Neues Forum für allgemeine und vergleichende Literaturwissenschaft 28 (Heidelberg: 2005).

6 Vgl. Mittelstraß J., „Kopernikanische oder Keplersche Wende?—Keplers Kosmologie, Philosophie und Methodologie", *Vierteljahrsschrift der Naturforschenden Gesellschaft in Zürich* 134 (1989) 197–215.

7 Kepler Johannes, *IOH. KEPPLERI MATHEMATICI OLIM IMPERATORII SOMNIVM, Seu OPVS POSTHVMVM DE ASTRONOMIA LVNARI*, ed. Ludwig Kepler (Sagan – Frankfurt a.M., Johannes Keplers Erben: 1634).

8 Kepler Johannes, *Der Traum, oder: Mond-Astronomie*, übers. H. Bungarten, ed. B. Langner (Berlin: 2011) 29. ‚[...] argumentum pro motu Terrae, seu solutionem potius objectionum ab vniversali contradictione gentis humanae desumptarum, moliri', Kepler Johannes, *Gesammelte Werke*, ed. Kepler-Kommission der Bayerischen Akademie der Wissenschaften, vol. 11.2: *Calendaria et Prognostica, Astronomica minora, Somnium*, ed. V. Bialas – H. Grössing (München: 1993) 315–438, hier 333.

Spekulationen sich nun bereits auf teleskopische Observationen stützen konnten, und die eigene lateinische Übersetzung eines wichtigen Prätextes für seine Erzählung: Plutarchs *De facie in orbe lunae*.[9] Kepler erörtert den gattungsmäßigen Ort seines Textes in einem Brief an Matthias Bernegger: Komplementär zu Campanellas *Civitas solis* als ‚Mondstaat' etikettiert, wird das Buchprojekt 1623 in die Nähe der Utopie gerückt und zugleich die Möglichkeit verworfen, ein lunares Staatswesen als Projektionsfläche für satirische Verhaltenskritik zu erfinden. Stattdessen kündigt Kepler—mit Hinweis auf die Schwierigkeiten, welche die *Utopia* Thomas Morus und das *Moriae encomion* Erasmus eingebracht hätten—die Fokussierung auf das ‚Feld der Philosophie' (d.h. der Naturforschung) an.[10] Sprich: Kepler interessiert sich weniger dafür, was die Mondbewohner tun, als dafür, was sie *sehen*. Das ist allerdings auch ein heißes Eisen.

Der Text hat die Form einer Traumerzählung, wie sie als Fiktionsrahmen von Astralreiseerzählungen durch Ciceros *Somnium Scipionis* etabliert war.[11] Daß Kepler den Traum als legitimatorisches Vehikel nutzte, ‚um sich selbst vom Inhalt des Textes zu distanzieren und seine eigene Verantwortung für die Argumentation zu reduzieren',[12] erscheint in Anbetracht der offenen und, insbesondere in den Endnoten, auch ironisch-streitbaren Parteinahme für das heliozentrische Weltmodell als wenig plausibel. Vielmehr ging es darum,

9 Romm J.S., „Lucian and Plutarch as sources for Kepler's *Somnium*", *Classical and Modern Literature* 9 (1989) 97–107.—Kepler erwog zudem, dem *Somnium* Lukians Raumreiseerzählung *Vera historia* als Anhang beizugeben. Vgl. den Brief Keplers an Bernegger vom 4. Dezember 1625 in Kepler Johannes, *Gesammelte Werke*, ed. W.v. Dyck – M. Caspar, vol. 18: *Briefe 1620–1630*, ed. M. Caspar (München: 1959) 143. Zur Lukian-Rezeption im *Somnium* vgl. Pantin I., „Kepler et Lucien: des voyages extraordinaires au *Ludus philosophicus*", in Ligota C. – Panizza L. (eds.), *Lucian of Samosata vivus et redivivus* (London – Turin: 2007) 115–127.

10 Vgl. den Brief Keplers an Bernegger vom 4. Dezember 1625: ‚Scripsit Campanella civitatem Solis, quid si nos Lunae? Anne egregium facinus Cyclopicos hujus temporis mores vivis coloribus depingere, sed cautionis causa terris cum tali scriptione excedere, inque Lunam secedere? Quanquam quid tergiversari juvabit? cum nec Morus in Utopia nec Erasmus in encomio Moriae, tuti fuerint, quin utrisque fuit apologiâ opus. Missam igitur penitus faciamus picem hanc politicam, nosque in amoenis Philosophiae viretis planè contineamus.' Kepler, *Gesammelte Werke* vol. 18, 143.

11 Vgl. speziell zu den frühneuzeitlichen Raumreiseerzählungen in Traumform Glomski J., „Science Fiction in the Seventeenth Century: The Neo-Latin Somnium and its Relationship with the Vernacular", in Tilg S. – Walser I. (eds.), *Der neulateinische Roman als Medium seiner Zeit. The Neo-Latin Novel in its Time* (Tübingen: 2013) 37–51.

12 Schirrmeister A., „Traum und Wissen in der Frühen Neuzeit. Erste Annäherungen", *Zeitsprünge* 5 (2001) 297–310, hier 307.

durch das ‚Traum'-Signal den (mehrdimensionalen) allegorischen Charakter des *Somnium* auszuweisen.[13]

Der Text schachtelt drei Erzählebenen ineinander: Auf der ersten Ebene berichtet ein ‚Kepler'-Ich davon, wie es in den Schlaf fällt und zu träumen beginnt:

> Als im Jahr 1608 Streit zwischen den Brüdern Kaiser Rudolf und Erzherzog Matthias entbrannt war, und man ihre Handlungen allgemein auf Beispiele zurückführte, die aus der Geschichte Böhmens herangezogen waren, wurde ich durch die allgemeine Neugier erregt und wandte mich der Lektüre böhmischer Literatur zu. Dabei stieß ich auf die Geschichte der Heldin Libussa, einer hochberühmten Zauberin; so kam es, dass ich eines Nachts, nachdem ich den Mond und die Sterne betrachtet hatte, behaglich im Bett lag und in tiefen Schlaf fiel. Und ich sah mich im Schlaf ein Buch durchlesen, das ich in Frankfurt auf der Messe erworben hattte, dessen Text folgender war:[14]

Auf der Ebene der Traumhandlung erzählt eine zweite Erzählinstanz, der Isländer Duracotus, wie es ihn in jungen Jahren, wegen einer Ungeschicklichkeit von der Mutter Fiolxhilde verstoßen, auf die dänische Insel Hven verschlägt, wo er von Tycho Brahe in der Astronomie unterwiesen wird. Nach Island zurückgekehrt, zeigt die Mutter des Duracotus, die sich auf die Magie versteht, großes Interesse an dem astronomischen Wissen ihres Sohnes und meint nun sterben zu können, weil dieser das Erbe ihrer Wissenschaft antreten werde. (Grob gesagt, die Allegorie eines wissenschaftlichen Paradigmenwechsels: der Überwindung ungelehrter Empirie, wie es Keplers vierte Anmerkung zum Text

---

13 Vgl. zur Einordnung des *Somnium* in die Tradition der Traumdichtung Swinford D., *Through the Daemon's Gate. Kepler's Somnium, Medieval Dream Narratives, and the Polysemy of Allegorical Motifs* (New York – London: 2006).—Vgl. zur allegorischen Dimension des Textes neben Swinford vor allem Bezzola Lambert L., *Imagining the Unimaginable: The Poetics of Early Modern Astronomy*, Internationale Forschungen zur Allgemeinen und Vergleichenden Literaturwissenschaft 58 (Amsterdam – New York: 2002), 66–105.

14 Kepler, *Der Traum, oder: Mond-Astronomie* 7. ‚Cum anno 1608 ferverent dissidia inter fratres Imp: Rudolphum et Matthiam Archiducem; eorumque actiones vulgo ad exempla referrent, ex historia Bohemica petita; ego publica vulgi curiositate excitus, ad Bohemica legenda animum appuli. Cumque incidissem in historiam Libussae Viraginis, arte Magica celebratissimae: factum quadam nocte, vt post contemplationem siderum et Lunae, lecto compositus, altius obdormiscerem: atque mihi per somnum visus sum librum ex Nundinis allatum perlegere, cuius erat tenor:' Kepler, *Gesammelte Werke* vol. 11.2, 321.

nahelegt.) Ihre Verbindung zur Dämonenwelt nutzt Fiolxhilde, um in einem magischen Ritual gemeinsam mit Duracotus den ‚Dämon aus Levania' anzurufen, der als dritte Erzählinstanz im Detail die astronomische Wahrnehmung der Mondbewohner referiert. Dieser Bericht bricht an einer Stelle plötzlich ab, an der von Regenfällen auf dem Mond die Rede ist, weil das ‚Kepler'-Ich der ersten Erzählebene durch den Regen geweckt wird, der an sein Fenster prasselt.—Keplers *Somnium* ist ein Buch über einen Traum, der von einem Buch handelt, in dem jemand erzählt, was ihm jemand erzählt hat. Die narrative Anlage des Textes folgt einer ‚linear structure of increasing „de-realisation" ',[15] wie Christian Schneider es nennt, und der Wahrheitsanspruch des *oraculum* bzw. der *visio*[16] vom Sonnensystem wird durch Tagesreste und Außenreize als profanen Auslösern von Traumgegenständen ironisiert.

Das Sprachrohr der lunaren Himmelsautopsie ist ein Dämon, dessen Gattung in der Lage ist, in den Schattenkegeln der Mond- und Sonnenfinsternisse zwischen Erde und Mond hin- und herzureisen (und dabei auch Menschen zu transportieren).[17] In den Noten zum Text ist dieser ‚Daemon ex Levania'[18] als Personifikation der Astronomie entschlüsselt, etwa in Anmerkung 55, in der die schattengebahnte Reise mit dem astronomischen Verfahren der Schattenmessung in Verbindung gebracht wird. Mit der Raumreisefähigkeit des Dämons ist die Astronomie als Wissenschaft behauptet, die zur Überwindung einer irdischen Bindung der Perspektive in der Lage ist. Durch den ‚interperspektivischen' Vergleich von Augenscheinlichkeiten auf Mond und Erde produziert der Text eine ‚transperspektivische' Evidenz des kopernikanischen Weltbildes (ein textrhetorisch raffiniert überspielter Zirkelschluß, da die lunare Empirie doch zuvor aus der heliozentrischen Hypothese extrapoliert wurde). Die *Interperspektivität* des Dämons wird auf der narrativen Ebene durch Momente eines unzuverlässigen Erzählens umgesetzt. Zum einen annonciert die Zwischenüberschrift, die dem Bericht von der lunaren

15 Schneider C., „Science as Science Fiction. Johannes Kepler's *Somnium* and the Poetics of Invention", in Bernhart T. – Mehne P. (eds.), *Imagination und Invention*, Paragrana. Beiheft 2 (Berlin: 2006) 259–268, hier 262.

16 Vgl. zur möglichen Kategorisierung des Astronomieberichts als *oraculum* (des Dämons als Autorität) bzw. als wahrheitsverbürgende *visio* Bezzola Lambert, *Imagining the Unimaginable* 131.

17 Vgl. Maus de Rolley T., „Voler avec le diable: démons et voyages célestes au tournant de XVIe et XVIIe siècles (Faust/Kepler)", in Holtz G. – Maus de Rolley T., *Voyager avec le diable. Voyages réels, voyages imaginaires et discours démonologiques (XVe–XVIIe siècles)* (Paris: 2008); ders., *Élevations: l'écriture du voyage aérien à la Renaissance* (Genf: 2011) 523–539.

18 Kepler, *Gesammelte Werke* vol. 11.2, 323.

Astronomie vorangestellt ist, den Dämon als ‚Daemon ex Levania' und läßt damit im Unklaren, ob er von dort herstammt oder ob er nach einer Reise von dort zurückgekehrt ist,[19] zum anderen ist der Bezug des ‚nos' und des ‚nobis' in der Rede des Dämons ambivalent: Zunächst wird damit nur die Gattung der Dämonen bezeichnet, ab der Aussage ‚Zwar ist der Anblick der Fixsterne für ganz Levania derselbe wie für uns'[20] (‚Etsi siderum fixorum aspectus tota Levania habet nobiscum eosdem')[21] aber eine *Wahrnehmungsgemeinschaft* des Dämons und der Menschen in Abgrenzung von der referierten Wahrnehmung der ‚Privolven' (d.h. der Bewohner der erdabgewandten Seite des Mondes) und der ‚Subvolven' (d.h. der Mondbewohner, die auf die Erde blicken).[22]

‚Volva' heißt die Erde auf dem Mond, weil die Erdrotation hier die ‚Kernsensation' bei ihrer Wahrnehmung ist.[23] Was die Subvolven aber auch sehen, wenn sie auf die Erde blicken, ist eine riesige Uhr:

> [Es] zeigt auch ihre Volva ihnen durch sich selbst die Stunden an. Obwohl man sie sich nämlich überhaupt nicht vom Ort bewegen sieht, vollführt sie an ihrem Platz eine Kreisbewegung um sich selbst, im Gegensatz zu unserem Mond, und zeigt nacheinander eine bewundernswerte Fülle verschiedener Flecken, die sich beständig von Osten nach Westen verschieben. Eine solche Umdrehung also betrachten die Subvolven, wenn dieselben Flecken wiederkehren, als eine Zeitstunde. Dem entspricht aber etwas mehr als ein Tag und eine Nacht bei uns: [...] Im Allgemeinen scheint diese Volva, was den oberen nördlichen Teil angeht, zwei Hälften zu haben, eine dunklere und mit zusammenhängenden Flecken bedeckte und eine leicht hellere. Als Trennung liegt zwischen beiden ein heller Gürtel zum Norden hin. Die Gestalt ist schwer zu erklären. Doch erkennt man im östlichen Teil etwas wie das Profil eines Menschen, in Höhe der Achseln abgeschnitten, der sich ein Mädchen zum Küssen heranzieht, das in ein langes Gewand gehüllt ist und mit nach hinten ausgestreckter

19 Vgl. Swinford, *Kepler's Somnium* 171.

20 Kepler, *Der Traum, oder: Mond-Astronomie* 13.

21 Kepler, *Gesammelte Werke* vol. 11.2, 324.

22 Vgl. Reiss T.J., *The Discourse of Modernism* (Ithaca – London: 1982) 158.

23 Kepler kommentiert in Note 89/90: ‚Aufgrund der Umdrehung also soll sie Volva genannt werden [von lat. volvere, umwälzen], und Subvolven oder Subvolanen diejenigen, welche die Volva sehen, Privolven die, welche des Anblicks der Volva beraubt sind [von lat. privare, berauben].' Kepler, *Der Traum, oder: Mond-Astronomie* 52. ‚A volvendo igitur Volva dicatur, et Subvolae vel Subvolvani, qui vident Volvam: Privolvae, qui sunt privati conspectu Volvae.' Kepler, *Gesammelte Werke* vol. 11.2, 343.

> Hand eine heranspringende Katze reizt. Doch der größere und breitere Teil des Fleckens springt ohne bestimmbare Form nach Westen vor.
>
> Die andere Hälfte der Volva besteht aus mehr hellerer Fläche und einem Fleck. Man könnte diesen das Bild einer Glocke nennen, die an einem Seil hängt und nach Westen geschwungen ist. Was darüber und darunter ist, kann man nicht identifizieren.[24]

Die Produktion einer Evidenz der Erdrotation stützt sich hier auf eine *evidentia*, in der die Wahrnehmung der Subvolven und der Erdenmenschen amalgamiert ist: Die Landmassen der Erde können von den Mondbewohnern nicht geographisch benannt werden, daher projizieren sie Bildassoziationen auf den Himmelskörper. Afrika wird zum Profil eines Menschen, Südamerika zu einer schwingenden Glocke. Aus der Perspektive des Lesers erscheinen die assoziativen Transformationen als Rätselbilder, die eine imaginative Rückübersetzung in ein Kartenbild fordern (was die Noten im Anhang durch Aufschlüsselung erleichtern). Doch gerade die Hürde der Imagination verstärkt die *evidentia*. Die entscheidende Rezeptionsanweisung für das Übersetzungsspiel der inneren Bilder gibt die Assoziation Europas als eines lang gewandeten Mädchens, das mit zurückgestreckter Hand (Britannien) eine Katze (Skandinavien) reizt. Das Bild modifiziert die in der Frühen Neuzeit geläufige Ikonographie der ‚Europa regina',[25] ein allegorisches Kartenbild, in dem die Frauenfigur auf

---

24 Kepler, *Der Traum, oder: Mond-Astronomie* 21. ‚[...] etiam sua ipsis Volva per seipsum distinguit horas. Etsi enim loco nequaquam moveri cernitur; intra tamen locum suum, contrà quam nostra Luna, gyratur, et admirabilem macularum varietatem successivè explicat, assiduè ab ortu in occasum translatis maculis. Una igitur talis revolutio, quando eaedem maculae redeunt, Subvolvanis habetur pro una hora temporali; aequat autem paulò quid ampliùs, quàm unum diem et unam noctem nostratem [...].

In universum Volva ista, quod superiorem Septentrionalem partem attinet, duas videtur habere medietates; alteram obscuriorem, et continuis quasi maculis obductam, alteram paulò clariorem, interfuso pro discrimine utriusque, cingulo lucido septentrionetenus. Figura difficilis est explicatu. Parte tamen orientaliori cernitur instar Protomes capitis humani, axillarum tenùs resecti, admoventis ad oscula puellulam cum veste longa, quae extenta retrorsum manu felem assultantem provocet. Major tamen et latior maculae pars sine evidenti forma versus occidentem procurrit.

In altera medietate Volvae, latiùs diffunditur splendor quàm macula. Effigiem diceres Campanae de fune dependentis et in occidentem jactatae. Quae suprà infraque sunt, assimilari nequeunt.' Kepler, *Gesammelte Werke* vol. 11.2, 328.

25 Vgl. Werner A.E., „Triumphierende Europa—Klagende Europa. Zur visuellen Konstruktion europäischer Selbstbilder in der Frühen Neuzeit", in Ißler R.A. – Renger A.B. (eds.), *Europa—Stier und Sternenkranz. Von der Union mit Zeus zum Staatenverbund*, Gründungsmythen Europas in Literatur, Musik und Kunst 1 (Göttingen – Bonn: 2009) 241–260.

einer von Südosten nach Nordwesten verlaufenden Achse liegt, ausgehend von Spanien als Krone und mit einem rechten italienischen Arm, der einen Reichsapfel hält, und einem linken skandinavischen Arm mit einem Szepter. Die Europa-Transformation in Keplers Text ermöglicht also sowohl einen Wiedererkennungseffekt beim Leser als auch das Bemerken einer Abweichung (Britannien statt Skandinavien als linker Arm)—eine Differenz, die *der anderen* (unwissenden) Wahrnehmung der Subvolven zugeschrieben werden kann.

Die Übersetzung der Erdkugel in einen Globus, d.h. in einen kartographischen Bildmodus, liefert für den Leser zum einen einen akzeptablen und funktionierenden Imaginationsrahmen (d.h. einen zeitgenössisch geläufigen Blick auf die Erde), zum anderen operiert die Bildtransformation mit ‚merkwürdigen' und also einprägsamen Assoziationen nach dem Schema der mnemonischen Bilderfindung, ohne dabei kartographische Darstellungslizenzen zu überschreiten.[26]

Ebenfalls als Uhr und als riesiger mathematischer Globus erscheint die Weltkugel in Francis Godwins *The Man in the Moone*.[27] Die Mondreiseerzählung[28] des Bischofs von Llandaff und Hereford, der im Gegensatz zu Kepler als wissenschaftsinteressierter Laie schrieb, war vermutlich nicht vor 1628 fertiggestellt[29] und wurde nach ihrer posthumen Publikation im Jahr 1638 europaweit zu einem

26 Man denke neben der ‚Europa regina' auch an die kartographische Allegorie des ‚Leo Belgicus'. Vgl. hierzu Michalsky T., *Projektion und Imagination. Die niederländische Landschaft der Frühen Neuzeit im Diskurs von Geographie und Malerei* (München: 2011) 109–113.

27 Godwin Francis, *THE MAN IN THE MOONE: OR A DISCOVRSE OF A Voyage thither BY DOMINGO GONSALES The speedy Messenger* (London, Joshua Kirton und Thomas Warren: 1638).

28 Vgl. zu den englischen Mondreisespekulationen des 17. Jahrhunderts (Godwin und Wilkins) Bachrach A.G.H., „Luna Mendax: Some Reflections on Moon-Voyages in Early Seventeenth-Century England", in Baker-Smith D. – Barfoot C.C. (eds.), *Between Dream and Nature: Essays on Utopia and Dystopia*, DQR Studies in Literature 2 (Amsterdam: 1987) 70–90; Roos A.M.E., *Luminaries in the Natural World: The Sun and the Moon in England, 1400–1720* (New York: 2001) 129–136 u. 153–156; Aït-Touati F., „La découverte d'un autre monde: fiction et théorie dans les œuvres de John Wilkins et de Francis Godwin", *Études Épistémè* 7 (2005) 15–31.

29 Vgl. zur Datierungsfrage McColley G., „The Date of Godwin's ‚Domingo Gonsales'", *Modern Philology* 35 (1937) 47–60; Empson W., „Godwin's Voyage to the Moon", in ders., *Essays on Renaissance Literature*, ed. J. Haffenden (Cambridge: 1993) vol 1: *Donne and the New Philosophy*, 220–254; Poole W., „The Origins of Francis Godwin's *The Man in the Moone* (1638)", *Philological Quarterly* 84 (2005) 189–210.

Erfolgsbuch.[30] Der Text, der Gattungselemente des pikarischen Romans,[31] der Reisebeschreibung und der Utopie verbindet, läßt seinen Protagonisten Domingo Gonsales, von einem Vogelgespann gezogen, ganz ungeplant auf den Mond reisen, wo er auf eine Gesellschaft moralisch vollkommener Wesen trifft. Der Erzählabschnitt, in dem der Raumflug geschildert wird, fällt insbesondere dadurch aus dem Rahmen, daß der Ich-Erzähler die im All beobachteten Phänomene nicht nur mit einem besonderen Exaktheitsgestus beschreibt, sondern auch mit Bezug auf die kosmologische Kontroverse wissenschaftlich erörtert; dabei wird, jedenfalls bezüglich der Frage der Erdrotation, eine kopernikanische Position eingenommen.[32] Als Quellen des kosmologisch-physikalischen Diskurses im *Man in the Moone* lassen sich u.a. William Gilberts *De magnete*[33] und die ‚Digression of Air' aus Robert Burtons *Anatomy of Melancholy*[34] bestimmen.

Das ‚Forschungsprogramm' der fiktionalen Reisebeschreibung wird bereits in der Vorrede in den Kontext der neuen Astronomie gestellt:

---

30 Vgl. zu den Übersetzungen und Fortschreibungen bis in das 18. Jahrhundert hinein Janssen A., „Wirkung eines Romans als Inspirationsquelle. Francis Godwins ‚The Man in the Moone' ", *Arcadia* 20 (1985) 20–46. Die deutsche Bearbeitung untersucht Kiesant K., „Science Fiction im Barock. *Der Fliegende Wandersmann nach dem Mond* (1659)", in Weber T.P. (ed.), *Science & Fiction II. Leben auf anderen Sternen* (Frankfurt a.M.: 2004) 41–60, 258f.

31 Vgl. Copeland T.A., „Francis Godwin's *The Man in the Moone*: A Picaresque Satire", *Extrapolation: A Journal of Science Fiction and Fantasy* 16 (1975) 156–163.

32 Vgl. Lawton H.W., „Bishop Godwin's *Man in the Moone*", *The Review of English Studies* 7 (1931) 25–55, hier 28–30; Johnson F.R., *Astronomical Thought in Renaissance England: A Study of the English Scientific Writings from 1500 to 1645* (Baltimore: 1937) 232–234; Janssen A., *Francis Godwins ‚The Man in the Moone'. Die Entdeckung des Romans als Medium der Auseinandersetzung mit Zeitproblemen*, Sprache und Literatur. Regensburger Arbeiten zur Anglistik und Amerikanistik 19 (Frankfurt a.M. – Bern: 1981) 92–102; Trinkner D., „Träume von Levania. Zur Rezeption neuer naturwissenschaftlicher Erkenntnisse am Beispiel von Galileis ‚Sidereus Nuncius' und Francis Godwins ‚The Man in the Moone' ", *Morgen-Glantz* 8 (1998) 227–254; Hutton S., „*The Man in the Moone* and the New Astronomy: Godwin, Gilbert, Kepler", *Études Épistémè* 7 (2005) 3–14; Aït-Touati, *Fictions of the Cosmos* 46–56.

33 Vgl. Hutton, „The man in the Moone and the New Astronomy" 8–11. Vgl. zur Kosmologie bei Gilbert Freudenthal G., „Theory of Matter and Cosmology in William Gilbert's *De magnete*", *Isis* 74 (1983) 22–37.

34 Vgl. Poole, „The Origins of Francis Godwin's *The Man in the Moone* (1638)". Vgl. zur Kosmologie in Burtons ‚Digression of Air' Barlow R.G., „Infinite Worlds: Robert Burton's Cosmic Voyage", *Journal of the History of Ideas* 34 (1973) 291–302.

> THou hast here an essay of *Fancy*, where *Invention* is shewed with *Judgment*. It was not the *Authors* intention (I presume) to discourse thee into a beleife of each particular circumstance. Tis fit thou allow him a liberty of conceite; where thou takest to thy selfe a liberty of judgment. In substance thou hast here a new discovery of a new *world*, which perchance may finde little better entertainment in thy opinion, than that of *Columbus* at first, in the esteeme of all men. Yet his than but poore espiall of *America*, betray'd unto knowledge soe much as hath since encreast into a vaste plantation. And the then unknowne, to be now of as large extent as all other knowne *world*.
>
> That there should be *Antipodes* was once thought as great a *Paradox* as now that the *Moon* should bee habitable. But the knowledge of this may seeme more properly reserv'd for this our discovering age: In which our *Galilæusses*, can by advantage of their spectacles gaze the Sunne into spots, & descry mountaines in the *Moon*. But this, and more in the ensuing discourse I leave to thy candid censure, & the faithfull relation of the little eye-witnesse, our great discoverer.[35]

Der Hinweis auf Galileis teleskopische Observationen der Sonnenflecken[36] und der Mondoberfläche affirmiert Erkenntnisse, die geeignet sind, die Fiktion durch Fakten zu initiieren (und zu plausibilisieren). Die Feststellung einer Erdähnlichkeit des Mondes lieferte das kopernikanische Planetenargument, daß die Erde nicht aus der schwersten und darum ruhenden Materie im All besteht, sondern offenbar physisch die Voraussetzungen eines Wandelsterns teilt. Zugleich erlaubt die Entdeckung von ‚mountaines in the Moon' als Indiz eines erdähnlichen Habitats aber auch die Projektion außerirdischen menschlichen Lebens auf den Mond.

Der Blick durch das Fernrohr liefert, als instrumentelle Verlängerung und ‚Versetzung' des Blicks in das All, das Modell der Raumreise. Die Observationen des *Sidereus nuncius* werden durch die imaginative Zeugenschaft des ‚speedy Messenger'[37] Domingo Gonsales übersetzt und fortgesetzt.[38] Gonsales, der

35 Godwin, *THE MAN IN THE MOONE* A3r –A4v.

36 Vgl. Biagoli M., „Picturing Objects in the Making: Scheiner, Galileo, and the Discovery of Sunspots", in Detel W. – Zittel C. (eds.), *Wissensideale und Wissenskulturen in der frühen Neuzeit / Ideals and Cultures of Knowledge in Early Modern Europe* (Berlin: 2002) 39–96.

37 Ebd. Titelblatt.

38 Vgl. zum Aspekt der Assoziierung von Fernrohr und Reise Trinkner, „Träume von Levania" 240f.

nicht zufällig auf den astronomischen Unterricht verweist, den er als Student in Salamanca genoß,[39] bemerkt auf seinem Raumflug

> three things very remarkeable: one that the further we went, the lesser the Globe of the Earth appeared unto us; whereas still on the contrary side the Moone shewed her selfe more and more monstrously huge.
>
> Againe, the Earth (which ever I held in mine eye) did as it were mask it selfe with a kind of brightnesse like another Moone; and even as in the Moone we discerned certaine spots or Clouds, as it were, so did I then in the earth. But whereas the forme of those spots in the Moone continue constantly one and the same; these little and little did change every hower. The reason thereof I conceive to be this, that whereas the Earth according to her naturall motion, (for that such a motion she hath, I am now constrained to joyne in opinion with *Copernicus*,) turneth round upon her owne Axe every 24. howers from *West* unto the *East*: I should at the first see in the middle of the body of this new starre a spot like unto a Peare that had amorsell bitten out upon the one side of him; after certaine howers, I should see that spot slide away to the *Eas*t side. This no doubt was the maine of *Afrike*.
>
> Then should I perceive a great shining brightnesse to occupy that roome, during the like time (which was undoubtedly none other then the great *Atlantick* Ocean). After that succeeded a spot almost of an Ovall form, even just such as we see *America* to have in our Mapps. Then another vast cleernesse representing the *West Ocean*; and lastly a medly of spots, like the Countries of the *East Indies*. So that it seemed unto me no other then a huge Mathematicall Globe, leasurely turned before me, wherein successively, all the Countries of our earthly world within the compasse of 24 howers were represented to my sight. And this was all the meanes I had now to number the dayes, and take reckoning of time.[40]

Wie in Keplers *Somnium* wird in Godwins Text der autoptische Befund—ein Arrangement von ‚Flecken' und lichtreflektierenden Flächen auf einer Kugel—im Rahmen einer kartographischen Topik transformiert. Auch findet sich das Verfahren der Bildassoziation: Das einprägsame ‚Schreckbild' Afrikas als ange-

---

39 Vgl. Godwin, *THE MAN IN THE MOONE* 53f.—Die Universität Salamanca nahm in ihrer Satzung von 1561 als erste Universität in Europa auch das kopernikanische Weltmodell auf. Vgl. Vernet J., „Copernicus in Spain", in Union Internationale d'Histoire et de Philosophie, Comité Nicolas Copernic (ed.), *Colloquia Copernicana I: Études sur l'audience de la Théorie héliocentrique*, Studia Copernicana v (Warschau: 1972) 271–279, hier 275.

40 Godwin, *THE MAN IN THE MOONE* 55–58.

bissener Birne wird zur Steigerung der *evidentia* aufgerufen, aber auch durch die geographische Benennung des Kontinents entschärft. Die dynamische sprachliche Taktgebung (,Then' und ,After that' und ,Then' und ,lastly') unterstützt die Evozierung der Erdrotation. Nach und nach kommen Formen in den Blick, die quasi mit einer Weltkarte abgeglichen werden. Eine Versammlung kleiner Flecken wird über die Feststellung einer *Ähnlichkeit* (,like the Countries of the *East Indies*') identifiziert. Das vorsichtige ,like' markiert eine sorgfältige Auswertung der Autopsie, bei der nichts vorschnell oder ungeprüft behauptet wird. Zuletzt bilanziert der Erzähler die Wahrnehmung der Erdrotation im Bild eines langsam gedrehten mathematischen Globusses; die Evidenz wird an die Rückübersetzung des Phänomens in eine ,verfügbare' Repräsentation der Erde und die leicht vorstellbare Hantierung mit ihr gebunden.

Im *Man in the Moone* steht Imagination gegen Imagination, oder anders gesagt: *evidentia* gegen behauptete Anti-*evidentia*. Die suggestive Übersetzung der Erde in einen sanft bewegten Globus wird mit einem ,irrsinnigen' Bild kontrastiert:

> O incredible thing, that those same huge bodies of the fixed stars in the highest orbe, whereof divers are by themselves confessed to be more then one hundreth times as bigge as the whole earth, should as so many nayles in a Cart Wheele, be whirled about in that short space [...].[41]

Das Argument dieser *Unvorstellbarkeit* des Sternenumlaufs im geozentrischen Weltbild ist weit verbreitet und findet sich auch in Gilberts *De magnete*,[42] einer Quelle für Godwins kosmologisches Wissen. Doch in der Erzählung wird, befördert durch den spezifischen Kontrast von leichter Drehung (der Globus) und Rasen (die Fixsternsphäre), die Unvorstellbarkeit ,erfahrbar' gemacht, genauer gesagt: auf ein Versagen der Rückübersetzung des Bildes vom Wagenrad in ein Bild der kosmischen Phänomene gesetzt.

Neben der Erdrotation wird auch das Planetenargument, und zwar in zwei Blickrichtungen, autoptisch bestätigt. Mit der Annäherung des Protagonisten an den Mond vollzieht sich ein Chiasmus: Auf der Ebene der wahrgenommenen Phänomene verwandelt sich der Mond zur Welt und die Erde zum Mond. Die Entdeckungsreise Richtung Mond ermöglicht die *eigentliche* Entdeckung in der Gegenrichtung. Domingo erkennt die alte Erde als ,new starre'. Die programmatische Verknüpfung von Entdeckungsreise und astronomischer Beobachtung in der Vorrede wird hier als Umkehrung des Blicks auf die Erde

41 Ebd. 58f.

42 Vgl. Gilbert William, *De magnete, magneticisque corporibus, et de magno magnete tellure* (London, Peter Short: 1600) 217.

realisiert. Die Erde kann als ‚starre', d.h. als etwas entdeckt werden, das äußerlich den Wandelsternen gleicht, und sie kann als ‚new', d.h. als bisher unbekannt entdeckt werden.

Weiter in den Kosmos hinein reicht die Autopsie aber nicht. Gonsales nimmt gegenüber der Frage, ob die Sonne eine Zentralstellung in der Welt innehabe, ausdrücklich eine neutrale Position ein, weil er im engen ‚experimentellen' Rahmen seiner Raumreise allein die Evidenz der Erdrotation gewinnen konnte.[43] Während die Forschung überwiegend dazu tendiert, die Nicht-Unterstützung des Heliozentrismus als *Meinung* des Textes zu verstehen,[44] ist es sinnvoller, die Zurückhaltung des Protagonisten mit Aït-Touati im Fiktionsrahmen des Textes als Ausweis der Glaubwürdigkeit des Zeugen zu interpretieren.[45] Gonsales kann den Umlauf der Erde um die Sonne nicht bezeugen und stellt dementsprechend auch keine Behauptung darüber auf.

Die Zentralstellung der Sonne kann aber Dyrcona bestätigen, die Hauptfigur von Cyrano de Bergeracs *L'Autre Monde*.[46] Der Roman, den Andreas Gipper als ‚Ausgangspunkt der modernen Literatur der Wissenschaftsvulgarisierung' bestimmt,[47] erschien posthum in zwei Teilen in den Jahren 1657 und 1662 und integriert burleske Handlung und ausführliche Reflexionen bzw. Dialoge über Themen der Physik, Chemie, Astronomie und Metaphysik.[48] Die Position Cyranos zur kosmologischen Kontroverse war eine kopernikanische.[49]

---

43 Vgl. Godwin, *THE MAN IN THE MOONE* 60.

44 So zuletzt Glomski, „Science Fiction in the Seventeenth Century" 45.

45 Aït-Touati, *Fictions of the Cosmos* 51.

46 *L'Autre Monde* (= der heute geläufige Manuskriptitel) erschien in zwei Teilen: (1.) Cyrano de Bergerac, Savinien, *HISTOIRE COMIQVE, PAR MONSIEVR DE CYRANO BERGERAC. Contenant les Estats & Empires de la Lune* (Paris, Charles de Sercy: 1657); (2.) ders., *LES NOVVELLES OEVVRES DE MONSIEVR DE CYRANO BERGERAC. CONTENANT L'HISTOIRE COMIQVE des Estats & Empires du Soleil, PLVSIEVRS LETTRES, ET AVTRES PIECES Diuertissantes* (Paris, Charles de Sercy: 1662).

47 Gipper A., *Wunderbare Wissenschaft. Literarische Strategien naturwissenschaftlicher Vulgarisierung in Frankreich. Von Cyrano de Bergerac bis zur Encyclopédie* (München: 2002) 45. Zu Cyranos *L'Autre Monde* vgl. 45–104.

48 Vgl. hierzu ausführlich Gipper, *Wunderbare Wissenschaft*, der plausibel macht, ‚daß es nicht ausreicht, einzelne Passagen des Romans als Bestandteile zeitgenössischer Theoriediskussion zu verstehen, sondern daß es gilt, ihren besonderen Status in einer epistemischen Struktur zu begreifen, für die Imagination und Experiment, Intuition und Demonstration, Spekulation und Erkenntnis, Mysterium und Mathematik noch keine getrennten oder gar gegensätzlichen Kategorien darstellen, sondern den gemeinsamen Raum der Wissenschaft.' (49).

49 Vgl. zu Kosmologie und Kopernikanismus bei Cyrano de Bergerac Loewenstein A.W., „Die naturphilosophischen Ideen bei Cyrano de Bergerac", *Archiv für Geschichte der*

Im ersten Teil des Romans konstruiert der Protagonist eine Flugmaschine, die von taugefüllten Kugelfläschchen angehoben wird, um zum Mond zu fliegen, weil er glaubt und beweisen will, ‚daß der Mond eine Welt ist wie diese hier, der die unsrige als Mond dient'[50] (‚que la lune est un monde comme celui-ci, à qui le nôtre sert de lune').[51] Der erste Flugversuch scheitert, beweist aber durch eben dieses Scheitern die Erdrotation, denn Dyrcona startet in Frankreich und landet schon nach kurzer Zeit in ‚la Nouvelle-France' (d.h. Canada),[52] weil sich die Erde unter ihm gedreht hat (was er aber nicht bemerkt). Dort diskutiert er mit dem französischen Gouverneur über kosmologische Fragen und kann gegen dessen geozentrischen Verweis auf die sinnliche Evidenz des Sonnenumlaufs (vorerst noch) keine Autopsie ins Feld führen, sondern vor allem rhetorische Argumente aufbieten, die sich aus der Voraussetzung einer *zweckmäßigen* Installierung der Sonne als zentraler Wärmequelle des Universums ableiten lassen—Argumente, wie das folgende:

> Denn es wäre gerade so lachhaft, zu glauben, der große leuchtende Körper drehe sich um einen Punkt, mit dem er nichts zu schaffen hat, als sich vorzustellen, wenn wir eine gebratene Wachtel sehen, man habe, um sie zu rösten, den Kamin um sie herum gedreht.[53]

Im zweiten Teil von *L'Autre Monde* wird komplementär zur dieser common sense-Argumentation eine Evidenz-Simulation geboten, die nun nicht mehr nur die Erdrotation betrifft, sondern in einer Reisebewegung über den Mond hinaus schließlich auch den Lauf der Erde um die Sonne und die Existenz weiterer Welten bestätigen kann:

Philosophie 16 (1903) 27–58, hier 51–57; Harth E., *Cyrano de Bergerac and the Polemics of Modernity* (New York – London: 1970) 80–111; Barbe C., „Cyrano: la mise à l'envers du vieil univers d'Aristote", *Baroque. Revue Internationale* 7 (1974) 49–70; Sick F., „Cyrano de Bergerac: Le Monde dans la perspective de l'Autre Monde", *Papers on French Seventeenth Century Literature* 21, Nr. 40 (1994) 65–80; Gipper, *Wunderbare Wissenschaft*, 67f., 90–97.

50 Cyrano de Bergerac Savinien de, *Reise zum Mond und zur Sonne*, übers. und ed. W. Tschöke (Frankfurt a.M.: 2004) 12.

51 Cyrano de Bergerac Savinien de, *Œuvres complètes. Edition critique*, ed. Alcover M., 3 vols. (Paris: 2000) vol. 1: *L'Autre Monde ou les États et Empires de la Lune. Les États et Emprires du Soleil. Fragment de Physique* 6.

52 Ebd. 13.

53 Cyrano de Bergerac, *Reise zum Mond und zur Sonne* 18. ‚Car il serait aussi ridicule de croire que ce grand corps lumineux tournât autour d'un point dont il n'a que faire, que de s'imaginer, quand nous voyons une alouette rôtie, qu'on a, pour la cuire, tourné la cheminée à l'entour.' Cyrano de Bergerac, *Œuvres complètes* vol. 1, 17.

Ich erkannte sehr deutlich, wie ich es früher bei meiner Fahrt zum Mond schon vermutet hatte, daß es tatsächlich die Erde ist, die sich von Westen nach Osten um die Sonne dreht, und nicht die Sonne um jene, denn ich sah auf Frankreich den Fuß des Stiefels von Italien folgen, darauf zogen das Mittelmeer, dann Griechenland, weiter der Bosporus, das Schwarze Meer, Persien, Indien, China und schließlich Japan vor dem Guckloch meines Kämmerchens vorüber, und einige Stunden nach meinem Aufstieg, als sich das Südmeer vorbeigedreht hatte, nahm das Festland von Amerika seinen Platz ein.

Ich erkannte deutlich all diese Umdrehungen und erinnere mich sogar, daß ich lange Zeit später Europa noch einmal auf dem Schauplatz auftauchen sah, aber da konnte ich die Staaten nicht mehr einzeln erkennen, weil ich schon zu hoch war. Ich ließ auf meinem Weg, mal links, mal rechts mehrere Welten wie die unsere zurück; sofern ich nur im mindesten in den Einfluß ihrer Sphären geriet, fühlte ich, wie ich mich ihnen zuneigte. Indessen überwand aber die reißende Kraft meines eigenen Aufstiegs diejenige solcher Anziehungskräfte.

Ich flog am Mond vorbei, der sich zu dieser Zeit zwischen Sonne und Erde befand, und ließ die Venus rechter Hand liegen. Aber anläßlich dieses Sterns hat die alte Astronomie so sehr gepredigt, die Planeten seien Sterne, die sich um die Erde drehen, daß die moderne wohl nicht wagen würde, daran zu zweifeln. Und dennoch stellte ich fest, daß ich die Venus während der ganzen Zeit, da sie diesseits der Sonne erschien, um die sie sich dreht, immer nur als Sichel sah; daß sich aber am Ende ihrer Umdrehung, in dem Maße, wie sie hinter die Sonne zog, ihre Sichelenden einander annäherten und ihr dunkler Leib sich wieder vergoldete. Dieser Wechsel von Licht und Dunkelheit zeigt ganz augenscheinlich, daß die Planeten wie der Mond und die Erde Weltkugeln ohne eigenes Licht sind, die nur zurückstrahlen können, was sie empfangen haben.

Tatsächlich machte ich beim weiteren Aufsteigen dieselbe Beobachtung beim Merkur. Darüber hinaus stellte ich noch fest, daß alle diese Welten noch andere kleine Welten haben, die sich um sie herum bewegen.[54]

---

54 Cyrano de Bergerac, *Reise zum Mond und zur Sonne* 177f. Je connus très distinctement, comme autrefois j'avais soupçonné en montant à la lune, qu'en effet c'est la terre qui tourne d'orient en occident à l'entour du soleil, et non pas le soleil autour d'elle; car je voyais en suite de la France, le pied de la botte d'Italie, puis la mer Méditerranée, puis la Grèce, puis le Bosphore, le Pont-Euxin, la Perse, les Indes, la Chine, et enfin le Japon, passer successivement vis-à-vis du trou de ma loge; et quelques heures après mon élévation, toute la mer du Sud ayant tourné laissa mettre à sa place le continent de l'Amérique.

Im Unterschied zu Keplers *Somnium* und Godwins *Man in the Moone* wird in *L'Autre Monde* nicht der Weg vom Phänomen zum Kartenbild genommen. Umgekehrt beginnt Dyrcona bei seiner Beschreibung mit einer Revue distinkter Länder und Reiche (Frankreich—Italien—Griechenland—Persien—Indien—China—Japan) und endet mit der Abfolge von Kontinenten (Amerika und dann noch einmal Europa). Der Bericht setzt so, zugleich mit der Erdrotation, die Reisebewegung weg von der Erde in Szene. Zuletzt erscheint Europa so klein ‚sur la scène', daß die einzelnen Staaten nicht mehr zu erkennen sind. Die Möglichkeit einer kartographischen Rekonstruktion kommt abhanden; die Erde ist nur noch ein ‚undefinierter' Himmelskörper unter anderen.

Die Anlage der Descriptio ist frei von Metaphern und Bild-Assoziationen. Während Kepler mit der ‚Europa regina' und Godwin mit der angebissenen Birne Afrika ein *Verfahren der Verfremdung* einsetzen, um das Wiedererkennen didaktisch zu nutzen, wählt Cyrano ein *Verfahren der Entfremdung* und läßt die Erde einfach nur unkenntlich werden.

Der ‚Protokollcharakter' der Passage wird nur von einem starken Bild unterbrochen, daß sich aber metareflexiv auf die Anlage des Textes selbst bezieht: Der astronomische Beobachtungsrahmen wird als ‚Schauplatz' (‚scène') bezeichnet—ein Hinweis darauf, daß Evidenz in ästhetisch kontrollierten Beobachtungssituationen produziert werden muß. Das geschieht in Experimenten und erst recht in Texten, die Gedankenexperimente vorführen.

Je distinguai clairement toutes ces révolutions, et je me souviens même que longtemps après je vis encore l'Europe remonter une fois sur la scène, mais je n'y pouvais plus remarquer séparément les états, à cause de mon exaltation qui devint trop haute. Je laissai sur ma route, tantôt à gauche, tantôt à droite, plusieurs terres comme la nôtre, où pour peu que j'atteignisse les sphères de leur activité, je me sentais fléchir. Toutefois, la rapide vigueur de mon essor surmontait celle de ces attractions.

Je côtoyai la lune qui pour lors se trouvait entre le soleil et la terre, et je laissai Vénus à main droite. Mais à propos de cette étoile, la vieille astronomie a tant prêché que les planètes sont des astres qui tournent à l'entour de la terre, que la moderne n'oserait en douter. Et je remarquai toutefois que, durant tout le temps que Vénus parut au-deçà du soleil, à l'entour duquel elle tourne, je la vis toujours en croissant; mais achevant son tour, j'observai qu'à mesure qu'elle passa derrière, [ses] cornes se rapprochèrent et son ventre noir se redora. Or cette vicissitude de lumières et de ténèbres montre bien èvidemment que les planètes sont, comme la lune et la terre, des globes sans clarté qui ne sont capables que de réfléchir celle qu'ils empruntent.

En effet, à force de monter, je fis encore la même observation de Mercure. Je remarquai de plus que tous ces mondes ont encore d'autres petits mondes que se meuvent à l'entour d'eux.' Cyrano de Bergerac, *Œuvres complètes* vol. 1, 212.

2 Inversion und Methode: Fiktionalitätsmarkierung und Perspektivenwechsel in Keplers *Somnium*

Im *Somnium* wird eine Evidenz der Erdrotation hergestellt, die aus irdischer Sicht die Vorstellung eines täglichen Umlaufs der Sonne widerlegt. Zugleich jedoch unterliegen die Mondbewohner wie die Erdbewohner dem falschen Augenschein, daß sich der Himmel um sie dreht. Der Text rekurriert auf diese Fehlwahrnehmung bereits im ersten Satz des überperspektivischen Dämons, wenn er—narratologisch gesprochen: mit interner Fokalisierung der Mondbewohner—formuliert: ‚50 000 deutsche Meilen entfernt liegt in der Tiefe des Äthers die Insel Levania.'[55] Der Satz scheint durch seinen Bezug auf das Entfernte allein aus irdischer Perspektive gedacht zu sein, doch das ‚Liegen' des Mondes widerspricht der Wahrnehmung auf der Erde. Kepler kommentiert in Note 54:

> Es [Levania] liegt nicht, sondern schwimmt eher, wenn wir das Gleichnis ‚Insel' berücksichtigen. Aber hier musste ich schon gemäß der Vorstellung sprechen, die man sich von der Wahrnehmung [dessen, der auf dem Mond ist] zu machen hat. Denn wer auf dem Mond wäre, der würde glauben, dass der Mond ganz und gar fest an einem Ort verharre.[56]

Die Amalgamierung der Perspektiven im Eingangssatz deutet eine Analogie an, die später im Klartext wiederholt wird: ‚Die Sonne umkreist die Berge. Levania scheint nämlich ihren Bewohnern genau so festzustehen und von den Gestirnen umkreist zu werden, wie uns [Menschen] die Erde.'[57] Keplers Kommentar 96 bindet an diesen Satz, der eine falsche Evidenz zum Gegenstand hat, die Kernhypothese seines *Somnium*:

> Hier offenbart sich die Hypothese des ganzen ‚Somnium', nämlich die Begründung für die Bewegung der Erde, oder eher die

55 Kepler, *Der Traum, oder: Mond-Astronomie* 11. ‚Quinquaginta millibus miliarium Germanicorum in aetheris profundo sita est LEVANIA insula:' Kepler, *Gesammelte Werke* vol. 11.2, 323.

56 Kepler, *Der Traum, oder: Mond-Astronomie* 43. ‚Non sita est, sed natat potiùs, si ad Insulae similitudinem respicimus. Sed hîc jam ad imaginationem visus loquendum fuit. Nam qui in Luna esset, omninò is Lunam stare loco fixam existimaret.' Kepler, *Gesammelte Werke* vol. 11.2, 339.

57 Kepler, *Der Traum, oder: Mond-Astronomie* 14. ‚Sol [...] montes circulo circumiens. Nec enim minùs Levania suis incolis immota stare videtur, currentibus astris, quàm Terra nostra nobis hominibus.' Kepler, *Gesammelte Werke* vol. 11.2, 324.

> Widerlegung der Begründung gegen die Bewegung der Erde, die auf der Sinneswahrnehmung aufgebaut ist.[58]

Im Kontext dieser Hypothese offenbart sich die Überperspektivität des Dämons sehr deutlich als Doppelstrategie. Sie dient nicht allein als Vehikel zur Produktion von Evidenz (die *evidentia* der Erdrotation), sondern sie vermittelt auch die Simulation von *fremder* falscher Evidenz (das Ruhen des Mondes im All), mit der die *eigene* falsche Evidenz ‚denkbar' und damit korrigierbar wird.

Die konträren Evidenzsimulationen des *Somnium* offenbaren eine grundsätzliche Ambivalenz des Sehens: Die Simulation der Augenzeugenschaft setzt auf die methodische Valenz von sinnlicher Empirie, während sie zugleich die Täuschbarkeit der Sinne ausstellt. Erst durch die ‚Verrechnung' der Sichtbarkeiten, wie sie sich von der Erde und vom Mond aus unterschiedlich darstellen, entsteht im Text die Evidenz des kopernikanischen Weltmodells. Was im engeren astronomischen Funktionsrahmen des Textes eine konkrete Hypothese stützen soll, impliziert eine generelle methodologische Weiterung. Keplers *Somnium* verhandelt ein methodisches Problem (die standortbedingten Mängel der Wahrnehmung bei der Bestimmung wahrer Sachverhalte) und zugleich dessen Lösung (Standort- bzw. Perspektivenwechsel als Korrektiv der unzureichenden Wahrnehmung).

Der Text präsentiert perspektivische Verunsicherungen und Sicherungen der Evidenz allerdings nicht allein auf der Ebene seines Sujets, er ‚thematisiert' sie auch auf der Ebene seiner literarischen Verfaßtheit. Das methodologische ‚Programm' des *Somnium*—Kritik an Schein-Evidenzen sowie Evidenzsicherung durch Perspektivenwechsel und Perspektivenumkehr—zeigt einen Widerschein in der *Methode des Textes*, genauer: in verschiedenen Fiktionalitätsmarkierungen und einer Motivik der Inversion, die für die *narrative* Ebene des Textes bereits gut untersucht ist.[59] Daher sollen hier drei textorganisatorische Ebenen in den Blick genommen werden, die bislang nur wenig Beachtung fanden: (1.) die Kommentierung des Textes in den Endnoten,

---

58 Kepler, *Der Traum, oder: Mond-Astronomie* 55. ‚En hypothesin totius somnij, argumentum scilicet pro motu Terrae, seu dissolutionem potius argumenti contra motum Terrae, ex sensu extructi.' Kepler, *Gesammelte Werke* vol. 11.2, 344. Vgl. auch die ausführlichere Behandlung der Grundthese in Note 135.

59 Vgl. Hallyn F., „Le *Songe* de Kepler", *Bibliothèque d'Humanisme et Renaissane. Travaux et Documents* 42 (1980) 329–347; Bozetto, „Kepler's ‚Somnium'; or, Science Fiction's Missing Link"; Hallyn F., *The Poetic Structure of the World. Copernicus and Kepler* (New York: 1990) 253–280, 349–354; Bezzola Lambert, *Imagining the Unimaginable*; Schneider, „Science as Science Fiction"; Swinford, *Through the Daemons's Gate*; Vogl J., „Robuste und idiosynkratische Theorie", *KulturPoetik* 7, 2 (2007), 249–258.

(2.) die Komplementarität von Mondastronomie (vorteleskopischer imaginativer Blick vom Mond) und angehängter Mondgeographie (teleskopischer Blick von der Erde) und (3.) die typographische Differenzierung von Textpassagen.

(1.) Das (manifeste) Buch vom Traum enthält nicht umsonst ein (imaginäres) Buch als Behälter der Traumbotschaft. ,The dream is presented here in the guise of a book to be read and annotated.'[60] Die 223 Endnoten zum *Somnium*, die um ein vielfaches länger als die Traumerzählung sind, integrieren die verschiedenen narrativen Ebenen des Textes[61] und systematisieren seine Aussagen.[62] Dabei werden nicht allein die fiktional dargebotenen astronomischen Observationen genauer ausgeführt, wissenschaftlich begründet und methodologisch reflektiert, es wird auch die literarische Genese des Textes thematisiert. Kepler bietet die Allegorese von Traumdetails sowie Rekurse auf intertextuelle Muster, Quellenbezüge, autobiographische Motive und Schreibanlässe. Inhalt und Form des Textes erhalten einen Rückhalt durch die Einheit des realen Autors, der sich in den Noten als Verursacher des *Somnium* ausstellt—und, im fiktionalen Rahmen des Buches, zugleich auf den Text als Ursache seines Endnoten-Diskurses reagiert.

Die Selbstkommentierung—selbst bereits strukturell ein Wechsel der Perspektiven—ermöglicht die programmatische Inszenierung von Inversionen. In den Noten kommt Kepler auf ein wichtiges Detail seiner Beschreibung der rotierenden Erdkugel zu sprechen.[63] Dort erscheinen die Landmassen als dunkel, die Wasserflächen als hell. In den Noten 154 und 155 wird die fiktionale *descriptio* mit Hinweis auf die Ergebnisse der teleskopischen Monderkundung in Galileis *Sidereus nuncius* korrigiert und ,faktualisiert':

> Galilei hat mich gelehrt, dass die Höhen und Schroffen des Mondes nicht als Flecken sichtbar seien, sondern als Helligkeit. Flächen aber, die sich in tiefer gelegenen Gegenden erstrecken, erschienen schwärzlich und nähmen das Aussehen von Flecken an. Auf dieselbe Weise also muss man von der Erdkugel annehmen, dass der Ozean und die zwischen dem

60 Reiss, *The Discourse of Modernism* 147.

61 Vgl. Poole W., „Kepler's *Somnium* and Francis Godwin's *The Man in the Moone*: Birth of Science-Fiction 1593–1638", in Houston C. (ed.), *New Worlds Reflected. Travel and Utopia in the Early Modern Period* (Surrey – Burlington 2010) 57–69, hier 66.

62 Vgl. Reiss, *The Discourse of Modernism* 144–148.

63 Siehe das entsprechende Zitat oben im Text.

> festen Land sich ausdehnenden Meere dunkel wirken. Die Kontinente aber und die Inseln erstrahlen hell vom Licht der Sonne.[64]

Kepler verzichtet darauf, den Text des *Somnium* gemäß den neuen Erkenntnissen zu revidieren; die Begründungen der weit ausgreifenden Note 154 wären dadurch ja keineswegs überflüssig geworden. Stattdessen konserviert er jedoch den Fehler der Beschreibung und nutzt die demonstrative Selbstkorrektur für ein Spiel der Inversionen. (1. Inversion:) Der vorteleskopische Blick auf die Erde wird korrigiert mittels des teleskopischen Blicks auf den Mond. (2. Inversion:) Durch die Distanzierung von Schlüssen, die er aus Beobachtungen zur Lichtreflexion in seiner *Astronomia pars Optica* gezogen hatte, gibt er das Exempel einer wissenschaftlichen *conversio* und zeigt damit eine methodologische Einstellung, die in der Auseinandersetzung um die Wahrheit des kopernikanischen Weltbildes generell vonnöten ist. (3. Inversion:) Zudem gibt die Korrektur des Kontrastverhältnisses zwischen Erde und Wasser (von dunkel/hell zu hell/dunkel) einen nachhaltigen imaginativen Impuls. Der Leser stellt sich zuerst das ‚falsche' Bild vor und erarbeitet sich sodann mithilfe des Kommentars das richtige; die Inversion läßt den Prozeß der *imaginatio* einen didaktischen Umweg nehmen und ist auf die gesteigerte Einprägsamkeit und Glaubwürdigkeit des Bildes berechnet. Gerade die Korrektur der falschen *evidentia* steigert die Evidenz—und läßt vergessen, daß beide Erdbilder (ob wahr, ob falsch) Fiktionen sind.

Die entscheidende Inversion des *Somnium* ist die oben behandelte ‚allegorische' Transformation der irdischen in eine mondgebundene astronomische Perspektive. Dieser Perspektivenwechsel wird allein möglich durch den Bericht des Dämons, der im letzten Absatz der Erzählung des Duracotus durch einen magischen Ritus der Fiolxhilde herbeigerufen wird:

> Die Mutter ging voraus bis zur nächsten Weggabelung, erhob ein Geschrei und brachte einige wenige Wörter hervor, mit denen sie ihre Bitte äußerte. Nachdem sie ihre Riten vollendet hat, kehrt sie zurück, gebietet mit ausgestreckter rechter Hand Schweigen und setzt sich neben mich. Kaum hatten wir das Haupt mit dem Gewand (wie es vereinbart war) verhüllt, da erhebt sich das Gekrächz einer stammelnden und dumpfen Stimme.

64 Kepler, *Der Traum, oder: Mond-Astronomie* 76f. ‚Docuit me Galilaeus, Edita Lunae et Aspera, non maculas esse, sed claritatem; fusa verò in depressas partes aequora, nigricare, macularumque speciem induere. Ad eundem igitur modum etiam de terrestri globo statuendum est, Oceanum et Maria Terris interfusa, obscuritatem induere; Continentes verò et Insulas, luce Solis eximiè resplendescere.' Kepler, *Gesammelte Werke* vol. 11.2, 355.

> Und sofort beginnt sie auf folgende Weise zu sprechen, aber in isländischer Sprache.[65]

Was hier als magischer Akt erscheint: die (binnenfiktionale) Installierung jener zentralen Erzählinstanz, welche die alternative astronomische Empirie übermittelt, wird in den Endnoten 44/46/47 mit einem Camera obscura-Experiment analogisiert, in dem die Inversion der Schriftlaufrichtung inszeniert wird:

> Auch das ist eine magische Zeremonie. [...] In meinen Prager Jahren führte ich oft auf besondere Art ein bestimmtes Experiment vor, wenn Zuschauer oder Zuschauerinnen zu mir kamen. Ich zog mich dafür jedes Mal zunächst von den plaudernden Leuten zurück in einen nahegelegenen Nebenraum meines Hauses, der zu meinem Zweck ausgewählt worden war, schloss das Tageslicht aus, passte ein ganz kleines Fenster in einen winzigen Rahmen ein und verhüllte eine Wand mit weißem Stoff. Sobald ich diese Vorbereitungen getroffen hatte, rief ich die Zuschauer. Das waren meine Zeremonien, meine Riten. Wollt ihr auch Zauber-Zeichen? Auf eine schwarze Tafel schrieb ich mit Kreide, was mir für die Zuschauer passend schien, in Großbuchstaben in umgekehrter Reihenfolge (Sieh da, der magische Ritus!), so wie die Juden schreiben. Diese Tafel hängte ich draußen im Tageslicht mit ihrer umgekehrten Reihenfolge der Buchstaben in die Sonne. Jetzt kommt's: Was ich geschrieben hatte, das erschien umgekehrt auf der weißbespannten Wand, in richtiger Reihenfolge, und wenn ein Windhauch draußen die Tafel bewegte, gerieten die Buchstaben drinnen an der Wand in entsprechend flatternde Bewegung.[66]

---

65 Kepler, *Der Traum, oder: Mond-Astronomie* 11. ‚[...] mater seorsim à me se recipiens in proximum bivium, et pauculis verbis, clamore sublato, enunciatis, quibus petitionem suam proponebat; caeremoniisque peractis, revertitur, praetensâ dextrae manus palmâ silentium imperans, propterque me assidet. Vix capita vestibus (vt conventum erat) involveramus; cum ecce screatus exoritur blaesae et obtusae vocis; et stàtim in hunc modum, sed idiomate Islandico, infit.' Kepler, *Gesammelte Werke* vol. 11.2, 323.

66 Kepler, *Der Traum, oder: Mond-Astronomie* 40f. ‚Haec quoque magica ceremonia: [...] In particulari observationis cujusdam praxi, quae mihi Pragae circa illos annos crebra erat, quoties me convenerunt spectatores spectatricesve; solitus ego sum prius ab illis colloquentibus me subducere in angulum domus proximum, ad hoc opus electum, diei lucem excludere, fenestellam aptare minutissimo ex foramine, parietem albo vestire; peractis ijs, advocare spectatores. Hae mihi caeremoniae, hi ritus; vultis et characteres? In tabella nigra, quae mihi videbantur apta spectatoribus, cretâ perscripsi, literis capitalibus, litera-

Der Kommentar läßt sich, mit seinem Bezug auf die Anrufungsszene, als poetologische Reflexion auf die Inversion als Verfahren lesen. Das sichtbare Ziel des berichteten Experiments: die Umkehrung (sprich Korrektur) der umgekehrten Schrift, ist eine Allegorie auf zweiter Stufe; sie allegorisiert die allegorische Anlage des Textes, der selbst Schrift ist und dessen Pointe ja darin besteht, auf der Sujetebene durch eine Perspektivenumkehrung einen falschen Augenschein zu korrigieren.

Wenn Kepler die ,umgekehrte' Schriftlaufrichtung des Hebräischen anführt, um seinen ,magischen' Schreibakt im Rahmen des Experiments vor Augen zu stellen, wird damit zugleich eine Beziehung zu Endnote 42 hergestellt, in welcher der Autor die Wahl des Namens ,Levania' für den Mond begründet:

> Mond heißt auf Hebräisch „Lebhana" oder „Levania"; ich hätte ihn auch „Selenitis" nennen können. Aber die hebräischen Wörter, dem Ohr fremder, empfehlen sich den geheimen Künsten durch stärkeren mystischen Klang.[67]

Hier wird zuvorderst auf die ,geheimwissenschaftliche' Alterität des Wortes abgehoben, doch mit Bezug auf die Assoziation von invertiertem Schreiben und hebräischem Schriftsystem, die in den Endnoten 44/46/47 als (wiederum kryptischer) Schlüssel für die Dechiffrierung der Namenswahl zur Verfügung gestellt wird, läßt sich ,Levania' auch als—*auf der Ebene der Schrift*, sprich: des Textes—*umgekehrter* Mond verstehen. Der Gegenstand also, an dem sich die allegorische Übertragung der irdischen Perspektive aufhängt: der Mond als invertierte Erde, wird mit einem Namen versehen, dessen Herkunft auf ein invertiertes Schreiben anspielt.

In Keplers Camera obscura-Experiment steht die verkehrte Schrift im Licht, die richtige erscheint im Dunkeln. Im besonderen astronomisch-methodologischen Zusammenhang des *Somnium* gewinnt die Camera obscura eine besondere Eignung zum allegorischen Signifikanten, weil sie als ,lichtbetriebener Inversionsapparat' sogar eine Korrektur der geläufigen Bildlichkeit von Erkenntnis (der emphatischen Formel vom *lux veritatis*) erlaubt. Die

---

rum figurâ retrò versa (en ritum magicum) vt Hebraica scribuntur; hanc tabellam foris sub dio, situ literarum everso suspendi in Sole; vt ita quae scripseram, ea introrsum ad album parietem pingerentur situ erecto, et si tabellam ventulus agitaret foris, literae intus motu vago parietem reciprocarent.' Kepler, *Gesammelte Werke* vol. 11.2, 338.

67 Kepler, *Der Traum, oder: Mond-Astronomie* 39. ,Luna Hebraeis est Lebhana, vel Levana, Selenitida potui indigetare; sed Hebraeae voces, remotiores ab auditu, majori superstitione commendantur in occultis artibus.' Kepler, *Gesammelte Werke* vol. 11.2, 337.

Evidenzproduktion wird hier nicht vorrangig dem Licht zugeordnet, sondern bedarf zusätzlich einer experimentell kalkulierten Schattenzone und einer *Umkehrung* des Licht-Bildes, das es transportiert.[68]

Die Verdunkelung als Rahmenbedingung einer gelingenden Herstellung von Evidenz begegnet in den Noten zum Text noch ein zweites Mal—wiederum mit Rekurs auf die Szene von der Herbeirufung des Dämons. Damit der Dämon erscheint, müssen Fiolxhilde und Duracotus die Häupter mit ihren Gewändern bedecken. Die Endnote 49 assoziiert dazu:

> Genau mit diesem Ritus (hui, auf wie magische Art magisch!) hatten wir ein wenig früher, als ich mein Buch in Angriff nahm, die Sonnenfinsternis im Jahre 1605 vom 2. bis zum 12. Oktober beobachtet. [...] Auf dem Sonnenplatz des Lusthauses in den Gärten des Kaisers fehlte uns nämlich ein verdunkelter Raum. Deswegen verhüllten wir die Köpfe mit unseren Mänteln und schlossen so das Tageslicht aus.[69]

Damit *die Stimme* des Dämons—die mondastronomische Fiktion—*gehört* werden kann, muß das *optische* Beobachtungsszenarium der Astronomie nachgeahmt und improvisiert werden, was wohl den Schluß erlaubt, daß die Erzählung *sehen* läßt—eine weitere (poetologische) Inversion, die ...

(2.) als Komplementarität von Mondastronomie und Mondgeographie auf der Dispositionsebene des Buches wiederkehrt: In der Rede des Dämons, die Kepler 1609, also vor dem Erscheinen von Galileis *Sidereus nuncius* (1610) verfaßte, sind die Mondbewohner, die in Löchern und Höhlen Schutz vor den extremen Umweltbedingungen des Mondes finden, quasi als ‚Troglodyten' konzipiert. In dem Brief an den Jesuiten Paul Guldin dagegen, der dem *Somnium* als mondgeographischer Anhang beigegeben ist, wird ihnen die Fähigkeit zum Städtebau zugeschrieben. Kepler deutet die von Galilei zuerst entdeckten Mondkrater als geschickt konstruierte monumentale Wall- und Grabenanlagen,

68 Vgl. generell zur zentralen Bedeutung der Schatten im *Somnium* Chen-Morris R., „Shadows of Instruction: Optics and Classical Authorities in Kepler's *Somnium*", *Journal of the History of Ideas* 66 (2005) 223–243. Die These von Chen-Morris lautet: ‚Kepler expects his reader not only to embrace Copernican theory but also to adopt a new ontology of vision comprising of instruments and shadows as positive elements' (225).

69 Kepler, *Der Traum, oder: Mond-Astronomie* 41. ‚Hoc ipso ritu (hui quàm magicè magico) observaveramus paulò prius, quàm libellum ego conciperem Eclipsin Solis, anno scilicet 1605 2./12. Octobris. [...] In solario enim domus deliciariae in hortis Caesaris camerâ destituebamur obscurâ; quare pallijs obnubentes capita, ita diei lucem arcuimus.' Kepler, *Gesammelte Werke* vol. 11.2, 338.

die jeweils eine urbane Einheit begrenzen. Wenn der Autor darauf verzichtet, die neuen Beobachtungsdaten und ihre Deutung in die Traumerzählung einzuarbeiten, ist damit nicht allein die literarische Integrität des Ausgangstextes geschützt. Kepler konstruiert vielmehr auch einen Kontrast von vorteleskopischer Imagination und teleskopischer Observation. Die Übersendung der Mondgeographie an Guldin wird als Dank für den Erhalt eines Fernrohrs von Niccolò Zucchi, einem jesuitischen Mitbruder des Adressierten, ausgewiesen. Durch den Einsatz des Instruments läßt sich die mentale Repräsentation von Städten auf dem Mond autoptisch legitimieren: ‚Wenn du den Geist auf die Mondstädte richtest, kann ich dir bestätigen, dass ich sie sehe.'[70]

Kepler imaginiert die Nutzungsweisen der Mond-Monumente; die rundumlaufenden Gräben etwa, die innen von den Wällen eingefaßt sind, dienen in Feuchtzeiten als schiffbare Kanäle, in Trockenzeiten als lebenswichtige Schattenzonen. Die Spekulationen bedürfen aber der zukünftigen Überprüfung:

> Das soll dir nach Art eines Problems vorgelegt sein. Es muss Stück für Stück mit den Phänomenen selbst bestätigt werden, die durch das Fernglas betrachtet werden können—wenn sie denn durch optische, physikalische und metaphysische Axiome mit den Schlussfolgerungen in Einklang zu bringen sind.[71]

Das Wechselspiel von Spekulation und Observation, das hier thematisiert wird, findet auf der textorganisatorischen Ebene des *Somnium*-Drucks seinen Niederschlag als ‚Schuß' und ‚Gegenschuß' zweier Blicke: ein imaginativer Blick vom Mond auf die Erde und ein teleskopischer von der Erde auf den Mond. Damit sind Standort und Modus der ‚Bildgebung' invertiert. Die instrumentelle Aufrüstung des Blicks verwandelt die Mondbewohner von Troglodyten zu Städtebauern (und allegorisiert durch diesen ‚Sprung' wiederum den Fortschritt in der Observationspraxis). Mit dem textgenetischen Nacheinander von Mondastronomie und Mondgeographie ist sowohl ein wissenschaftsgeschichtlicher Schritt (die Einführung des Teleskops) ausgestellt als auch eine metholodogische Komplementarität von Imagination und

---

70 Kepler, *Der Traum, oder: Mond-Astronomie* 102. ‚Ad Oppida Lunaria si mentem afferas, tibi me videre comprobabo.' Kepler, *Gesammelte Werke* vol. 11.2, 368.

71 Kepler, *Der Traum, oder: Mond-Astronomie* 104. ‚Haec problematis in morem tibi proposita sunto, membratim comprobanda ex ipsis apparentijs, perspicillo detectis; si eae per Axiomata Optica, Physica, Metaphysica ad collectiones istas accommodentur.' Kepler, *Gesammelte Werke* vol. 11.2, 369.

Observation, sprich auf der Textebene: von narrativer/rhetorischer *evidentia* und instrumenteller Evidenz—welche wiederum das imaginative Spiel antreibt. Observationen erzeugen Imaginationen, die Obversationen nach sich ziehen, die neue Imaginationen erzeugen: ‚Aber das ist unterhaltsam etc.'[72]

(3.) Der Druck des *Somnium* von 1634 weist ein typographisches Markierungssystem auf,[73] das in das Spiel der Fiktionalitäts- und Perspektivitätsreflexion einbezogen ist. Den größten Schriftgrad teilen die Rede des Dämons und der mondgeographische Anhang. Beide Passagen sind gleichrangig als inhaltlicher ‚Kern' der Publikation ausgewiesen. Die Anmerkungen Keplers zur Mondastronomie, Mondgeographie und zu Plutarchs *De facie in orbe lunae* im Anhang stehen, auch das ist geläufig, im kleinsten Schriftgrad. Bemerkenswert ist allerdings die typographische Binnendifferenzierung der Traumerzählung: Der einleitende Absatz, in dem die autobiographische Stimme Keplers die äußerlichen Rahmenbedingungen des Traumes darlegt, erscheint in einer mittleren Schriftgröße, die folgende Erzählung des Duracotus im gleichen Schriftgrad, jedoch in einer Kursive, die auch in der angehängten Plutarch-Edition Anwendung findet. Hier entsteht eine visuelle Allianz zwischen den Textpassagen: Die Kursivierung assoziiert die traumallegorische Erzählung des Duracotus mit dem Plutarch-Text; der fiktionale Rahmen gewinnt an Legitimität durch die Autorität des fiktionalen Intertextes im Anhang.

Wichtiger ist jedoch die typographische Markierung der dreistufigen Staffelung der Erzählinstanzen: das ‚Kepler'-Ich (mittlerer Schriftgrad recte), Duracotus (mittlerer Schriftgrad kursiv), der Dämon aus Levania (größter Schriftgrad recte); der Stimmen- respektive Perspektivenwechsel wird *augenfällig*. Am Ende der Traumerzählung jedoch ‚versagt'—ob kalkuliert oder nicht[74]—das typographische Markierungssystem der Erzählung. An der Stelle, an der das ‚Kepler'-Ich aus dem Traum erwacht und ‚zu sich selbst zurückkehrt',[75] wäre, nach der binnentypographischen Logik des Druckes, der mittlere Schriftgrad des ersten Absatzes zu erwarten gewesen.

72 Kepler, *Der Traum, oder: Mond-Astronomie* 104. ‚Sed haec ludicra sunt, etc.' Kepler, *Gesammelte Werke* vol. 11.2, 369.

73 Im folgenden kann ich die relevanten Schriftgradabstufungen lediglich relativ bestimmen; der exakte Schriftgrad ließ sich nicht feststellen, da mir nicht der Originaldruck, sondern ein Digitalisat zur Verfügung stand.

74 Ich betrachte hier den Umstand der mangelnden, weil unmöglichen typographischen Autorisation des posthumen Drucks durch Kepler als nicht maßgeblich und gehe, mit einem rezeptionsästhetischen Ansatz, vom Faktischen der Seite aus.

75 Vgl. Kepler, *Der Traum, oder: Mond-Astronomie* 26; Kepler, *Gesammelte Werke* vol. 11.2, 331.

Stattdessen endet die Erzählung mit dem Schriftgrad, in dem die Rede des Dämons gesetzt ist. Die typographische Performanz der Seite ‚entlarvt' den Perspektivenwechsel als literarisches Spiel; die zunächst optisch getrennten Instanzen erscheinen amalgamiert. Während auf der Textebene die dämonische Stimme der Astronomie abbricht und das autobiographische ‚Kepler'-Ich sich selbst auf dem Bett wiederfindet, sticht dem Leser ins Auge, daß hier wohl tatsächlich *einer* (ein dritter) *durchgängig gesprochen* hat: der ‚reale' Kepler als Produzent des Textes.

Die Reflexionen auf den fiktionalen Status des *Somnium* sind zahlreich. Die konsequenteste Ironisierung des Textes findet sich in Endnote 50, die auf *den Klang* der dämonischen Stimme Bezug nimmt, d.h. auf den medialen Status jener Erzählinstanz, durch welche die mondastronomische Beobachtungssituation vermittelt wird. Die Stimme des Dämons ist ein Krächzen, stammelnd und dumpf. Kepler merkt an:

> Ich halte es nicht für unmöglich, mit verschiedenen Instrumenten sowohl die einzelnen Vokale als auch die Konsonanten zur Nachahmung der menschlichen Sprache hervorzubringen. Was jedoch dabei herauskommen wird, dürfte einem Getöse und Gekrächze ähnlicher sein als einer lebendigen Stimme. Ich glaube auch, dass in solcher Mechanik einige Fallen für Abergläubische und Leichtgläubige stecken, so dass sie manchmal glauben, Dämonen sprächen zu ihnen, während doch die Technik magische Gaukelei erzeugt. [...]
>
> Doch kommt mir hier die angenehme Erinnerung an Matthias Seiffart seligen Angedenkens in den Sinn, einen Schüler Tycho Brahes, der auch nach dessen Tod noch bei seinen Erben blieb. Er verbrachte drei Monate damit, die Ephemeride des Mondes nach Tychos Regeln für ein Jahr zu berechnen. Dessen Stimme war solchen Tönen ganz ähnlich. Ihn befiel eine melancholische, wahnhafte Krankheit, die keinen Platz zum Scherzen ließ und schließlich in Tod durch Wassersucht endete.[76]

---

76 Kepler, *Der Traum, oder: Mond-Astronomie* 42. ‚Non impossibile existimo, varijs instrumentis singulas tàm vocales, quàm consonantes ad imitationem loquelae humanae repraesentare. Id tamen, quicquid futurum est, strepitui et screatui quàm vivae voci similius erit. Et in hac mechanica puto nonnullas esse collocatas insidias superstitiosis et credulis, vt quandoque existiment, sibi loqui daemonas, cùm praestigias magicas ars imitetur. [...] Offert se tamen hîc mihi jucunda memoria Matthiae Seiffarti bonae memoriae Symmystae à Tychone Braheo haeredibus suis relicti, qui tres menses insumpsit in computanda ex praeceptis Tychonicis Ephemeride Lunae in annum unum: cui equidem

Die Maschinenspekulation thematisiert vordergründig die auf Unwissenheit beruhende dämonische Wirkung, welche durch artifizielle Simulationen von Lebendigem entstehen könnte. Der Bezug der Note auf die zentrale narrative Gelenkstelle des *Somnium* legt jedoch nahe, die ‚abschweifende' Assoziation auch als ironisches poetologisches ‚Geständnis' zu lesen. Die Stimme des Dämons, die ja zuletzt eine maskierte Kepler-Stimme ist, wird als Artefakt ausgewiesen, hinter dem Kepler als der ‚Maschinist des Textes' steht. Zugleich ähnelt das Krächzen der Stimme eines ‚echten' Astronomen, der sich als Ephemeridenproduzent ausgerechnet der tabellarischen Positionsbestimmung des Mondes widmete und mit dem daher ein wichtiges Motiv des *Somnium* angespielt ist: die Bewegung des Mondes (= der Erde), auf dem postiert man keine Bewegung wahrnimmt. Dieser Astronom aber wird wahnsinnig.—So wird die ‚Glaubwürdigkeit' der dämonischen Stimme zweimal assoziativ unterlaufen, denn sie ähnelt einer maschinellen Simulation, die eine wahnhafte Wahrnehmung hervorbringen kann, und sie ähnelt der Stimme eines Wahnsinnigen.

Natürlich ist die Stimmen-Verkettung von Kepler, Dämon, Wahnmaschine und wahnsinnigem Astronomen keinesfalls als Zurücknahme des Textes auf inhaltlicher Ebene zu verstehen. Hier wird vielmehr das im *Somnium* durchgespielte Prinzip der Entlarvung falschen Glaubens auch noch auf der Ebene der Selbstthematisierung des Textes betrieben. Wie der Dämon eine (explizite) Allegorie der ‚richtigen' Astronomie ist und die durch ihn berichtete Wahrnehmung der Mondbewohner eine Allegorie der irdisch-menschlichen Wahrnehmung, so fungiert die literarische Anlage des *Somnium* als ‚textperformative' Allegorisierung der Methode.[77]

vox erat verbis hisce non absimilis, morbus etiam obtigit melancholicus et phreneticus, in quo non erat locus ludendi, quique tandem in Hydropem lethalem exijt.' Kepler, *Gesammelte Werke* vol. 11.2, 338.

77 Zum Verhältnis von Epistemologie und allegorisch-literarischer Faktur des *Somnium* vgl. insbesondere auch Paxson, J.J., „Kepler's Allegory of Containment, the Making of Modern Astronomy, and the Semiotics of Mathematical Thought", *Intertexts* 3, 2 (1999), 105–123; Chen-Morris R., „Shadows of Instruction"; Schneider, „Science as Science Fiction"; Vogl, „Robuste und idiosynkratische Theorie".

3 Selenozentrische Aporie: Ein Wissenschafts-Prozeß in Cyrano de Bergeracs *L'Autre Monde*

Im ersten Teil von *L'Autre Monde* imaginiert Cyrano de Bergerac eine andere Welt, in der geleugnet wird, daß es andere Welten geben kann: In einer Prozeß-Episode, die sehr deutlich als satirische „Kontrafaktur"[78] des Prozesses gegen Galilei angelegt ist,[79] wird der Ich-Erzähler Dyrcona, den die Mondbewohner für eine Art von Tier halten, auf Betreiben der Priester vor Gericht gestellt, weil er die Einzigartigkeit des Mondes als Welt leugnet und den Mond einen Mond nennt. Dyrconas Verteidiger erreicht, daß die Frage nach dem biologischen Status seines Mandanten—Tier oder Mensch—bei der juristischen Entscheidungsfindung Berücksichtigung findet. Im folgenden wird in den Blick genommen, wie der Prozeß *anthropozentrische* Gewißheiten zweier (religiös kontrollierter) wissenschaftlicher Diskurse—der Kosmologie und der Zoologie—in einen methodischen Widerspruch münden läßt. Um den ‚Selenozentrismus' zu retten, muß die mondübliche Tier/Mensch-Unterscheidung aufgegeben werden.—Cyrano greift bei seinem wissenschaftskritischen Spiel irdische Anthropozentrismen auf (den aufrechten Gang, die exklusive Sprachfähigkeit als Zeichen der exklusiven Vernunft, die exklusive Stellung im Weltall) und verunsichert ihre Geltung durch Inversionen.[80]

Der thematische Schwerpunkt der Forschung zu den Bezügen Cyranos auf zeitgenössische Epistemologie und ‚naturwissenschaftliche' Wissensbestände lag bislang auf der Kosmologie (Kopernikanismus, Vielheit der Welten) und der (epikureischen) Theorie der Atome in ihrer Formulierung und Vermittlung durch Gassendi, dessen Pariser Vorlesungen Cyrano besuchte.[81] Kaum verfolgt

---

78 Guthke, *Der Mythos der Neuzeit* 178.

79 Vgl. ebd. 174, 178f.; Sick, „Cyrano de Bergerac", 75f.

80 Vgl. zur erkenntniskritischen Stoßrichtung der Inversionsmotive Gipper, *Wunderbare Wissenschaft* 98.

81 Zu den naturwissenschaftlichen (verstanden im modernen Sinn) und philosophischen Positionen bzw. Quellen Cyranos, vornehmlich in *L'Autre Monde*, vgl. Loewenstein, „Die naturphilosophischen Ideen bei Cyrano de Bergerac"; Jordan L., [Einleitung zu] *Savinien de Cyrano Bergerac's L'Autre Monde ou Les Estats et Empires de la Lune. Nach der Pariser und der Münchener Handschrift sowie nach dem Drucke von 1659*, ed. L. Jordan, Gesellschaft für Romanische Literatur 23 (Dresden: 1910) 7–97, hier 43–97; Alcover M., *La pensée philosophique et scientifique de Cyrano de Bergerac*, Histoire des idées et critique littéraire 109 (Genève: 1970); Darmon, J.-Ch., „Cyrano de Bergerac et les images de la nature", in Biet Ch. (ed.), *L'idée de nature au début du XVIIe siècle*, Littératures classiques 17 (Paris: 1992) 153–175; Darmon J.-Ch., „L'épicurisme et les fables du monde: remarques sur Gassendi et Cyrano", in Beugnot B., *La notion de „monde" au XVIIe siècle*, Littératures classiques

wurden indes die Spuren der frühneuzeitlichen Tierseelendiskussion,[82] speziell auch der cartesianischen bête-machine-Theorie,[83] die immer wieder im Text begegnen.[84] Und wenig verfolgt wurde das satirische Spiel mit

---

22 (Paris: 1992) 87–125, hier 111–125; Darmon J.-Ch., *Philosophie épicurienne et littérature au XVIIe siècle. Études sur Gassendi, Cyrano de Bergerac, La Fontaine, Saint-Evremond*, Perspectives littéraires (Paris: 1998) 211–261.—Forschungsliteratur zur Kosmologie bei Cyrano de Bergerac ist in Fußnote 49 angegeben.

82 Überblicke zur französischen Tierseelendiskussion der Frühen Neuzeit bieten Boas G., *The Happy Beast in French Thought of the Seventeenth Century* (Baltimore: 1933); Seban J.-L., „La question de la psychologie de la faune dans la pensée théologique française du crépuscule de la Renaissance au mitan des Lumières", in Niewöhner F. – Seban J.-L. (eds.), *Die Seele der Tiere*, Wolfenbütteler Forschungen 94 (Wiesbaden: 2001) 201–247.

83 Zu Descartes' Position im Tierseelendiskurs vgl. Heintel E., „Tierseele und Organismusproblem im cartesianischen System", *Wiener Zeitschrift für Philosophie, Psychologie und Pädagogik* 3 (1950) 73–121; Cottingham J., „‚A Brute to the Brutes?': Descartes' Treatment of Animals", *Philosophy. The Journal of the Royal Institute of Philosophy* 53 (1978) 551–559; Sutter A., *Göttliche Machinen. Die Automaten für Lebendiges bei Descartes, Leibniz, La Mettrie und Kant* (Frankfurt a.M.: 1988) 64–78; Gontier Th., „Les animaux-machines chez Descartes: modèle ou réalité?", *Corpus* 16/17 (1991) 3–16; Harrison P., „Do Animals Feel Pain?", *Philosophy. The Journal of the Royal Institute of Philosophy* 66 (1991) 25–40; Harrison P., „Descartes on Animals", *The Philosophical Quarterly* 42 (1992) 219–227; Steiner G., „Descartes on the Moral Status of Animals", *Archiv für Geschichte der Philosophie* 80 (1998) 268–291; Boros G., „Fictum Brutum? Über die ambivalente Tierseelenlehre des Descartes", in Niewöhner F. – Seban J.-L. (eds.), *Die Seele der Tiere*, Wolfenbütteler Forschungen 94 (Wiesbaden: 2001) 181–202; Wild M., *Die anthropologische Differenz. Der Geist der Tiere in der frühen Neuzeit bei Montaigne, Descartes und Hume*, Quellen und Studien zur Philosophie 74 (Berlin – New York: 2006) 135–210.—Speziell zur *Wirkungsgeschichte* der cartesianischen bête-machine-Theorie im 17. und 18. Jahrhundert vgl. Rosenfield L.C., „Un chapitre de l'histoire de l'animal-machine (1645–1749)", *Revue de Littérature Comparée* 17 (1937) 461–487; Rosenfield L.C., „Pardies and the Cartesian Beast-Machine", *Publications of the Modern Language Association* 53 (1937) 763–773; Rosenfield L.C., *From Beast-Machine to Man-Machine. Animal Soul in French Letters from Descartes to La Mettrie*, new and enlarged edition (New York: 1968); Shugg W.: „The Cartesian Beast-Machine in English Literature (1663–1750)", *Journal of the History of Ideas* 29 (1968) 279–292; Guichet J.-L., „Les ambiguïtés de la querelle de l'âme des bêtes dans la seconde moitié du XVIIe siècle: L'exemple du *Discours de la connaissance des bêtes* d'Ignace-Gaston Pardies", in Gontier, T. (ed.), *Animal et animalité dans la philosophie de la Renaissance et de l'Âge classique* (Louvain – Paris – Dudley, MA: 2003) 59–75.

84 Eine Einordnung Cyranos in die durch Montaigne begründete frühmoderne ‚Theriophilie', die auf eine Verunsicherung des Anthropozentrismus bzw. der menschlichen *superbia* durch den Aufweis von Vernunft und Moralität im Tierreich zielt, findet sich in Boas, *Happy Beast* 142–145. Boas behandelt in diesem Zusammenhang lediglich den Strafprozeß, der gegen Dyrcona im Sonnenreich der Vögel unter dem Anklagepunkt

Leitkriterien, Argumenten und Topoi der ‚anthropozentrierten' Tier/Mensch-Unterscheidung,[85] welches *L'Autre Monde* leitmotivisch durchzieht. Damit klar wird, was für den Erhalt der religiös-anthropozentrischen Kosmologie im ‚Fall Dyrcona' schließlich geopfert wird, soll hier zunächst in gebotener Ausführlichkeit die Topik dieser Differenz behandelt werden. Was ist für die Außerirdischen—unter der Bedingung ihrer anderen Gestalt—ein Mensch, was ein Tier? Die Frage läßt sich verfolgen am Leitfaden der taxonomischen und praktizierten Vertierung (‚brutification')[86] Dyrconas, oder anders gesagt: der ‚naturgeschichtlichen Karriere' (Mensch—Affe—‚wilder Mensch'—Monstrum—Papagei—Vogel Strauß—Mensch), die der Außermondische auf dem Mond durchläuft. Dort trifft Dyrcona riesenhafte Wesen—gestaltet nach Art der Erdenmenschen, doch auf allen Vieren gehend—, die ihn an die Sirenen, Faune und Satyrn aus den Ammenmärchen seiner Kindheit erinnern. Die Alteritätserfahrung ist beiderseitig: Dyrcona wird mit einem solchen Spottgelächter bedacht, daß er glaubt, selbst ein Monstrum geworden zu sein (‚quasi être devenu monstre').[87] In eine Stadt getragen, wo man ihn für eine Art von Affen hält, gerät der Held in die Menagerie eines Gauklers, der ihn ausstellt und dazu ‚dressiert', Purzelbäume zu schlagen und Grimassen zu schneiden.

Nach einiger Zeit wird Dyrcona einem Artgenossen als vermeintliches Weibchen zugesellt; das ‚petit animal'[88] der Königin, mit dem der Franzose gepaart werden soll, entpuppt sich als ein Spanier, der von Vögeln auf den Mond getragen worden sei: eine Anspielung auf die Hauptfigur Domingo

---

geführt wird, der Erzähler sei ein Mensch. Den ersten Hinweis auf eine anticartesianische Position in der Tierseelendebatte, die im Roman erkennbar sei, gibt Rosenfield, *Beast-Machine* 114–118; vgl. hierzu ferner Barbe, „Cyrano: la mise à l'envers du vieil univers d'Aristote". Die bislang ausführlichste Analyse des Verhältnisses von Mensch und Tier in *L'Autre Monde* bietet Harth, *Cyrano de Bergerac and the Polemics of Modernity* 156–184. In dem Kapitel ‚Satirical Relativism: Man and Society' werden die Inversionen des Tier/Mensch-Verhältnisses im Roman allerdings nicht als Beitrag zur Tierseelendebatte interpretiert, sondern—im Rahmen der Topik der ‚Verkehrten Welt'—als geläufiges satirisches Mittel der Darstellung menschlichen Denkens und Handelns.

85 Zum topischen Spektrum der Tier/Mensch-Differenz in der Frühen Neuzeit vgl. Münch P., „Die Differenz zwischen Mensch und Tier. Ein Grundlagenproblem frühneuzeitlicher Anthropologie und Zoologie", in Münch P. – Walz R. (eds.), *Tiere und Menschen. Geschichte und Aktualität eines prekären Verhältnisses* (Paderborn – München – Wien – Zürich: 1998) 323–347; Münch P., „Verwandtschaft oder Differenz? Zur Theorie des Mensch/Tier-Verhältnisses im 17. Jahrhundert", in Lehmann H. – Trepp A.-Ch. (eds.), *Im Zeichen der Krise. Religiosität im Europa des 17. Jahrhunderts* (Göttingen: 1999) 517–535.

86 Cyrano de Bergerac, *Œuvres complètes* vol. 1, 76.

87 Ebd. 51.

88 Ebd. 75.

Gonsales aus Francis Goodwins *The Man in the Moone*. Cyranos Gonsales wird inmitten einer Horde von Affen gehalten, die mit Halskrausen und kurzen Hosen ausstaffiert sind. Es stellt sich heraus, daß der Affenstatus, der dem Spanier zugeschrieben wird, sich einem kulturell bedingten Fehlschluß verdankt: Als der Spanier in die Hände der Königin gefallen sei, habe sie ihn aufgrund seines Habits für einen Affen gehalten, ‚weil sie in diesen Land hier die Affen von ungefähr nach spanischer Art kleiden'.[89]

Der Syllogismus ‚Affen tragen spanische Tracht—das Fremde trägt spanische Tracht—das Fremde ist ein Affe' ist formallogisch einwandfrei, übersieht jedoch den für den Leser offensichtlichen Fehler, daß die klassifikatorische Bestimmung des Spaniers als Affe auf der Ineinssetzung eines akzidentiellen kulturellen Merkmals mit einem wesentlichen natürlichen Merkmal beruht. Der Phänotyp des Affen ist ein Kulturprodukt, Ergebnis einer Denaturierung, die in einem weiteren Schritt begrifflich und taxonomisch ‚naturalisiert' wird. Als ‚Affe' wird (nur noch?) erkannt, was durch die ‚zweite Natur' des Tieres geprägt ist, ein kostümierter Spaßmacher zu sein.

Dyrcona und der Spanier dienen eine Weile als Zootiere; immer wieder wird das falsche Weibchen von seinen Haltern, König und Königin, abgetastet, ob es nicht trächtig geworden sei. Unterdessen hat der Franzose die Fähigkeit erworben, die (zwei) Sprache(n) der Mondbewohner zu verstehen und bruchstückhaft zu sprechen.[90] Die sensationelle Sprachkompetenz des Tieres provoziert

89 Cyrano de Bergerac, *Reise zum Mond und zur Sonne* 59. ‚[...] à cause qu'ils habillent, par hasard, en ce pays-là, les singes à l'espangnole [...]' Cyrano de Bergerac, *Œuvres complètes* vol. 1, 76.—Dyrcona antwortet dem Spanier mit einer Bemerkung, welche dessen zoologische Klassifizierung als Affe ironisch bestätigt, indem die kulturelle Überformung als rückgebunden an seine Natur angedeutet wird. Dyrcona sinngemäß: Man habe wohl keine lächerlichere und also der Spaßmacherfunktion der Affen dienlichere Kleidung finden können als das spanische Habit. Die implizite These lautet: Wie die spanische Kleidung beim Affen seine lächerliche Natur erst vollständig zum Ausdruck bringt, so läßt sie auch beim Spanier auf eine lächerliche Naturanlage schließen. Die satirische Bestätigung der Grenzverwischung zwischen Spanier und Affe, die den Franzosen natürlich eindeutig auf der Seite der Menschen positioniert, läßt sich als eine Variation des frühneuzeitlichen Topos vom natürlichen Gegensatz zwischen Spaniern und Franzosen lesen. Vgl. hierzu Garcia Carlos., *La oposición y conjunción de los dos grandes luminares de la tierra o La antipatia de Franceses y Españoles (1617)*, ed. Bareau M. (Edmonton: 1979).

90 Die Sprache der Mondmenschen zerfällt sozialhierarchisch in zwei fundamental verschiedene Formen: die stumme Ganzkörpergestik des einfachen Volkes und die auf Tonhöhenunterschieden beruhende Sprache der Vornehmen. Cyranos Erfindung der ‚höheren' Musiksprache rekurriert auf Godwins Modell einer Notensprache in *The Man in the Moone*. Vgl. hierzu Davies H.N., „Bishop Godwin's ‚Lunatique Language' ", *Journal of the Warburg and Courtauld Institutes* 30 (1967) 296–316; Knowlson J.R., „A Note on Bishop

im Mondreich die These, man habe es bei den beiden Tieren der Königin mit ‚deux hommes sauvages'[91] zu tun, deren Winzigkeit und Unfähigkeit, sich auf die verkümmerten Vorderbeine zu stützen, durch einen—modern gesprochen—genetischen Mangel ihrer Väter und das karge Nahrungangebot der Wildnis erklärbar sei.

Hier nun treten *die Priester* des Landes auf den Plan. Der ‚naturwissenschaftlich' begründete Versuch, die fremden Wesen auf der *scala naturae* zu (wenn auch verkümmerten) Menschen aufsteigen zu lassen bzw. die Mangelmutante wenigstens im Zwischenreich des *homme sauvage* anzusiedeln,[92] wird sogleich als ketzerisch zurückgewiesen. Die Priester okkupieren den zoologischen Diskurs und beanspruchen für sich, d.h. für die Theologie die alleinige Kompetenz zur Definition der Tier/Mensch-Grenze. Die radikalere Fraktion unter den Priestern bestimmt Dyrcona und den Spanier als ‚monstres'[93]—und unterläuft damit die Frage nach der Klassifikation der fraglichen Tiere mit der impliziten These ihrer Nicht-Klassifizierbarkeit; als Monstren sind sie weder Tier noch Mensch und quasi außerhalb der *scala naturae* angesiedelt. Die genauen Argumente der Tier/Mensch-Unterscheidung müssen im konkreten Fall somit gar nicht erst erörtert werden. Auch die gemäßigteren Priester wollen eher noch ihren ‚animaux domestiques' das ‚privilège de l'humanité, et de l'immortalité' zugestehen mit dem Argument, daß diese Tiere im eigenen Land geboren sind im Gegensatz zu einem ‚bête monstrueuse',[94] dessen genaue Herkunft nicht bestimmt werden kann. Selbst die ‚revolutionäre' Position, welche die *humanité* von Tieren als Gedankenexperiment durchspielt, grenzt die umstrittenen Tiere als monströs von den ‚humanisierten' Haustieren ab, deren Phänotyp und Verhalten kulturell überformt, ja häufig sogar kulturell produziert ist. Die Menschendiener aus der Tierwelt sind tendenziell menschlich, weil sie in ihrer Funktion von Menschen ‚verursacht' werden.

Im strengeren Sinne in den Zuständigkeitsbereich der Priester gehören zwei ihrer *theologischen* Argumente: (1.) die physikotheologische Interpretation der Vierfüßigkeit der Mondmenschen und (2.) die kontrastive Exegese der

Godwin's *Man in the Moone*: The East Indies Trade Roote and a ‚Language' of Musical Notes", *Modern Philology* 65, 4 (1968) 357–361.

91 Cyrano de Bergerac, *Œuvres complètes* vol. 1, 88.

92 Einen historischen Überblick über Quellen und Theorien zum Phänomen des ‚wilden Menschen' bietet Blumenthal P.J., *Kaspar Hausers Geschwister. Auf der Suche nach dem wilden Menschen*, erweiterte Ausgabe (München – Zürich: 2005).

93 Cyrano de Bergerac, *Œuvres complètes* vol. 1, 88.

94 Ebd.

zweibeinigen und der vierbeinigen Körperhaltung als Ausdruck unterschiedlicher religiöser Praktiken.

(1.) Die Vierfüßigkeit der Mondmenschen verweise auf die Providenz des Schöpfers, der dem Menschen als dem wertvollsten Geschöpf einen besonders sicheren Stand habe geben wollen:

> Wir hier, wir laufen auf allen vieren, weil Gott sich bei einem solch kostbaren Wesen nicht auf eine Haltung verlassen wollte, die weniger Sicherheit böte. Er fürchtete, andernfalls könnte dem Menschen etwas zustoßen; weshalb er selbst Sorge trug, ihn auf vier Pfeiler zu setzen, daß er nicht falle. Da er es aber für unter seiner Würde hielt, sich mit der Verfertigung dieser beiden Viecher zu befassen, überließ er sie den Launen der Natur, welche, den Verlust von so Geringfügigem nicht achtend, sie nur auf zwei Pfoten stützte.[95]

Das physikotheologische Interpretationsprinzip, die funktionale Perfektion der Natur an die Providenz Gottes zu binden, ist zwar nicht strikt antihierarchisch, zielt jedoch tendenziell auf die Integration auch der ‚niedrigsten' Wesen unter dem Gesichtspunkt eines Gesamtplans der Schöpfung. In der anthropozentrischen Argumentation der Priester dagegen wird die kreative Zuwendung Gottes skaliert. Die Providenz bemißt sich am (unterstellten) Wert der Kreatur. Daher wird, nach Meinung der Priester, auf die Konstruktion des Menschen alle ‚statische' Sorgfalt verwendet, während andere Geschöpfe vernachlässigt werden und die unwürdigsten Kreaturen als Laune der Natur (‚caprice de la nature') sogar aus der zweckorientierten göttlichen Schöpfung herausfallen können.[96] Die *monstrositas*,[97] die Dyrcona und dem Spanier zugeschrieben wird, beweist die Nichtbeteiligung Gottes bei der Entstehung dieser Tiere.

---

95 Cyrano de Bergerac, *Reise zum Mond und zur Sonne* 69f. ‚Nous autres, nous marchons à quatre pieds, parce que Dieu ne se voulut pas fier d'une chose si précieuse à une moins ferme assiette. Il eut peur qu'il arrivât fortune de l'homme; c'est pourquoi il prit lui-même la peine de l'asseoir sur quatre piliers, afin qu'il ne pût tomber; mais [dédaignant] de se mêler de la construction de ces deux brutes, il les abandonna au caprice de la nature, laquelle, ne craignant pas la perte de si peu de chose, ne les appuya que sur deux pattes.' Cyrano de Bergerac, *Œuvres complètes* vol. 1, 88.

96 Vgl. Adamowsky N. – Böhme H. – Felfe R. (eds.), *Ludi naturae. Spiele der Natur in Kunst und Wissenschaft* (München: 2010), hier speziell den Beitrag von Fischel A., „Groteske Spiele—Monster als Werke einer spielenden Natur" (161–176).

97 Zur Interpretation und Klassifikation der *monstra* in der Frühen Neuzeit vgl. Park K. – Daston L.J., „Unnatural Conceptions: The Study of Monsters in Sixteenth- and Seventeenth-Century France and England", *Past and Present* 92 (1981) 20–54; Park K. –

(2.) Die vierbeinige Körperhaltung wird von den Priestern zudem als Naturanlage zu einer spezifischen Art der Kontemplation gedeutet. Der Kopf sei tief gesenkt ‚in Betrachtung der Dinge, deren Herren wir sind'.[98] Der ‚Dingbezug' des auf die Erde gerichteten Blicks soll eine meditative Betrachtung der Welt provozieren, über die der Mensch herrscht. Der Blick in den Himmel, zu dem die zweibeinigen Tiere genötigt sind, sei dagegen der Blick der Unglücklichen, die nach ihrem Schöpfer suchen, um sich zu beklagen und Hilfe zu erflehen. Der natürliche Unterschied zwischen Zweibeinern und Vierbeinern wird als Gegensatz des Gottesbezugs, als Gegensatz von verzweifelter Klage und meditativer Inbesitznahme der Gottesgaben, motiviert.[99]

Die Gangart dient in der Argumentation der Priester als entscheidendes Kriterium der Tier/Mensch-Differenz. Dyrcona bestreitet die Triftigkeit dieses Differenzkriteriums; kurz nach seiner Gefangennahme reflektiert er über den Nutzen des vierfüßigen Ganges:

> Und wirklich, beim späteren Nachsinnen darüber fand ich diese Haltung gar nicht so sehr ungereimt, wenn ich daran dachte, daß unsere Kinder, solange sie nur von der Natur angeleitet werden, auf allen vieren laufen und sich erst auf zwei Beinen erheben durch die Sorglichkeit ihrer Ammen, die sie in kleinen Laufwagen aufrichten und ihnen Riemen

---

Daston L.J., *Wunder und die Ordnung der Natur 1150–1750* (Berlin: 2002) 205–252; Bates A.W., „Good, Common, Regular, and Orderly: Modern Classifications of Monstruos Births", *Social History of Medecine* 18 (2005) 141–158.

98 Cyrano de Bergerac, *Reise zum Mond und zur Sonne* 70. ‚pour contempler les biens dont nous sommes seigneurs'. Cyrano de Bergerac, *Œuvres complètes* vol. 1, 89.

99 Die Argumentation der Priester verkehrt satirisch das auf Ovid gründende geläufige Lob des aufrechten menschlichen Ganges, das Montaigne in der Abhandlung der Tier/Mensch-Differenz in seiner *Apologie de Raimond Sebond* referiert—und relativiert: ‚Der Vorzug, welchen die Poeten darinnen finden, daß wir aufrechts gehen, und im Himmel, unsern Ursprung, schauen,
Pronaque cum spectent animalia caetera terram,
Os homini sublime dedit, coelumque videre
Iussit, et erectos ad sidera tollere vultus.
ist ein wahrhaft poetischer Vorzug. Es giebet verschiedene kleine Thiere, die vollkommen gen Himmel sehen: und die Kameele und Strause scheinen mir noch einen höhern und geradern Hals, als wir, zu haben.' Montaigne Michel de, *Essais* [*Versuche*] *nebst des Verfassers Leben nach der Ausgabe von Pierre Coste ins Deutsche übersetzt von Johann Daniel Tietz* (Neudruck Zürich: 1992) vol. 2, 103.—Vgl. generell zur Geschichte des ‚aufrechten Gangs' als anthropologisches Differenzkriterium und auch speziell zum Topos der aufrechten Himmelsbetrachtung Bayertz K., *Der aufrechte Gang. Eine Geschichte des anthropologischen Denkens* (München: 2012).

> anbinden, um zu verhindern, daß sie auf alle viere fallen, in die einzige Ruhestellung, zu der die Gestalt unserer Masse neigt.[100]

Nach Maßgabe der Natur, die kleinen Kinder beweisen es, tendiert auch der irdische Mensch zum vierfüßigen Gang, und nur eine maschinengestützte Denaturierung (durch Laufwagen und Riemen) im Rahmen der menschlichen Ontogenese erzwingt die Fortbewegung auf zwei Beinen. Der Habitus des aufrechten Ganges, auf der Erde ein Hauptkriterium der Zugehörigkeit zur menschlichen Art, erscheint Dyrcona bei genauerer Überlegung als akzidentielles Unterscheidungsmerkmal zwischen Mensch und Tier. Die Differenz kann in dieser Sichtweise nicht ‚wesentlich' sein, wenn sie durch kulturelle Maßnahmen hergestellt wird. Der Unterschied ist sinnvoll nur bestimmbar als ein Unterschied der ‚Naturen' von Mensch und Tier.

Die Zweibeinigkeit, die Dyrcona als aktiv betriebene Denaturierung wertet, stellt sich den Priestern als lediglich erlittene Denaturierung dar. Die Tiere der Königin sind nicht allein Zweibeiner, sie sind unvollkommene Zweibeiner:

> ‚Selbst den Vögeln', behaupteten sie, ‚wurde nicht solches Unrecht angetan wie diesen da, denn sie erhielten wenigstens Federn, um der Schwäche ihrer Füße abzuhelfen und sich in die Lüfte zu schwingen [...].'[101]

Der Flug der Vögel wird nicht als die besonders elaborierte Fortbewegungsart gewertet, die sie ist, sondern nur als Notbehelf eines Tieres, das schwache Beine hat. Die Tiere der Königin wirken im Vergleich wie gerupfte Vögel; ihre Vorderbeine sind gänzlich unbrauchbar. Dyrcona, das sprechende Tier der Königin, wird von den Priestern schließlich konsequenterweise zum gerupften Papagei (‚perroquet plumé')[102] erklärt, vom Spanier getrennt und—‚artgerecht'—in einen Vogelkäfig gesperrt. Die Umwidmung des fremden Tieres

---

100 Cyrano de Bergerac, *Reise zum Mond und zur Sonne* 44. ‚Et en effet, rêvant depuis sur ce sujet, j'ai songé que cette situation de corps n'était point trop extravagante, quand je me suis souvenu que nos enfants, lorsqu'ils ne sont encore instruits que de Nature, marchent à quatre pieds, et ne s'élèvent sur deux que par le soin de leurs nourrices, qui les dressent dans de petits chariots et leurs attachent des lanières pour les empêcher [de choir] sur les quatre, comme la seule assiette où la figure de notre masse incline de se reposer.' Cyrano de Bergerac, *Œuvres complètes* vol. 1, 52.

101 Cyrano de Bergerac, *Reise zum Mond und zur Sonne* 70. ‚«Les oiseaux mêmes, disaient-ils, n'ont pas été si maltraités qu'elles, car au moins ils ont reçu des plumes pour subvenir à la faiblesse de leurs pieds, et se jeter en l'air [...].' Cyrano de Bergerac, *Œuvres complètes* vol. 1, 89.

102 Ebd. 89.

vom Affen zum Papagei hat den (unausgesprochenen) Vorteil, das Skandalon der Sprechfähigkeit zu mildern: Ein sprechender Affe erzwingt entweder den Schluß, daß es sich bei ihm tatsächlich um einen Menschen handelt, oder den, daß auch Tiere über eine dem Menschen analoge Ratio verfügen können. Der Papagei wird dagegen als vernunftloser Sprechapparat begriffen, der Laute ‚aufzeichnet' und reproduziert, ohne sie zu verstehen.[103]

Nach kurzer Zeit lernt Dyrcona durch den Unsinn, der ihm täglich vorgesprochen wird, die Sprache der Mondmenschen so perfekt zu sprechen, daß er zu einem geistvollen Erzähler und Unterhalter avancieren kann:

> Schon weidete man sich in den Gesellschaften nur noch an der Artigkeit meiner lustigen Einfälle, und die Wertschätzung für meinen Geist ging so weit, daß sich die Geistlichkeit gezwungen sah, einen Bescheid zu erlassen, wonach es verboten war zu glauben, ich besäße Vernunft, mit einem ganz ausdrücklichen Geheiß an alle Personen welchen Standes und Ranges auch immer, sich vor Augen zu halten, was an Geistreichem ich auch vollbringe, nur der Instinkt sei, der es mich anstellen ließe.[104]

Die außerordentliche Sprechfähigkeit des Papageis provoziert eine Glaubensspaltung (‚accroc de religion')[105] im Land. Trotz der angedrohten Exkommunikation glaubt eine wachsende Fraktion an die Vernunft des fremden Wesens und fordert die Untersuchung der Streitfrage durch ein neutrales

---

103 Vgl. Boehrer B., „‚Men, Monkeys, Lap-Dogs, Parrots, Perish All!' Psittacine Articulacy in Early Modern Writing", *Modern Language Quarterly* 59 (1998) 171–193. Boehrer untersucht den Wandel von der mittelalterlichen Bewunderung der Papageiensprache zu ihrer frühneuzeitlichen Marginalisierung als ‚index of mindlessness and inferiority' (ebd. 186). Die Papageiensprache, aufgefaßt als ‚model of empty, labile, and mechanical articulacy' (ebd. 181), diente in der Frühen Neuzeit als ‚marker not of rarity or excellence or wisdom but as the emblem of an amusing ontological inferiority' (ebd. 182) und wurde in satirischen Kontexten zur sozialen Diskriminierung verwendet—etwa zur Behauptung der intellektuellen Defizienz von Dienerfiguren. Im schlimmsten Fall ließ sich der Vergleich von Menschen mit Papageien dazu nutzen, im Rahmen der europäischen Expansion die Versklavung von Indianern zu rechtfertigen (vgl. ebd. 176f.).

104 Cyrano de Bergerac, *Reise zum Mond und zur Sonne* 71. ‚Déjà les compagnies ne s'entretenaient plus que de la gentillesse de mes bons mots, et l'estime qu'on faisait de mon esprit vint jusque-là que le clergé fut contraint de faire publier un arrêt, par lequel on défendait de croire que j'eusse de la raison, avec un commandement très exprès à toutes personnes de quelque qualité et condition qu'elles fussent, de s'imaginer, quoi que je pusse faire de spirituel, que c'était l'instinct qui me le faisait faire.' Cyrano de Bergerac, *Œuvres complètes* vol. 1, 90.

105 Ebd. 91.

Richtergremium. Bei einem Philosophieexamen reproduziert Dyrcona (in der Annahme, es könne ihm hier nutzen) artig aristotelische Grundsätze—und disqualifiziert sich dadurch, denn Aristoteles gilt den Mondmenschen als Vertreter einer irrationalen Denkweise, in der die Tatsachen der Philosophie angepaßt werden statt umgekehrt. Die Untersuchungskommission beschließt daher einstimmig, Dyrcona sei kein Mensch, sondern vielleicht ‚irgendeine Art Strauß, in Anbetracht, daß ich den Kopf wie diese aufrecht trage'.[106]

In einer zweiten Befragung, die um Themen der Physik kreist, kommt es zu dem Streitpunkt, ob die Welt von Ewigkeit her besteht (mondisches Weltwissen) oder nach der Genesis in einem sechstägigen Schöpfungsakt entstanden sei (Dyrconas irdisches Weltwissen). Durch die Arroganz eines Mondgelehrten provoziert, läßt sich Dyrcona in gespielter Allianz mit Aristoteles zu der provokativen Aussage hinreißen, die ‚andere Welt' sei nur ein Mond:

> ‚Aber', sagten alle zu mir, ‚Ihr seht hier Erde, Wälder, Flüsse und Meere, was wäre denn das alles?'
>
> ‚Einerlei', gab ich zurück, ‚Aristoteles versichert, dies sei nur der Mond, und wenn Ihr in den Schulen, in denen ich studierte, das Gegenteil behauptet hättet, würde man euch ausgepfiffen haben'.[107]

Die (dummerweise) gespielte Dummheit ist für die philosophische Untersuchungskommission ein Grund zum Lachen.[108] Die Unsinnigkeit der Aussage liegt auf der Hand und scheint damit einmal mehr die angenommene Vernunftlosigkeit des Untersuchungsobjektes zu beweisen. Cyrano läßt seinen Protagonisten, der eigentlich zum Mond aufgebrochen war, um dessen Weltartigkeit zu beweisen, nun trotz der sinnlichen Evidenz dieser Weltartigkeit die kosmologische Gegenthese vertreten—eine doppelte Inversion: nämlich eine Verkehrung des experimentell-imaginativen Ansatzes des Textes durch den ‚irrationalen' Positionswechsel Dyrconas. Indem sich die Erzählerfigur mit

106 Cyrano de Bergerac, *Reise zum Mond und zur Sonne* 72. ‚quelque espèce d'autruche, vu que je portais comme elle la tête droite'. Cyrano de Bergerac, *Œuvres complètes* vol. 1, 92.

107 Cyrano de Bergerac, *Reise zum Mond und zur Sonne* 77. ‚« Mais, me dirent-ils tous, vous y voyez de la terre, des forêts, des rivierès, des mers, que serait-ce donc tout cela?—N'importe, repartis-je. Aristote assure que ce n'est que la lune, et si vous aviez dit le contraire dans les classes où j'ai fait mes études, on vous aurait sifflées. »' Cyrano de Bergerac, *Œuvres complètes* vol. 1, 97.

108 Diese Reaktion greift motivisch die Eingangssituation von *L'Autre Monde* auf, in welcher Dyrcona von seinen Freunden ausgelacht wird, weil er behauptet, der Mond sei eine Welt und unsere Welt deren Mond. Vgl. Cyrano Cyrano de Bergerac, *Reise zum Mond und zur Sonne* 12; Cyrano de Bergerac, *Œuvres complètes* vol. 1, 6.

einem ‚irren' theoretischen Phantasma gegen die vom Leser akzeptierte fiktionale (Mond-)Welt stellt (die Dyrcona gleichzeitig selbst erzählend produziert), wird diese eben nicht aufgehoben, sondern *beurteilt* und als ‚das Faktische' im Gegensatz zum ‚Nicht-Faktischen' bestimmt. Die Fiktion funktioniert nicht nur, unbeschadet des Widerspruchs, weiter (das ist ohnehin das Wesen der Fiktion), sie wird *im Denkprozeß* des Lesens als *verifiziert* betrachtet. Der Figur Dyrcona kommt an dieser Stelle die paradoxe Doppel-Funktion zu, die Leugnung der Möglichkeit außerirdischer Welten durch erzählte ‚Fakten' zu widerlegen und im gleichen Moment als Exempel für die Irrationalität dieses Leugnens zu dienen. Ihr Argumentieren mit Aristoteles ist pragmatisch dumm auf der Ebene der Erzählhandlung und erkenntnistheoretisch dumm auf der Ebene der Wirklichkeit, aber konzeptionell schlau auf der Ebene des Textes.

Die Gelehrten in der Kommission nehmen die Abqualifizierung des Mondes mit Humor, doch die nunmehr wieder auf den Plan tretenden Priester[109] reagieren auf die damit verknüpfte Aufwertung der Erde todernst und initiieren einen Inquisitionsprozeß gegen Dyrcona:

> Die Priester indessen erhielten Nachricht davon, daß ich gewagt hatte zu sagen, der Mond sei eine Erde, von der ich komme, und ihre Welt sei nur ein Mond. Sie glaubten, dies liefere ihnen einen gerade ausreichenden Vorwand, mich zum Wassertod verurteilen zu lassen (das ist die Art und Weise, die Atheisten auszurotten).'[110]

Problematisch ist für die Priester weniger die offensichtlich sinnlose Aussage, der Mond sei keine Welt, als vielmehr die Behauptung, der Himmelskörper, den man hier für den Mond halte, sei eine Weltkugel. Die These, mit der die Einmaligkeit der göttlichen Schöpfung infrage gestellt ist, erfüllt in den Augen der Priester den Tatbestand der Ketzerei. Der religiöse ‚Selenozentrismus' trifft sich mit der irdischen theologischen Frontstellung gegen die Vorstellung

109 So in der hier benutzten kritischen Textausgabe von Alcover, welche die handschriftliche Überlieferung berücksichtigt; in der Erstausgabe von 1657 ist weniger spezifisch von ‚anderen, sich noch mehr als die ersten ereifernden Gelehrten' die Rede: ‚autres sçauans plus emportez que les premiers' (Cyrano de Bergerac, *Histoire Comique* 109).

110 Cyrano de Bergerac, *Reise zum Mond und zur Sonne* 78. ‚Les prêtres, cependant, furent avertis que j'avais osé dire que la lune était un monde dont je venais, et que leur monde n'était qu'une lune. Ils crurent que cela leur fournissait un prétexte assez juste pour me faire condamner à l'eau: c'était la façon d'exterminer les athées.' Cyrano de Bergerac, *Œuvres complètes* vol. 1, 98.

von einer ‚Pluralität der Welten'.[111] Eine solche Vielheit der Welten bedeutete nicht nur eine (letztlich konsequenzlose) Entwürdigung der besonderen *nobilitas*, die sich die Gattung Mensch zurechnete,[112] sie war auch mit erheblichen theologischen Problemen, vor allem im Bereich der Sünden- und Erlösungstheologie, verbunden. Die Existenz extraterrestrischer ‚Menschen' implizierte (überkonfessionell) die Frage, ob man sich mit ihnen in einer heilsgeschichtlichen Gemeinschaft befindet. ‚Undenkbar' war die Vorstellung, daß die anderen Menschen frei von der Erbsünde und ohne Wissen um Jesus Christus zu Auferstehung und ewigem Leben bestimmt sein könnten. Undenkbar war aber auch, wie etwa Melanchthon in seinen *Initia doctrinae physicae* ausführt,[113] daß der eine und einzige Sohn Gottes, der *einmal* in unserer Welt geboren, gestorben und auferstanden ist, seine Erlösungstat auf verschiedenen Planeten wiederholen sollte. Die Annahme einer Vielheit der Welten provozierte die unerträgliche Frage: ‚Was Jesus Christ to be seen as a planet-hopping Savior in the new cosmology?'[114]

Der Inquisitionsprozeß gegen Dyrcona rekurriert auf den Prozeß der römischen Inquisition gegen Galilei. Obgleich sich dieser an Spekulationen über menschenähnliche Wesen auf anderen Himmelskörpern niemals beteiligte, war die theologische Brisanz der galileischen Observationen aber allein schon durch den Aufweis der physischen Ähnlichkeit von Erde und Mond

111 Vgl. zur Geschichte des kosmologischen Konzepts einer Vielheit der Welten McColley G., „The Seventeenth Century-Doctrine of a Plurality of Worlds", *Annals of Science* 1 (1936) 385–430; Dick S.J., „The Origins of the Extraterrestrial Life Debate and its Relation to the Scientific Revolution", *Journal of the History of Ideas* 41 (1980) 3–27; ders., *Plurality of Worlds. The Origins of the Extraterrestrial Life Debate from Democritus to Kant* (Cambridge: 1982); Guthke, *Der Mythos der Neuzeit*.—Zur Lehre von der Vielheit der Welten in *L'Autre Monde* vgl. Gipper, *Wunderbare Wissenschaft* 90–97.

112 Vgl. Rossi P., „Nobility of Man and Plurality of Worlds", in Debus A.G. (ed.), *Science, Medicine and Society in the Renaissance* (New York: 1972) vol. 2, 131–162. Auch ein Kopernikaner wie Kepler bemühte sich noch, die Vorstellung einer Mehrheit der Welten auf ein Sonnensystem zu beschränken und die Mittelstellung der Erde zwischen den Himmelskörpern als Argument für die Sonderstellung des Menschen zu konstruieren, dessen Planet als privilegierter Observations-, d.h. Kontemplationsort seine jährlichen Runden um die Sonne macht (vgl. ebd. 138–144).—Zur frühneuzeitlichen Reflexion der Stellung des Menschen im Kosmos vgl. ferner Danielson D.R., „Poeticizing the Earth in the Copernican Cosmos", in Mahlmann-Bauer B. (ed.), *Scientiae et artes. Die Vermittlung alten und neuen Wissens in Literatur, Kunst und Musik*, Wolfenbütteler Arbeiten zur Barockforschung 38 (Wiesbaden: 2004) vol. 1, 399–409.

113 Vgl. Melanchthon Philipp, *Initia doctrinae physicae* (Wittenberg, Johann Lufft: 1550), Bl. 43v.

114 Dick, *Plurality of Worlds* 89.

gegeben. Dementsprechend warnte der Jesuit Giovanni Ciampoli seinen Freund Galilei in einem Brief aus dem Jahr 1615, mit seinen Worten vorsichtig zu sein. Die Behauptung der Ähnlichkeit von Erde und Mond könnte zu der Unterstellung führen, der Astronom nehme menschliches Leben auf dem Mond an—ein möglicher Anstoß zu Diskussionen über die Herkunft der Mondmenschen: wie sie von Adam abstammen könnten, ob sie von Noahs Arche herkämen und sonstige Traumgespinste.[115] Tatsächlich diente der empirische Nachweis einer physischen Vergleichbarkeit von Erde und Mond ja nur dazu, die Wahrscheinlichkeit des heliozentrischen Weltmodells zu erhöhen, wurde doch die geozentrische Annahme der ruhenden Mittelstellung der Erde u.a. durch die Schwere ihrer Materie begründet. War der um die Erde kreisende Mond als von gleicher Materie ausgewiesen, fiel damit ein wichtiges (aristotelisches) Argument gegen die kopernikanische Theorie.—Relevant ist in diesem Zusammenhang ein Detail im kosmologischen Streit Dyrconas mit den Gelehrten. Als dieser die Welthaltigkeit des Mondes in Frage stellt, wird er gefragt, was Erde, Wälder, Flüsse und Meere denn sonst seien, wenn nicht eine Welt. Die Widerlegung Dyrconas beruft sich also überhaupt nicht auf die ebenfalls offenkundige Existenz vernunftbegabter Wesen auf dem Mond, sondern auf dessen physische Geographie (die allenfalls als potentielles Habitat für Leben präsent ist). Ins Spiel kommt mithin allein jene Kategorie von Fakten, um die es Galilei im Rahmen seines Beweisinteresses ging. Die Untersuchungskommission vertritt quasi den Beobachtungsansatz des Astronomen. Damit kommt es im Wechsel von der philosophischen Untersuchung zur priesterlichen Inquisition zu einem Chiasmus der personalen Repräsentationen: Die Auseinandersetzung zwischen dem Beobachter Galilei (autoptischer Ansatz der Mond-Philosophen) und dem Antikopernikanismus (Dyrcona als aristotelischer Simulant) invertiert zur Auseinandersetzung zwischen dem ‚Ketzer' Galilei (nunmehr Dyrconas Rolle) und der Kirche.

Der Ketzerprozeß verläuft schließlich anders, als es die Priester geplant hatten. Der Verteidiger[116] vermag den Angeklagten zu retten, indem er den entscheidenden Anklagepunkt mit der immer noch virulenten Frage verknüpft, *was* ‚cet homme, ce singe ou ce perroquet'[117] denn nun tatsächlich ist. Ist er ein Mensch, so der Anwalt in seinem Plädoyer, müsse ihm Gedanken- und

115 Vgl. Ciampolis Brief an Galilei vom Februar 1615 in Galilei Galileo, *Opere*, ed. A. Favaro, 20 vols. (Florenz: 1902) vol. 12, 145–147.

116 Es ist der ‚Dämon des Sokrates', der schon zuvor im Text als philosophischer Dialogpartner und Helfer Dyrconas eingeführt wird.

117 Cyrano de Bergerac, *Œuvres complètes* vol. 1, 99.

Glaubensfreiheit zugestanden werden. Gehöre er aber zur ,catégorie des bêtes'[118] dürfe man ihn nicht verurteilen, erstens, weil seine Aussagen als Ergebnis einer Instinkthandlung durch die Natur und damit letztlich durch Gott autorisiert seien, und zweitens, weil vernünftige Menschen und unvernünftige Tiere keine gemeinsame Ethik besitzen und daher keine Rechtsgemeinschaft bilden könnten:

> Denn setzt den Fall, er sei ein Tier ohne Vernunft, wie vernünftig wäre es dann von Euch, wenn Ihr ihn anklagt, gegen sie verstoßen zu haben? Er hat gesagt, der Mond sei eine Welt; nun handeln die vernunftlosen Tiere aber nur nach einem Naturinstinkt; folglich ist es die Natur, die so spricht, und nicht er. Nun anzunehmen, diese weise Natur, die die Welt erschuf und den Mond, wisse selbst nicht, was er ist, und Ihr, die Ihr davon nur so viel Kenntnis habt, wie Ihr von der Natur erhieltet, wüßtet es mit größerer Bestimmtheit, das zu glauben wäre recht lächerlich. Wenn nun aber gar die Leidenschaft Euch dazu bewegen sollte, Euch von Euren ersten Prinzipien loszusagen und Ihr dafürhieltet, daß die Natur die Tiere gar nicht leitete, so errötet zum allermindesten über die Unruhe, die der Eigensinn eines Tieres über Euch gebracht hat. Ja wahrhaftig, Ihr Herren, wenn Ihr einen Mann im verständigen Alter träft, der auf einen Ameisenhaufen ein wachsames Auge hätte, nur um bald einer Ameise, die ihre Gefährtin zu Fall gebracht, eine Maulschelle zu geben, bald eine andere hinter Schloß und Riegel zu bringen, die ihrer Nachbarin ein Körnchen Getreide raubte, eine dritte wiederum vor Gericht zu ziehen, die ihre Eier im Stich ließ, hieltet Ihr den nicht für verrückt, daß er Dinge besorgte, die zu sehr unter seiner Würde sind, und Anspruch darauf erhöbe, Tiere der Vernunft zu unterwerfen, die damit gar nicht umgehen können? Wie also, hochwürdige Oberpriester, würdet Ihr es wohl nennen, daß Euch soviel an den Grillen dieses kleinen Tieres liegt? Gerechte, ich habe gesprochen.[119]

118 Ebd.

119 Cyrano de Bergerac, *Reise zum Mond und zur Sonne* 79f. ‚[...] car supposez qu'il soit animal sans raison: quelle raison vous-même avez-vous de l'accuser d'avoir péché contre elle? Il a dit que la lune était un monde; or les brutes n'agissent que par un instinct de nature; donc c'est la nature qui le dit et non pas lui. De croire maintenant que cette savante nature qui a fait et la lune et ce monde-ci ne sache elle-même ce que c'est, et que vous autres, qui n'avez de connoissance que ce que vous en tenez d'elle, le sachiez plus certainement, cela serait bien ridicule. Mais quand même, la passion vous faisant renoncer à vos premiers principes, vous supposeriez que la nature ne guidât point les brutes, rougissez à tout le moins des inquiétudes que vous causent les cabrioles d'une bête. En vérité, Messieurs,

Hält man Dyrcona tatsächlich für ein Tier, ist man mit dem Problem konfrontiert, daß sein instinktives Sprechen seinen Grund in Gott als einer zwecksetzenden Instanz *außerhalb* des Tieres haben muß. Wenn also die tierische Sprechmaschine, welche die Priester abschalten wollen, von Gott angeworfen wurde, können seine Laute nicht zwecklos—und im schlimmsten Fall auch nicht restlos sinnlos sein; das Tier müßte vielleicht sogar als Medium einer autorisierten Botschaft angehört werden, durch die Gott sein Weltmodell erklärt.

Das Argument der (heteronomen) Instinktwahrheit des Tieres impliziert zugleich, daß allein der (autonom denkende) Mensch im Irrtum sein kann. Der Prozeß richtet sich unversehens gegen die Ankläger, indem nicht eine grundsätzliche Kompetenz, sondern ein grundsätzlicher Mangel die anthropologische Differenz begründet. Um die Aussagen Dyrconas überhaupt als Irrlehre einstufen zu können, muß er zum Menschen erklärt werden.

Die Argumentation des Verteidigers zieht eine heikle Güterabwägung der Anthropozentrismen mit sich. Beharrt man auf der Annahme eines animalischen repektive monströsen Status des Angeklagten, riskiert man die Geltung des eigenen kosmolgischen Weltmodells. Die Entscheidung fällt zugunsten des ‚größeren' Anthropozentrismus. Der König der Mondmenschen erklärt, der Angeklagte sei fortan als Mensch zu betrachten. Dyrcona wird freigelassen und die drohende Hinrichtung zur Pflicht gemildert, die ketzerischen Lehren öffentlich zu widerufen. Leichter ist es offenbar, die Exklusivität des Humanen für die innerweltliche *monstrositas* Dyrconas zu öffnen, als die Gleichheit bzw. potentielle *superioritas* außerirdischer Menschen zu akzeptieren.

Mit der Entscheidung, Dyrcona (wieder) zum Menschen zu machen, ist zugleich ein gefährlicher Präzedenzfall geschaffen, denn mit ihr geht die *morphologische Homogenität* als eindeutiges Kriterium der Abgrenzung des vernünftigen Menschen von den unvernünftigen Tieren verloren. Im konkreten Fall Dyrconas wird der Anthropozentrismus noch um den Preis der Eingemeindung des zuvor Anderen gerettet, aber schon deutet sich—ganz im

---

si vous rencontriez un homme d'âge mûr qui veillât à la police d'une fourmilière, pour tantôt donner un soufflet à la fourmi qui aurait fait choir sa compagne, tantôt en emprisonner une qui aurait dérobé à sa voisine un grain de blé, tantôt mettre en justice une autre qui aurait abandonné ses œufs, ne l'estimeriez-vous pas insensé de vaquer à des choses trop au-dessous de lui, et de prétendre assujettir à la raison des animaux qui n'en ont pas l'usage? Comment donc, vénérables pontifes, appellerez-vous l'intérêt que vous prenez aux cabrioles de ce petit animal? Justes, j'ai dit.' Cyrano de Bergerac, *Œuvres complètes* vol. 1, 99f.

Sinne der Stellung Cyranos zur Tierseelendebatte—die Notwendigkeit an, die Vernunft auch in anderen Gattungen und Arten für möglich zu halten.

Nach dem Prozeß verkehrt sich alles noch einmal: Dyrcona wird wieder zum Papagei und der im Urteil des Königs geforderte Widerruf gerät zum Widerruf des Widerrufs. Auf einem Wagen wird der Verurteilte herumgefahren, um auf den fünf großen Plätzen der Stadt auszurufen:

> ‚Untertanen, ich erkläre, daß dieser Mond hier kein Mond ist, sondern eine Erde, und daß diese Erde dort keine Erde ist, sondern ein Mond. Das ist, was die Prieser für gut erachten, das Ihr glauben sollt.‘[120]

Die ideologische Sprachkontrolle der Priester ist mangelhaft: ‚cette lune‘ und ‚ce monde de là-bas‘ werden im Widerruf grammatisch gerade als das bestätigt, was sie nicht sein sollen. Zugleich wird *die Wiederholung der Worte*—im doppelten Sinne: als Nachsprechen von Worten und als (fünfmal) wiederholter Akt—zum Comeback Dyrconas als Papagei. Am Ende verliert die ideologische Botschaft ihren Sinn.

Selective Bibliography

Aït-Touati F., *Fictions of the Cosmos: Science and Literature in the Seventeenth Century* (Chicago – London: 2011).

Bayertz K., *Der aufrechte Gang. Eine Geschichte des anthropologischen Denkens* (München: 2012).

Bezzola Lambert L., *Imagining the Unimaginable: The Poetics of Early Modern Astronomy*, Internationale Forschungen zur Allgemeinen und Vergleichenden Literaturwissenschaft 58 (Amsterdam – New York: 2002).

Boas G., *The Happy Beast in French Thought of the Seventeenth Century* (Baltimore: 1933).

Campbell M.B., *Wonder & Science. Imagining Worlds in Early Modern Europe* (Ithaca, N.Y.: 1999).

Campe R., „Vor Augen Stellen. Über den Rahmen rhetorischer Bildgebung“, in Neumann G. (ed.), *Poststrukturalismus. Herausforderung an die Literaturwissenschaft*, Germanistische Symposien. Berichtsbände 18 (Stuttgart – Weimar: 1997) 208–225.

120 Cyrano de Bergerac, *Reise zum Mond und zur Sonne* 80. ‚«Peuple, je vous déclare que cette lune ici n'est pas une lune, mais un monde; et que ce monde de là-bas n'est point un monde, mais une lune. Tel est ce que les prêtres trouvent bon que vous croyiez.»‘ Cyrano de Bergerac, *Œuvres complètes* vol. 1, 100.

Cyrano de Bergerac, Savinien, *HISTOIRE COMIQVE, PAR MONSIEVR DE CYRANO BERGERAC. Contenant les Estats & Empires de la Lune* (Paris, Charles de Sercy: 1657).

Cyrano de Bergerac, Savinien, *LES NOVVELLES OEVVRES DE MONSIEVR DE CYRANO BERGERAC. CONTENANT L'HISTOIRE COMIQVE des Estats & Empires du Soleil, PLVSIEVRS LETTRES, ET AVTRES PIECES Diuertissantes* (Paris, Charles de Sercy: 1662).

Cyrano de Bergerac, Savinien, *Œuvres complètes. Edition critique*, ed. Alcover M., 3 vols. (Paris: 2000) vol. 1: *L'Autre Monde ou les États et Empires de la Lune. Les États et Emprires du Soleil. Fragment de Physique.*

Cyrano de Bergerac, Savinien, *Reise zum Mond und zur Sonne*, übers. und ed. W. Tschöke (Frankfurt a.M.: 2004).

Dick S.J., *Plurality of Worlds. The Origins of the Extraterrestrial Life Debate from Democritus to Kant* (Cambridge: 1982).

Gipper A., *Wunderbare Wissenschaft. Literarische Strategien naturwissenschaftlicher Vulgarisierung in Frankreich. Von Cyrano de Bergerac bis zur Encyclopédie* (München: 2002).

Godwin Francis, *THE MAN IN THE MOONE: OR A DISCOVRSE OF A Voyage thither BY DOMINGO GONSALES The speedy Messenger* (London, Joshua Kirton und Thomas Warren: 1638).

Guthke K.S., *Der Mythos der Neuzeit. Das Thema der Mehrheit der Welten in der Literatur- und Geistesgeschichte von der kopernikanischen Wende bis zur Science Fiction* (Bern – München 1983).

Hallyn F., *The Poetic Structure of the World. Copernicus and Kepler* (New York: 1990).

Janssen A., *Francis Godwins ‚The Man in the Moone'. Die Entdeckung des Romans als Medium der Auseinandersetzung mit Zeitproblemen*, Sprache und Literatur. Regensburger Arbeiten zur Anglistik und Amerikanistik 19 (Frankfurt a.M. – Bern: 1981)

Kepler Johannes, *Gesammelte Werke*, ed. Kepler-Kommission der Bayerischen Akademie der Wissenschaften, vol. 11.2: *Calendaria et Prognostica, Astronomica minora, Somnium*, ed. V. Bialas – H. Grössing (München: 1993) 315–438.

Kepler Johannes, *IOH. KEPPLERI MATHEMATICI OLIM IMPERATORII SOMNIVM, Seu OPVS POSTHVMVM DE ASTRONOMIA LVNARI*, ed. Ludwig Kepler (Sagan – Frankfurt a.M., Johannes Keplers Erben: 1634).

Kepler Johannes, *Der Traum, oder: Mond-Astronomie*, übers. H. Bungarten, ed. B. Langner (Berlin: 2011).

Münch P., „Die Differenz zwischen Mensch und Tier. Ein Grundlagenproblem frühneuzeitlicher Anthropologie und Zoologie", in Münch P – Walz R. (eds.), *Tiere und Menschen. Geschichte und Aktualität eines prekären Verhältnisses* (Paderborn – München – Wien – Zürich: 1998) 323–347.

Niewöhner F. – Seban J.-L. (eds.), *Die Seele der Tiere*, Wolfenbütteler Forschungen 94 (Wiesbaden: 2001).

Rosenfield L.C., *From Beast-Machine to Man-Machine. Animal Soul in French Letters from Descartes to La Mettrie*, new and enlarged edition (New York: 1968).

Rößler H., *Die Kunst des Augenscheins. Praktiken der Evidenz im 17. Jahrhundert*, Wissenschaftsgeschichte 2 (Berlin – Münster – Zürich: 2012).

Swinford D., *Through the Daemon's Gate. Kepler's Somnium, Medieval Dream Narratives, and the Polysemy of Allegorical Motifs* (New York – London: 2006).

Wild M., *Die anthropologische Differenz. Der Geist der Tiere in der frühen Neuzeit bei Montaigne, Descartes und Hume*, Quellen und Studien zur Philosophie 74 (Berlin – New York: 2006).

Zittel C., *Theatrum Philosophicum. Descartes und die Rolle ästhetischer Formen in der Wissenschaft*, Wissenskultur und gesellschaftlicher Wandel 22 (Berlin: 2009).

CHAPTER 8

Cosmology after Copernicus: Decentralisation of the Sun and the Plurality of Worlds in French Engravings

Lucía Ayala

Summary

Was the central position of the Sun the pivotal factor par excellence in the beginning of Early Modern astronomy? What role did France play in this process? And what role did Copernicus's contribution play in this context? The following pages focus the attention on crucial images around the large-scale structure of the universe to discuss some of its milestones in Modern history. We will consider the period from the colonisation of heliocentrism by the *Sun* King to the consolidation of the *new system*, i.e., the plurality-of-worlds model. Some steps in the unfolding of this process are considered, especially the influence of Descartes in shaping the modern complex structure of the cosmos. The idea of a plurality of worlds succeeds as the main general cosmic model after Copernicus', which is indispensable for understanding the further development of astronomy.

1 The Sun King: Astronomy under Political Control

In the frontispiece of a Dutch re-edition of the *Éloges des académiciens de l'Académie Royale des Sciences* [Fig. 8.1], the French artist Bernard Picart (1673–1733) gave a visual overview of the state of the sciences in the Paris of Louis XIV (1638–1715).[1] The enlightened engraver and designer depicted Minerva driving the search for knowledge and truth that should encourage the scientific endeavours. She emanates her wisdom from the sun-head at her chest, which illuminates the entire scene. In the room, several female figures personify the sciences—Botany, Medicine, Astronomy, Mechanics, Geometry, Chemistry

1 Fontenelle Bernard Le Bovier de, *Éloges des académiciens de l'Académie Royale des Sciences morts depuis l'an 1699* (The Hague, van der Kloot: 1731). The previous editions of Paris (1708, 1709, 1714, 1717, 1719–1722, 1724) and the Netherlands (Amsterdam, 1709) did not include Picart's frontispiece.

 | DOI 10.1163/9789004281127_010

FIGURE 8.1 Picart Bernard, allegorical representation of the Academy of Science in Paris. Frontispiece of Fontenelle's *Éloges des académiciens de l'Académie Royale des Sciences morts depuis l'an 1699* (The Hague, van der Kloot: 1731). Engraving, Berlin. © Staatsbibliothek zu Berlin—Preußischer Kulturbesitz.

and Algebra. They occupy themselves in preparing the busts and glories of the deceased academicians, while Fame, at the upper part of the room, spreads their achievements for all eternity. In the background, each branch of knowledge labels a shelf containing its corresponding key literature, displaying on the top its main instruments. It represents a kind of 'scientific still-life', a common genre at the time.

In this image, Picart gave an impression of the state of the Academy and its main successes, by means of its published books and its celebrities. It shows the most relevant names related to the institution, principally the scientists themselves (whose busts, very realistic portraits showing individual personalities, are placed on pedestals displayed in the room). Moreover, it also includes two famous rulers who are spatially distinguished from the rest: Tsar Peter I (recently died in 1725; for that reason, the sculpture is still in the process of assemblage) and, of course, Louis XIV (who died earlier, in 1715, and whose bust is already finished but still being transported to its final place). The French king's bust is placed at the very centre of the composition, and its size is notably bigger than the others.

As the authoritarian sovereign stands out against the rest of the people, so does Astronomy, who is situated at the most privileged position among all sciences. Its shelf is at the centre of the room and, consequently, it is the only one showing its name and instruments frontally. This fact implies that one can see them in their entirety and without perspectival distortion. In other words, in this room—a condensed metaphor of the Academy—Astronomy is visually presented as the leading science.

However, according to this image, its leadership does not necessarily lie in its scientific relevance, but rather in its political role. The Sun King is superimposed over the space reserved for this branch of knowledge in the background, which frames his bust. Moreover, his effigy leans towards the woman representing Astronomy, a position visually explained due to its instability while being carried. As a result of this intricate composition, the monarch fixes his eyes on Astronomy who is in turn the only figure in the image looking at him. This reciprocal gaze centring the engraving reveals, on the one hand, the special attention this science attracted and, on the other, the relationship of dominance, imposition and intimidation exerted by the king over it. The royal sculpture casts a shadow over Astronomy's face, preventing her to be properly illuminated by the light of wisdom that emanates from Minerva. Furthermore, we should note that Minerva does not bear on her chest the head of Medusa, her most common sign, but an image formally very similar: a sun-head, the symbol *par excellence* of Louis XIV.[2] Within the context of the Parisian Academy, she does not symbolise wisdom and erudition, but rather the 'Truth' imposed by the Sun King. While Minerva guides the sciences following the instructions of the monarch, he takes himself care of Astronomy, dominating her by means of his intimidating presence. As we will see, Louis XIV's threatening gaze, stressed here by Picart, resulted in the leadership of Copernicanism as the main cosmic model during his reign.

This image was printed fourteen years after the death of the sovereign, during Picart's voluntary exile in the protestant Amsterdam.[3] It represents one of the most revealing analyses of the interplay between sciences and politics under Louis XIV.

This powerful (though subtle) representation embodies the basis for the present text. The following pages deal with the political tension behind the spread of Copernicanism[4] at the end of the seventeenth century, to explain how the appropriation of heliocentrism by the Sun King lead to a certain misapplication of astronomical theories.[5] Once the conception of the universe transcended the 'conservative' persistence of heliocentrism, the cosmic structure, finally emancipated from political pressure, could be freely shaped in modern terms. In this process, the idea of a plurality of worlds[6] played an essential and decisive role.

2 In the thesis frontispiece engraved in 1707 for Monsieur Brillon de Jouy, who defended Cartesian ideas, Picart correctly depicted Minerva with the head of Medusa on her chest. See Leiden University Library: PK–P–106.523. Even if Medusa has been associated with the Sun (one cannot look closely at either of them without being injured), this image directly alludes to Louis XIV; the sun–head is not a *mere* Sun, but the symbol of his monarchy.

3 In Amsterdam, Picart could develop his intellectual and artistic freedom. See his biography and the political difficulties he suffered when he tried to leave France in *Impostures innocentes, ou Recueil d'estampes* [...] *accompagné d'un Discours sur les Préjugés de certains Curieux touchant la Gravûre. Par Bernard Picart, dessinateur et graveur: avec son Eloge historique, et le catalogue de ses ouvrages* (Amsterdam, Chez la Veuve de Bernard Picart: 1734).

4 We should note that we are not considering here the multiple dimensions of Copernicus's work. Rather, we will only focus on the most popular aspect: the central position of the Sun in the Solar System. The general idea of heliocentrism was the most influential concept in the contexts we are tackling, namely the plurality of worlds and popular science during the 17–19th centuries.

5 For a clear political use of the principles of heliocentrism by the Sun King, see, for instance, Lorrain's book, mentioned in the next section. For a historical study on this topic, see Hutchinson K., "Towards a Political Iconology of the Copernican Revolution", in Curry P. (ed.), *Astrology, Science and Society. Historical Essays* (Suffolk: 1987) 95–141.

6 Apart from the many editions of Fontenelle's *Entretiens sur la pluralité des mondes* appeared in the course of the centuries, the bibliographical references on the plurality of worlds is

2 The Perversion of Copernicanism in France: The Absolute Power of the Sun

In 1707, Pierre Le Lorrain, known as the Abad de Vallemont (1649–1721), who cultivated his intellect amongst the gardens of Versailles, published *La Sphére du Monde, selon l'Hypothèse de Copernic*.[7] The text was aimed at presenting, attesting to and consolidating the new world order according to the heliocentric theory, promoting its truth against the previous out-dated systems. In order to facilitate the understanding of the differences between the models, five visualisations are attached: three diagrams showing respectively the systems of Ptolemy (circa 90–168 AD), Tycho Brahe (1546–1601) and Copernicus, and two Copernican armillary spheres, one of them more elaborated, and signed by Bernard Picart as designer and I.B. Scotin as engraver. The models by Ptolemy and Tycho Brahe are depicted in a common way: schematic diagrams of concentric circles with lineal abstract representations of the stars, the planets and the orbits, including the signs of each planet to identify them. However, the Copernican system [Fig. 8.2] significantly stands out from them. In this image, the abstract diagram has been inserted into a cloud-shaped environment,[8] while a smiling head with frizzy hair replaces the Sun. The main title is placed at the top, similarly than in the other images of the book. In this case, it is entitled 'Le monde selon l'hypothése de Copernic'. However, a further denomination has been added below: 'Le systéme du Monde au moment de la Naißance de Louis Le Grand le 5. de Septembre á u. heures 30. minutes. du matin 1638. Pol. 48°.55'.' In other words, to show the Copernican system, the universe is presented exactly as it was arranged when Louis XIV was born.[9] The scientific

extensive from the last years of the seventeenth century until the nineteenth century. See, for instance, an early example in Christiaan Huygens, *Kosmotheoros* (The Hague: 1698), translated from the original Latin into English as *The Celestial Worlds discover'd* (London: 1698); a great late example is Camille Flammarion, *La pluralité des mondes habités* (Paris: 1862).

7 Lorrain Pierre Le, *La Sphére du Monde, selon l'Hypothèse de Copernic, présentée au Roy: Décrite, démontrée, & Comparée avec les Sphéres & les Systèmes de Ptolomée, & de Tyco–Brahé* (Paris, Prosper Marchand: 1707).

8 Picart might have been inspired by the engraving for the first edition of the *Entretiens sur la pluralité des mondes* by Fontenelle [Fig. 8.6] for making this image. Visually, it is also similar to the geocentric representation published in Mallet Allain Manesson, *Description de l'Univers* (Paris, Denis Thierry: 1683) vol. I.

9 Although it was not signed, its author is quite probably Picart. This image and the armillary sphere signed by Picart are of a similar artistic complexity, notably different from the rest, which in turn are very similar to each other. His style is recognisable, for example, in

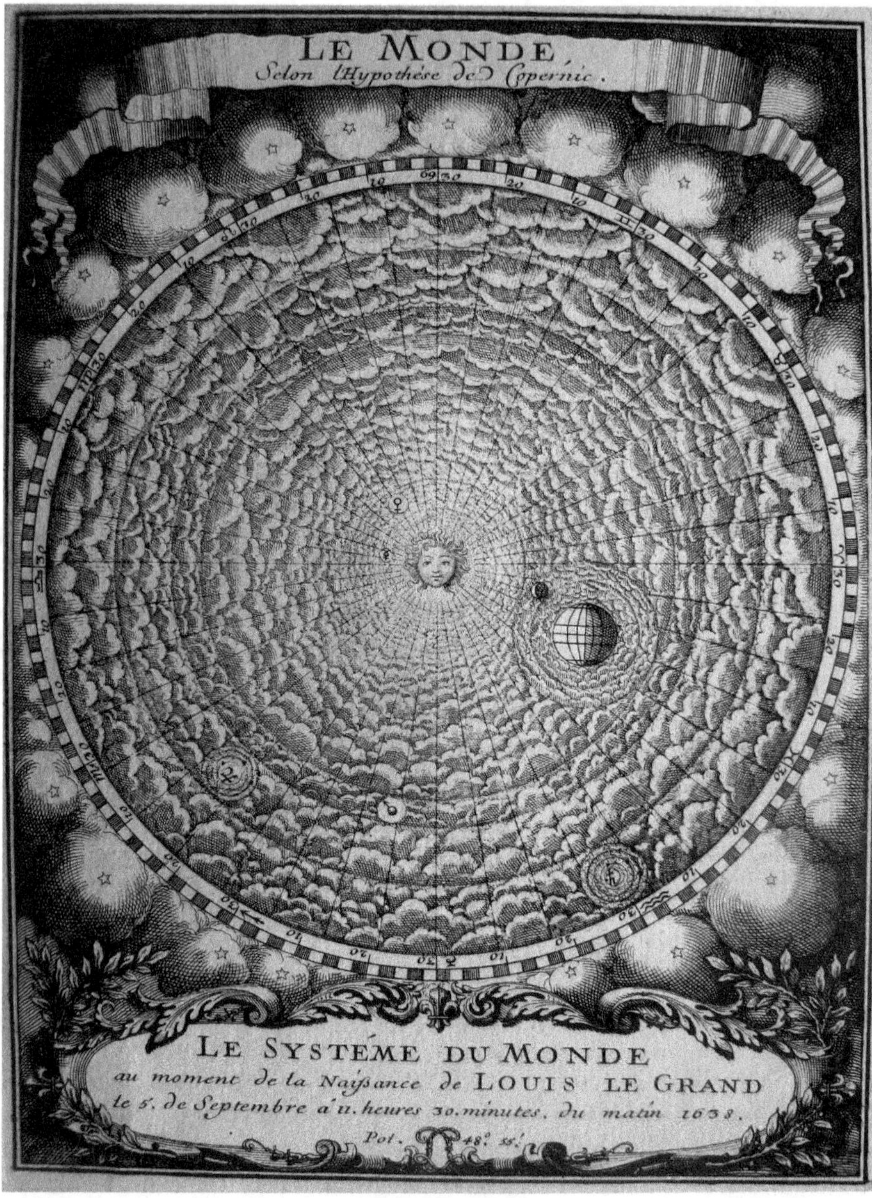

FIGURE 8.2 Picart Bernard, "Le monde selon l'hypothése de Copernic", illustration in Le Lorrain's *La Sphére du Monde, selon l'Hypothèse de Copernic, présentée au Roy: Décrite, démontrée, & Comparée avec les Sphéres & les Systèmes de Ptolomée, & de Tyco-Brahé* (Paris, Prosper Marchand: 1707). © Bibliothèque Nationale de France.

diagram is therefore mixed with a natal chart of the king, including an outer ring with the zodiacal signs positioned according to the moment of his birth.

The Sun King undertook many endeavours to convince others of his own conviction: that natural forces legitimised his power, Copernicanism being his major proof. This idea was not fortuitous: it happened coincidentally that when he was born, the heavenly bodies were in fact very specially arranged in an unusually good combination. Tommaso Campanella (1568–1639) and other astrologers who were in charge of elaborating the horoscope of the king stated that the Sun was at the Midheaven (the most important angle and whose sign will identify the person), followed by Leo with ascendant in Capricorn.[10] Therefore, the main signs marked his birth: Leo gave him the task of ruling the world, and the Sun was his alter ego, a metonymic figure to be identified with. According to this 'fact', it was clear that his power came from the universal, natural order and that he must perform the important task he was appointed for. Following this conviction, his absolute domination was a must, ratified thanks to Copernicus: all heavenly bodies revolve around the Sun, who imposes from its central position control over them. As a consequence, Louis XIV did not only make a traditional use of the Sun as a typical symbol of monarchic power; moreover, he wanted to replace the Sun, to take over its role. His natal chart was good evidence to legitimate it. This state of things induced the King to carry out an intense programme of propaganda. Thus, his natal chart was spread, being engraved many times by means of a small plate probably designed by Sébastien Leclerc (1637–1714) [Fig. 8.3].[11] This political

the dotted surface of the faces. In addition, the publisher of the book, Prosper Marchand (1678–1756), was Picart's closest friend. He also emigrated with the artist to Amsterdam and wrote his biography.

In 1707, when the book was published, Picart still lived in Paris. One year later he travelled to Holland for the first time. From the time he settled in the Netherlands, his artistic production did not praise the king again as in this image; on the contrary, his later images concerning Louis XIV display a succinct and incisive criticism of him.

10 The natal charts made at this time can be corroborated with the modern ones, easily available on the Internet. Jean Racine (1639–1699), who was appointed as the historiographer of the king, selected Campanella's version of the natal chart above the other versions because of his exaltation of the Sun. See Grenet M., *La passion des astres au XVIIe siècle. De l'astrologia à l'astronomie* (Paris: 1994) 133. Campanella suggested the king should make real his utopian ideas about *The city of the Sun.*

11 The identification of the designer is difficult to assert because I found the same engraving in work collections of several artists; however, Leclerc actually singed the engraving, while the others reproduced it anonymously. The close relation of Leclerc with astronomy (he worked for Cassini at the Parisian Observatory) helps to postulate this attribution.

FIGURE 8.3 D'Olivar Juan (engraver) after Leclerc Sébastien (designer), a version of Louis XIV's natal chart (second half of the seventeenth century). Engraving, Paris. © Bibliothèque Nationale de France.

strategy made it possible for many people to have in their hands the astral explanation of the king's authority. The same idea was applied, for instance, to the construction of Vincenzo Coronelli's (1650–1718) famous celestial globe. Rather than being a neutral representation of the stars, this huge sphere shows figurative representations of the constellations exactly as they were arranged in the king's natal chart.[12] Actually, the golden line of the ecliptic crosses the globe perpendicularly, showing the Sun over the roaring Leo.

12 Coronelli's celestial and terrestrial globes, made between 1681 and 1682, have been reconstructed and are shown in one of the halls of the French National Library (BNF) François Mitterrand in Paris.

The identification of the monarch with the Sun was performed *ad nauseam*. His public performances are well known. For instance, Louis appeared in the court ballet *La nuit* (1653), written by Isaac de Benserade (1613–1691), acting as the Sun [Fig. 8.4]. He came on stage in a chariot, dressed with golden clothing, a flame-like crown, trimmings everywhere resembling sunbeams, and even an image of the Sun on his chest. This performance was the apotheosis of his solar enactment, a direct way of transmitting his message—or, even more, of self-satisfying his insistent heliocentric demands.

In this context, it is easy to understand that the Paris of Louis XIV accepted heliocentric ideas with no censorship and considered Copernicus as a kind of new prophet. The appropriation of heliocentrism by the Sun King goes beyond the poetic metaphors; it explicitly asserts his rightful place in the universe. Thus, the preface of the above-mentioned book by Le Lorrain contains the following statement:

> Indeed, SIRE, you are at the centre of this kingdom as the sun, according to the hypothesis of Copernicus, is at the centre of the universe. The respectable serenity, which is always seen glowing on your noble face, in situations when Philosophy would be disconcerted, shows us how religion has elevated your sentiments above nature; and to us you are a vivid image of this perfect repose, as Copernicus represents the sun inscribed from the centre of the world out to its circumference, whose powerful virtue makes the Earth and all the celestial globes move. Is your state not ordered like the visible world?[13]

As we can see in these words, in Copernicus's hypothesis was found the political legitimation of the monarchy. Astronomical theories became more appropriate for Louis XIV's purposes than the strictly political treatises. Due to the special signification of his natal chart, astrology, placed in a space between the natural and the symbolic orders, acted as a master of ceremonies, mixing

13 'En efet, SIRE, vous êtes au centre de ce Royaume ce que le Soleil, selon l'Hypothèse de Copernic, est au centre du Monde. La respectable sérénité, que l'on a toûjours vûe reluire sur vôtre auguste visage, dans des conjonctures, où la Philosophie se seroit déconcertée elle-même, nous montre, combien la Religion a élevé vos sentimens au dessus de la Nature; & nous est une vive image de ce parfait repos, où Copernic représente le Soleil, imprimant du centre du Monde jusqu'à sa circonférence, cette vertu puißante, qui fait mouvoir & la Terre & tous les Globes célestes. Vôtre Etat n'est-il pas réglé comme le Monde visible?' Lorrain, *La Sphére du Monde*, Preface.

FIGURE 8.4 Gissey Henry, custom design for Louis XIV acting as the Sun in the court ballet *La nuit* (1653). Coloured drawing. © Bibliothèque Nationale de France.

together astronomy and politics.[14] A book in principle devoted to the explanation of the newest astronomy, like Le Lorrain's, became thus a manifestation of political propaganda.

One of the most pernicious consequences of the Sun King's reign was this colonisation of the sky. The monarch not only surrounded himself with a number of variants of the solar symbol (architecturally manifested in Versailles, iconographically spread by the sun-heads insistently repeated in all kinds of decorations, etc.)—moreover, he interpreted the universe in political terms. During Louis XIV's time, to learn about Copernicanism meant to know the grounds of the monarch's absolute power. 'You are in the centre of your kingdom just as the Sun, according to Copernicus's hypothesis, is placed at the centre of the World', said Le Lorrain.

3 Monarchical Heliocentrism as Idolatry

Picart, one of the sharpest minds of the Enlightenment, again provides us with a subtle but forceful visual statement regarding the critical reception of Louis XIV's appropriation of the Sun. In the frontispiece for *Cérémonies et coutumes religieuses de tous les peuples du monde*[15] [Fig. 8.5], he depicted all religions known at this time, including references from Asia, Africa, and America, showing everything with an unusual sense of respect. This image displays the diversity of cultures and civilizations on Earth. In this crowd of peoples, rites, creeds, and races, Picart also depicted three groups under a tree, notably smaller than the rest, on the left side of the composition: it represents the idolaters, the most primitive practices. One of these groups is dancing around a goat stuck on a post, another is seated around a bonfire, and the third is worshiping in theatrical gestures a niche resembling a tomb with a sun-head. The Parisian artist, who would eventually emigrate to Amsterdam in search of freedom, indirectly criticises the cult of the Sun promoted by Louis XIV in this detail, by identifying it with idolatrous practices. While the king tried to impose his new creed based on Copernicus, for Picart it became indeed an irrational cult. Through this small detail, Picart reveals how, under the Sun King, politics and religion (their symbols and arguments) were mixed up to manipulate

14 Nonetheless, the Parisian society in the course of Louis XIV's lifetime became disappointed with respect to astrology, especially after a scandal involving some magicians and astrologers who were eventually convicted. See Grenet, *La passion des astres* 136–159.

15 Picart Bernard, *Cérémonies et coutumes religieuses de tous les peuples du monde*, 9 vols. (Amsterdam, Jean Frederic Bernard: 1723–1743).

FIGURE 8.5 Picart Bernard, frontispiece of *Cérémonies et coutumes religieuses de tous les peuples du monde* (Amsterdam, Jean Frederic Bernard: 1723–1743), detail. Engraving. © Department of Special Collections, Charles E. Young Research Library, UCLA.

the human will, like in the primitive tribes—like these people who, although worshiping a bright and shinny sun-head, are still in the shadows, as the image explicitly shows.

4 The New System: A Universe Beyond Copernicus

Reinforced by Picart's artistic suggestions, we see how, during the reign of the Sun King, astronomical ideas were strongly manipulated for political ends in France. Accordingly, the critical responses to his absolute monarchy involved the support of an astronomical model different from Copernicus', allowing science to push forward in a new direction. A change of times was imminent.

The book *Entretiens sur la pluralité des mondes,* by Bernard Le Bovier de Fontenelle (1657–1757),[16] was conceived as a compendium of astronomy until

16 Fontenelle Bernard Le Bovier de, *Entretiens sur la pluralité des mondes* (Paris, Veuve C. Blageart: 1686). For a complete documentation of the images and the visual significance of the book, see my doctoral thesis. It gives a comprehensive analysis of each image,

then, explaining and popularising a complementary paradigm to Copernicus yet during Louis XIV's lifetime, namely the plurality of worlds. The book was published in 1686 *'avec privilege du Roy'*. Certainly the king was pleased with its defence of Copernicanism but underestimated the already unstoppable revolutionary potential of the new complex universe it describes. This popular science book, presented in the form of a series of conversations between two fictional characters, a man and a woman, included in its first edition an engraving made by the Spanish artist Juan d'Olivar (1641–1692) [Fig. 8.6]. He depicted a universe composed of many similar systems arranged around stars. This engraving is an early figurative representation of the plurality-of-worlds model.

Each substructure is a planetary system including planets, satellites and their respective orbits, and is framed by a cloud-shaped environment. D'Olivar, who worked in the artistic circle of the court, knew quite well the solar symbol of the king, since he reproduced it a number of times in his engravings. On this occasion he used the same icon but subversively: he depicted the star at the centre of each system as a sun-head but, paradoxically, he did not use this symbol for our Sun, in the middle of the composition. In contrast, our Sun was presented as an astronomical object: a sphere with dots on its surface, emulating scientific observations. Taking into account the political connotations of a sun-head commonly assumed at that time, two main consequences could be extrapolated in view of this image. First, Louis XIV was dethroned from the Sun. Second, and even more important, the relevance of our star (namely the king) became radically relativised since thousands of similar stars were shining in the cosmos, ruling their respective areas of influence. Times had already changed. Heliocentrism was not the ultimate explanation.

One of the clearest descriptions of this new world order was given in a later English edition of Fontenelle's book:

> This infinite number of worlds is called, to distinguish it from the rest, the *new system*, which is the same as the Copernican, in regard to the situation of the sun and the planets revolving round him. But whereas the Copernican hypothesis supposes the firmament of the fixed stars to be the bounds of the universe, and to be placed at equal distance from its centre the sun; the *new system* suposeth there may be many other sys-

---

especially the engraving of the first edition. Lucia Ayala, *Las imágenes de los 'Entretiens sur la pluralité des mondes' de Bernard Le Bovier de Fontenelle. Visualizaciones astronómicas de una pluralidad de mundos/Die Bebilderung der 'Entretiens sur la pluralité des mondes' von Bernard Le Bovier de Fontenelle. Astronomische Visualisierungen einer Vielzahl von Welten* (Granada: 2011).

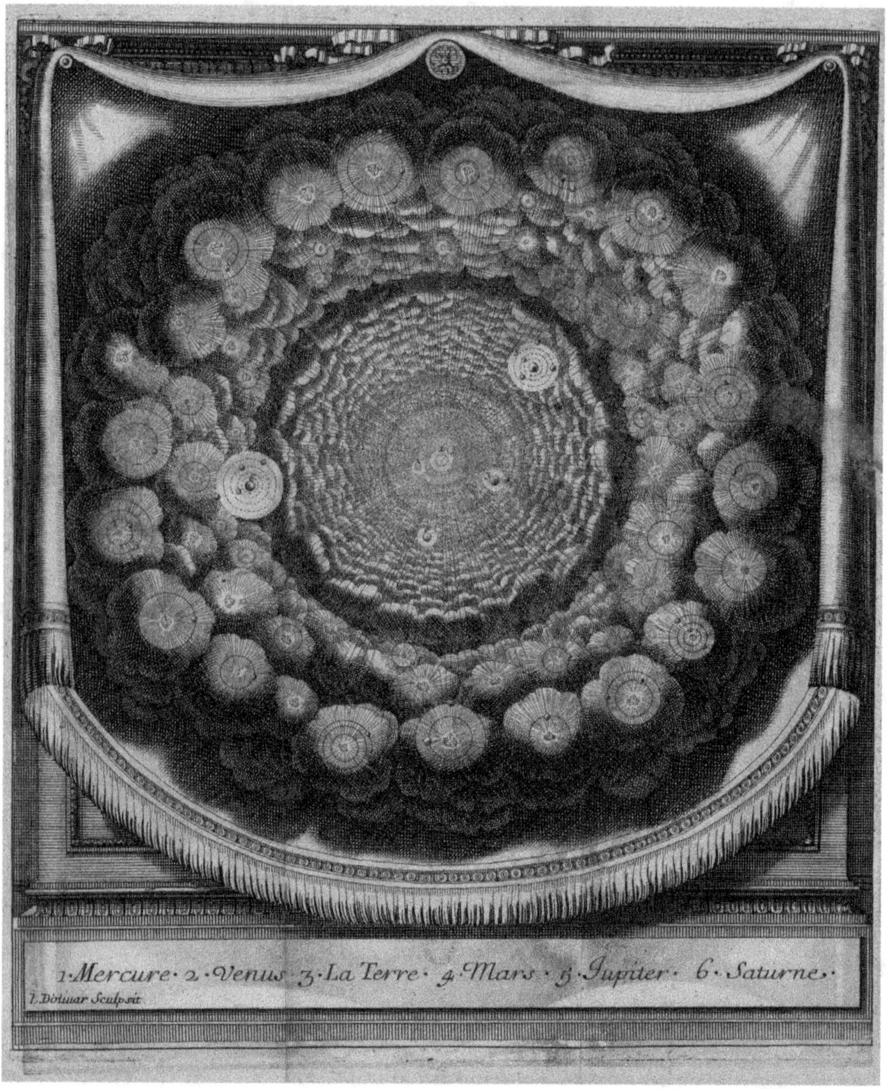

FIGURE 8.6 D'Olivar Juan, depiction of the universe in the first edition of Fontenelle's *Entretiens sur la pluralité des mondes* (Paris, Veuve C. Blageart: 1686). Engraving. © Bibliothèque Nationale de France.

> tems of suns and planets, besides this in which we reside; namely, that every fixed star is a sun.[17]

At the time of this quote, by the middle of the eighteenth century and after the convulsions of the previous century, these ideas, previously revolutionary, were already consolidated. The conclusion was clear: they implied the establishment of a *new model*, a defined paradigm that surpasses Copernicanism. Thus, heliocentrism was understood as a tiny component within the huge complexity of the universe. Copernicus was right, but his achievements were placed on a radically lower level of relevance. So we see, in the introduction to a previous English edition of the *Entretiens*, how the key idea emphasised was the large size of the universe:

> We can perceive no bounds of the vast expanse in which natural causes operate; nor can we fix any border, or termination of the Universe [...]. The objects, which we commonly call great, vanish when we contemplate the vast body of the *Earth*: the terraqueous globe itself is soon lost in the *Solar System*, being in some parts, seen as a *Planet*, or distant Star; and, in a great part of the System unknown, or visible only at rare times to vigilant observers, assisted, perhaps, with instruments, like our telescopes: the *Sun* itself dwindles into a Star [...]. Other Suns illuminate other Systems, where our Sun's rays are unperceived: but all these also are swallowed up in the vast expanse of the UNIVERSE.[18]

This complex cosmos is closer to our current picture of the universe than heliocentrism is. As a matter of fact, the plurality-of-worlds model should be placed as one of the main triggers of contemporary astronomy. Copernicus gave a preliminary, crucial, and partial step, opening the door, whereas a decisive and further-reaching revolutionary step was made by the theoreticians of the plurality of worlds. They rethought the entire structure of the universe

17 Italics added by the author. Fontenelle, *Entretiens* (Dublin, Peter Wilson: 1761) 152.

18 Fontenelle, *Entretiens* (London, Robert Withy: 1760) 36. We cannot forget in this context the famous text by Koyré explaining that the idea of an infinite universe was the main factor for the new paradigm. Koyré's book provides a key approach to understand the background of these processes, which are materialised by the authors of the plurality-of-worlds model. While Koyré tells us that the existing space was understood to be possibly infinite, the authors of the plurality of worlds tell us how could be the actual structure of this space. Koyré A., *From the Closed World to the Infinite Universe* (Baltimore, Md.: 1957).

in quite right terms, in spite of the (technological, observational, theoretical) limitations of science at that time. Inferring the ideas mainly by reasoning and basic physics rather than deriving them from direct observations or advanced science, many authors managed to understand that the universe goes further beyond the Solar System.[19] In the eighteenth century, these ideas were internationally accepted, and many people were aware of being in a another era, characterised by a *new system* different from the Copernican.

5 Vortices and the Plurality of Worlds: Descartes's Role in the Consolidation of Western Modernity and Its Later Disregard

The colonisation of the Sun by Louis XIV was in the long run counterproductive for his purposes. It paved the way for the enthusiastic reception of the idea of the plurality of worlds, what compelled the Sun to give up its throne. The spread and consolidation of the plurality-of-worlds model was especially triggered by the influence of the Cartesian system of vortices. Within this system, the idea of an infinite cosmos or, at least, a huge, complex, and diverse universe, was a fundamental premise.[20] Within the context of the large-scale structure of the universe, with no doubt the French natural scientist and philosopher[21] outlined the most innovative model of his time. Thanks to his contribution, France led European cosmology[22] by the middle of the seventeenth century.

---

19 For a better understanding of the historical role of the plurality of worlds, see Ayala L., "Worlds and Systems in Early Modern Europe", in Way M.J. – Hunter D. (eds.), *Origins of the Expanding Universe: 1912–1932* (San Francisco: 2013) 193–204.

20 Descartes carefully avoided defending the infinitude of the universe: 'Mas afin que cette infinité ne nous empefche & ne noue embarrasse point', he said, it is preferable to consider that the universe extends itself 'jusques à une distance *indéfinie*'. He prudently refers to a undefined rather than to an infinite space. Descartes René, *Le Monde. Traité de la lumière/El Mundo. Tratado de la luz*, bilingual edition, trans. S. Turró (Barcelona: 1989 [1664]) 98–100.

21 Descartes is usually regarded as philosopher. His role as natural scientist has not always been properly considered. See Zittel's claim of a deeper consideration of the multiple dimensions of his work. Zittel C., *Theatrum Philosophicum. Descartes und die Rolle ästhetischer Formen in der Wissenschaft*, Wissenskultur und gesellschaflicher Wandel 22 (Berlin: 2009) especially 13–16.

22 Many authors use the term 'cosmogony' in this context. I consider 'cosmology' more accurate. 'Cosmogony' may create confusion, since it usually refers to the origin of the Solar System. Within the plurality-of-worlds tradition, the theoreticians referred to the whole universe, even if sometimes they described the formation of the Solar System as a case

Fontenelle's *Entretiens* was a key factor for the popularisation of his theories. Through the hundreds of editions published of this book, the Cartesian cosmos reached many cities, languages and debates in the course of the centuries, calling the attention of the new bourgeoisie in addition to the scientific circles. Because of Fontenelle's engaging style, the *Entretiens* were a useful tool to easily understand the structure and dynamics of the cosmos. Together with the system of vortices, the plurality of worlds, also described in the book, became a common topic in intellectual conversations, not only for the scientific elites, but also for the emergent middle-class.

Thanks to the mediation of Fontenelle, cosmological Cartesianism was the driving force for justifying and spreading the plurality-of-worlds model at first—however, it turned out to be its weak point in the long term. When Newton (1642–1727) published his *Pincipia Mathematicæ*,[23] Great Britain could impose its hegemony, becoming the new leading European country in science. The scientific community bent over backwards to extol its new prophet. Thus, in the English edition of the *Entretiens* mentioned above, the Cartesian system is discredited in favour of

> the great Philosopher of this age, the most ingenious and incomparable Mr. Newton [...] so that, the notion of a Vortex being ruined, the whole Cartesian System must, of necessity, fall to the ground; and that World, whose Origination he pretended to have deduced from mechanical Principles, must be a wild chimera of his own imagination.[24]

Meanwhile, many French enlightened thinkers, including Fontenelle, still defended Descartes's worldview[25] over Newton's. For them, the first represented a purely scientific way of conceiving the universe, since Descartes's explanations were based on physical forces and the nature of matter, rather than on 'mysterious forces' that could be perceived closer to religious notions.

study to be extrapolated to the entire space. 'Cosmology' refers to the origin and evolution of the universe, what is exactly the case here.

23 Newton Isaac, *Philosophiæ naturalis principia mathematica* (London, John Streater – Samuel Smith: 1687).

24 Fontenelle, *Entretiens* (London, Robert Withy: 1760) 49–51.

25 Descartes René, *Le Monde, ou le Traité de la lumière* [...] (Paris, Théodore Girard: 1664). It was the first outline of his physical theories. See also the later, more detailed elaboration of his ideas in Descartes, *Principia Philosophiæ* (Amsterdam, Lowijs Elzevier: 1644). Let us remember that the publication of this second book was determined by Galileo's condemn by the Inquisition, which made him express his ideas much more carefully.

In spite of that, many scientists started to prove that Newton's gravity actually worked, so that English science finally won the battle for power.

The ideas of a plurality of worlds survived even after Descartes was generally overcome. Presented as an independent idea not exclusively linked to the vortices, it was still the focal point of many treatises on astronomy until the nineteenth century.

Contemporary history of science, when tracing the sources of our understanding of the universe, usually revives Newton, Maxwell and other scientists, considering them the 'right' path. However, besides these achievements and successes, it is important to recall the 'wrong' tendencies, since they include certain aspects that are fundamental to understand the further scientific developments. This is the case of the Cartesian system of vortices. Descartes was branded as an inventor of chimeras, and Fontenelle as a proto-science-fiction writer. Nevertheless, the current conceptions about the large-scale structure of the universe represent the latest consequence of a long tradition whose early steps can be traced, at least, up to Descartes. The system of vortices, on the one hand, and the plurality of worlds, on the other, are early key steps for a scientific understanding of the macrostructure of the universe.[26] Before them, the universe was generally assumed to be a limited space only encompassing the Solar System and an outer sphere of fixed stars. These authors managed to change our conception on the structure of the universe, from them on understood as the vast space we still consider today.[27]

6 A New Visual Paradigm for the Complex Universe

The plurality-of-worlds model was shaped through theories, texts, and images, its visual dimension being a key component. The ideas around the complex

26 We should note that these ideas were not new. Ancient Greek philosophers, Nicholas of Cusa, or Giordano Bruno, for instance, had already stated that the universe (probably) goes further beyond the boundaries of the Solar System. Descartes's particular significance lies in giving detailed scientific explanations of this idea, and also in consolidating it as a solid and precise paradigm (even if it involved many misunderstandings and mistakes, as later scientists pointed out). After him, it became an ongoing, non-stop and widespread tradition.

27 Cosmology before the seventeenth century presents many different tendencies. We are speaking about mainstream ideas. In the second half of the twentieth century, cosmology entered in a new scientific paradigm regarding the structure of the universe: the so-called cosmic web. Mutatis mutandis, we can find some similarities in the basic ideas behind this new paradigm and the Cartesian notions. See Ayala L. – Forero-Romero J.E., "Visualising matter and cosmologies: A Transhistorical Example", *Column* 7 (2011) 76–82.

structure of the cosmos generated a new set of images that evolved in time. Roughly speaking, we can detect three phases in the visual shaping process of the plurality of worlds:

1) The prevalence of heliocentrism. The first depictions of the plurality of worlds placed the Sun at the centre of the universe. At first stages, Copernicus's model still had a stark impact in the definition of the cosmos.
2) The influence of the vortices. In a second step, the Solar System was represented as an anonymous component among many similar others, following Descartes's theories.
3) The final autonomy of the *new system*, conceived and defined as an independent model.

Let us look at some examples from each tendency.[28]

Figure 8.6, engraved by d'Olivar for the first edition of Fontenelle's *Entretiens*, is one of the major examples of the strong presence of heliocentrism in the first stages. This fact is not a surprise, taking into account that Fontenelle's main pedagogical goal was to present heliocentrism and the vortex model as the two leading principles in astronomy. He clearly described how the Sun is placed at the centre of the Solar System. Fifty years after Galileo was condemned for publicly asserting it, finally heliocentrism could be proudly spread.

Apart from the explicit emphasis on heliocentrism present in the text, there is an additional reason to understand why d'Olivar conferred such a prominent role to our Sun. This reason, an even more satisfactory explanation, lies in a visual level. A series of broken lines representing concentric circles are still recognizable among the figurative motifs. These circles are actually the orbits of planets and satellites. Extracting the lines from the background texture, we find a perfect diagram of the Solar System, similar to those commonly used in scientific texts at that time. They reveal how the process of production of this image began with a schematic diagram of our system. Given that Fontenelle was in touch with the scientists at the Paris Observatory, it seems quite plausible that he might have asked them for a diagram to be used as starting point. That way, its scientific grounds would be guaranteed. Then, the artist was commissioned to design the rest of the image. Similarly to the text,

28 An overview of the images depicting the universe as a plurality of worlds can be seen in Ayala L., "The Universe in Images: an Art-Historical Approach to the Plurality of Worlds", in Nicolas Campion – Ron Sinclair (eds.), *The Inspiration of Astronomical Phenomena* VIII (Wales, Sophia Centre Press: in press). The text collects the main visual sources within this astronomical model. These references have been put together for the first time.

in the engraving observations and scientific facts were enhanced by artistic imagination, resulting in a harmonious composition. Thus, as consequence of the process of image production (which began with a traditional heliocentric diagram), our Sun turned out to rule the universe from its centre.

A third level of explanation was mentioned above: this Copernican representation of the plurality of worlds matched the Sun King's expectations, who approved the publication of the book and might have seen it as an additional component of his political propaganda.

It is remarkable that in a passage of the text the main character says, after praising Copernicus's merits: 'The *Sun* is in the centre of the Universe, and there remains immoveable'.[29]

From a visual and a philosophical point of view, the combination of this premise with a more complex universe was easy to solve, since it did not imply changing the previous notions or diagrams related to the Copernican universe. Instead, the surrounding 'worlds' were added to the known Solar System, practically without altering its structure or function.

With the passing of time and the spread of the plurality-of-worlds model, people soon realised one of its basic consequences: in a universe much larger than the Solar System, our Sun is merely a star, a tiny heavenly body in an immense space. Therefore, it makes no sense assuming its superiority in the cosmic order. At this point, we should take up Descartes again.

As mentioned above, the popularisation of the plurality of worlds began intrinsically bound to the system of vortices, principally due to Fontenelle's fame. He was not the first in dealing with the vortices or a universe beyond the Solar System, but he was indeed the one who made it better known to the public. Fontenelle, who was a convinced Cartesian thinker until his death,[30] became the reference of the plurality-of-worlds model for many later authors.

Descartes's model entailed a core question: a conception of the general structure of the universe radically distant from Copernicus's world. The theoretical dimension of his ideas, that is, the existence of many autonomous and interconnected systems, changed the course of astronomy. Descartes's macrovision represented a step further in notions about the large-scale structure of the universe, a field that was radically transformed after him. He triggered a crucial change in our way of visually and theoretically conceiving the universe as a whole, a change that paved the way to modern notions in astronomy.

29 Fontenelle, *Entretiens* (London, Robert Withy: 1760) 19.

30 At the end of his long life, when he was 95 years old and Newtonianism was widely accepted, Fontenelle published a book in defense of the Cartesian vortices. Fontenelle, *Théorie des tourbillons cartésiens avec des réflexions sur l'attraction* (Paris, Guérin: 1752).

Picart's contribution to the *Entretiens* is one of the clearest examples of this new tendency in which heliocentrism is not the main question anymore to define the universe. Rather, it is about a decentralised model with the stamp of Descartes. Among the images Picart engraved for the edition of 1728,[31] we find a depiction of the macrostructure of the cosmos [Fig. 8.7]. This image is a striking version of the Cartesian universe close to the original source but freely (and figuratively) interpreted.[32] Picart knew very well the images of the cosmos made by Frans van Schooten the Younger for Descartes's books.[33] Owing to the fact that Fontenelle constantly mentioned Descartes in his text, and given that Picart always worked as close to the original sources as possible,[34] for depicting the universe he followed van Schooten's main visual parameters regarding the structure of the universe. Therefore, Picart's universe visually stresses a decentralised and complex structure, being our Sun and Solar System blended anonymously with the other elements. Heliocentrism was therefore integrated into a larger state of things, where the Sun lost its position at the centre of the universe and each vortex had its own central body. Copernicus's revolution was in this moment finally surpassed by another, further-reaching revolution. In this context, Picart's image represents the transition to the *new system* as such, through the clues given by Descartes.

In a third stage, the plurality of worlds reached its maturity and became independent from the dominance of the Sun and the vortices. Notwithstanding, the decentralised complex structure proposed by Descartes remained—and has remained, actually, until now.

---

31 Fontenelle, *Entretiens* (The Hague, Chez Gosse et Néaulme: 1728).

32 It is important to note that d'Olivar did not have in mind the images from Descartes's books when he figured out his universe. So we cannot say that the first engraving of the *Entretiens* is a visualisation of Descartes's vortices, even though the text addresses it. The Spanish engraver alluded to other references, specifically taken from his closest artistic circle in Paris. See, for instance the stage designs by Torelli (1658–1709) for the opera *Le nozze di Peleo e di Theti* (1654), BNF Col. Henin fol. 45, or the tribute to the Rostaing family engraved by Jean Lepautre (1618–1682), who supposedly was d'Olivar's uncle, entitled *Trophée médallique, maritime at naval...*, BNF Ed. 42a, fol. 103. For further information see Ayala, *Las imágenes de los 'Entretiens sur la pluralité des mondes'* 92–97.

33 For the attribution of the images in Descartes's cosmological books to Frans van Schooten the Younger, see Zittel, *Theatrum Philosophicum.* For a detailed analysis of the images from *Le Monde* and the *Principia*, see especially the chapter 5.

Picart owned in his personal library several editions of Descartes's books. See the records of his library made by his wife after Picart's death in Bernard Jacques, *Catalogue de livres curieux* (Amsterdam: 1733).

34 See, for instance, Paola von Wyss-Giacosa, *Religionsbilder der frühen Aufklärung. Bernard Picarts Tafeln für die Cérémonies et Coutumes religieuses de tous les Peuples du Monde* (Wabern: 2006).

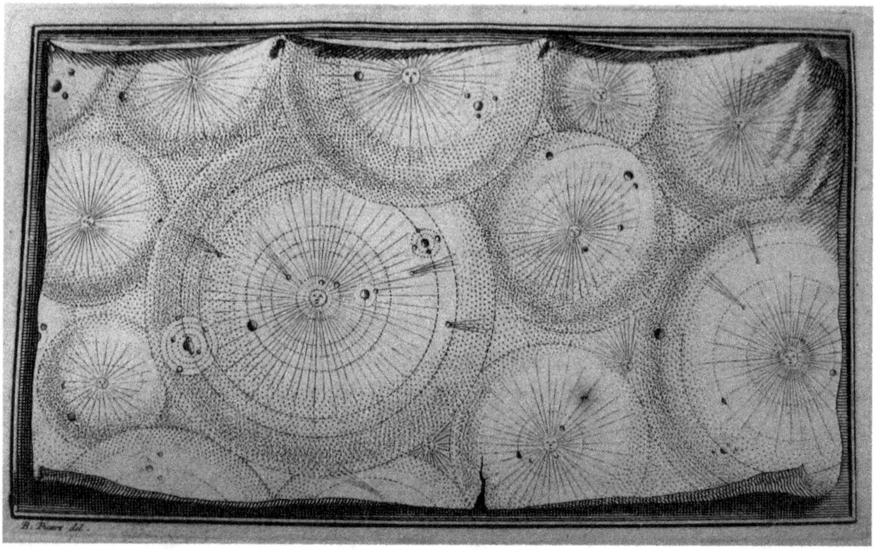

FIGURE 8.7 Picart Bernard, depiction of the universe in Fontenelle's *Entretiens* (The Hague, Chez Gosse et Néaulme: 1728). Image taken from the 1743 reprint. Engraving. © Staatsbibliothek zu Berlin—Preußischer Kulturbesitz.

A great example of the emancipation of the *new system* was published by Pierre Louis Moreau de Maupertuis (1698–1759).[35] Particularly, I am referring to a new edition of his *Discours sur les differentes figures des astres* published the same year that he was elected director of the Academy of Science in Paris, which included a frontispiece engraved by Guillaume Dheulland after Charles Maugein [Fig. 8.8].[36] This image can be taken as a prime example of how the outlines of Descartes's cosmic structure endured over time, even in a work aimed at defending Newton's theories. The frontispiece highlights the presence of comets, one of the main arguments against the vortices. However, the general composition follows the standard principles of the French natural scientist: similar neighbouring systems cover the space on an equal basis. This diagrammatical representation of the cosmos involved a model that had forgotten Copernicus's parameters and evolved in another direction. In

35 Let us remember that Maupertuis established the shape of the Earth as an oblate spheroid after his famous expedition to Lapland. See, for instance, Mary Terrall, *The Man who Flattened the Earth* (Chicago – London: 2002).

36 Maupertuis Pierre Louis Moreau de, *Discours sur les differentes figures des astres* (Paris, Martin, Coignard & les Freres Guerin: 1742) [1st edition without this frontispiece: Paris, Imprimerie Royale: 1732].

FIGURE 8.8 Dheulland after Maugein, depiction of the universe. Frontispiece of Maupertuis' *Discours sur les differentes figures des Astres* [...] (Paris, Martin, Coignard & les Freres Guerin: 1742). Colored engraving. © Bibliothèque Nationale de France.

Maugein's engraving the Sun is not placed at the centre of the cosmos, nor do the vortices define its constitutive essence. Beyond Copernicus and Descartes, a new universe consisting of many systems arranged by gravitational forces in a decentralised general composition was shaped. These basic features of the large-scale structure were an early starting point for the contemporary framework. Nevertheless, the scientific community is still arguing many issues around the details and physics of the general structure.[37]

7 Conclusions: An Appeal for a Continuous History of the Structure of the Universe

The significance of the French *siècle des lumières* and its immediately preceding period in shaping Early Modern science is well known in general. Nevertheless, from an astronomical point of view, it still remains to be positioned at the right place—actually, at two places at least.

In the first place, the structural concepts implicit in the plurality-of-worlds model should be considered at the same level of relevance as universal gravitation, with regard to their roles in the process that paved the way for Western Modern astronomy. The contributions made from both contexts (the French and the English) were indispensable milestones of a complete new state of things. Gravity and the complex universe with many systems are two sides of the same coin.

In the second place, French scientists took up Copernicus's baton in reshaping the entire structure of the universe. In fact, it was the beginning of a long process that reaches up until the present. However, several historical facts (such as the merciless appropriation by Louis XIV of Copernicanism or the English supremacy imposed after Newton) have blackened a proper regard for a continuous, well-defined tradition coming from these achievements. Strictly speaking, this is true from a historiographical perspective; during the eighteenth and nineteenth centuries there were indeed many authors actively and consciously taking part in restructuring a *new system*.

When trying to recompose the history of European astronomy and cosmology, the criteria are usually mixed up and, therefore, misused. We recognise a sequence of models with respect to the structure of the universe: Ptolemy, Copernicus, Tycho Brahe, Descartes. But then a gap is suddenly introduced and the sequence continues in another direction following Newton. Although

37 Let us remember once more that currently astronomy and cosmology are within a new theoretical paradigm far beyond the plurality of worlds, namely the era of the cosmic web.

this path was of course crucial, it belongs to a complementary process that ran parallel to the evolution of the theories on the general arrangement of the cosmos, which certainly kept going on after Descartes.

History of cosmology, especially with regard to the theories around the large-scale structure of the universe, should be continuously traced until now. To fill the historiographical gap concerning the evolution of theories on the macrostructure, we should consider the plurality-of-worlds tradition. It was a defined paradigm just after Descartes's vortices and highly influenced by it (though not limited to its scope). It involved the most relevant set of theories after Copernicanism in relation to the structure of the universe, becoming therefore an indispensable source, a *sine qua non* for understanding the shaping process of Western (Early) Modern cosmology.[38]

New scenarios, protagonists and references should come into play for a more precise comprehension of our recent past. In this context, art history and image analysis provide research tools that, complementary to other approaches, can give an answer to unsolved misunderstandings in the history of European astronomy and cosmology.

Selective Bibliography

Ayala L., *Las imágenes de los 'Entretiens sur la pluralité des mondes' de Bernard Le Bovier de Fontenelle. Visualizaciones astronómicas de una pluralidad de mundos / Die Bebilderung der 'Entretiens sur la pluralité des mondes' von Bernard Le Bovier de Fontenelle. Astronomische Visualisierungen einer Vielzahl von Welten* (Granada: 2011).

Ayala L., "The Universe in Images: an Art-Historical Approach to the Plurality of Worlds", in Nicolas Campion—Ron Sinclair (eds.), *The Inspiration of Astronomical Phenomena VIII* (Wales, Sophia Centre Press: in press).

Ayala L., "Worlds and Systems in Early Modern Europe", in Way M.J. – Hunter D. (eds.), *Origins of the Expanding Universe: 1912–1932* (San Francisco: 2013) 193–204.

Ayala L., Forero-Romero Jamie E., "Visualising Matter and Cosmologies: A Transhistorical Example", *Column 7* (2011) 76–82.

38 The main historiographical references on the plurality of worlds emphasize the issue of extraterrestrial life. The core question was, however, the new conception of the space, in which the presence of life in other planets was a kind of side effect, but not the main factor. For the classical consideration of the plurality of worlds, see Dick S.J., *Plurality of Worlds. The Origins of the Extraterrestrial Life Debate from Democritus to Kant* (Cambridge – New York – Melbourne: 1982) and Crowe M.J., *The Extraterrestrial Life Debate, 1750–1900: The Idea of a Plurality of Worlds from Kant to Lowell* (Cambridge: 1988).

Crowe M.J., *The Extraterrestrial Life Debate 1750–1900. The Idea of a Plurality of Worlds from Kant to Lowell* (Cambridge: 1986).

Descartes René, *Principia Philosophiæ* (Amsterdam, Lowijs Elzevier: 1644).

Descartes René, *Le Monde. Traité de la lumière / El Mundo. Tratado de la luz*, bilingual edition, trans. Salvio Turró (Barcelona: 1989 [1664]).

Dick S.J., *Plurality of Worlds. The Origins of the Extraterrestrial Life Debate from Democritus to Kant* (Cambridge: 1982).

Flammarion C., *La pluralité des mondes habités* (Paris: 1864).

Fontenelle Bernard Le Bovier de, *Entretiens sur la pluralité des mondes* (Paris, Veuve C. Blageart: 1686).

Fontenelle Bernard Le Bovier de, *Entretiens* (The Hague, Chez Gosse et Néaulme: 1728).

Fontenelle Bernard Le Bovier de, *Éloges des académiciens de l'Académie Royale des Sciences morts depuis l'an 1699* (The Hague, van der Kloot: 1731).

Fontenelle Bernard Le Bovier de, *Entretiens* (London, Robert Withy: 1760).

Fontenelle Bernard Le Bovier de, *Entretiens* (Dublin, Peter Wilson: 1761).

Grenet M., *La passion des astres au XVIIe siècle. De l'astrologia à l'astronomie* (Paris: 1994).

Guthke K.S., *Der Mythos der Neuzeit. Das Thema der Mehrheit der Welten in der Literatur- und Geistesgeschichte von der kopernikanischen Wende bis zur Science Fiction* (Bern – Munich: 1983).

Hentschel K. – Wittmann A.D. (eds.), *The Role of Visual Representation in Astronomy: History and Research Practice* (Thun – Frankfurt am Main: 2000).

Hutchinson K., "Towards a Political Iconology of the Copernican Revolution", in Curry P. (ed.), *Astrology, Science and Society. Historical Essays* (Suffolk: 1987) 95–141.

Koyré A., *From the Closed World to the Infinite Universe* (Baltimore, Md.: 1957).

Lorrain Pierre Le, *La Sphére du Monde, selon l'Hypothèse de Copernic, présentée au Roy: Décrite, démontrée, & Comparée avec les Sphéres & les Systèmes de Ptolomée, & de Tyco- Brahé* (Paris, Prosper Marchand: 1707).

Maupertius Pierre Louis Moreau de, *Discours sur les differentes figures des astres* [...] (Paris, Martin, Coignard & les Freres Guerin: 1742).

Newton Isaac, *Philosophiæ naturalis principia mathematica* (London, John Streater – Samuel Smith: 1687).

Picard Bernard, *Cérémonies et coutumes religieuses de tous les peuples du monde*, 9 vols. (Amsterdam, Jean Frederic Bernard: 1723–1743).

Picard Bernard, *Impostures innocentes, ou Recueil d'estampes* [...] *accompagné d'un Discours sur les Préjugés de certains Curieux touchant la Gravûre. Par Bernard Picart, dessinateur et graveur: avec son Eloge historique, et le catalogue de ses ouvrages* (Amsterdam, Chez la Veuve de Bernard Picart: 1734).

Whewell W., *Of the plurality of worlds: an essay* (London: 1853).

Zittel C., *Theatrum Philosophicum. Descartes und die Rolle ästhetischer Formen in der Wissenschaft*, Wissenskultur und gesellschaflicher Wandel 22 (Berlin: 2009).

PART 3

New Astronomers: Biographical Transformations

CHAPTER 9

Timid Mathematicians vs. Daring Explorers of the Infinite Cosmos: Giordano Bruno, Literary Self-Fashioning and *De revolutionibus orbium coelestium*[1]

Sergius Kodera

Abstract

This article examines Giordano Bruno's literary strategies in imagining, assimilating and appropriating heliocentrism to his infinitist and vitalistic cosmology and metaphysics. A close reading of a short passage from the Latin epic *De Immenso* (1591) examines how Bruno creates an emblematic image of himself that not only bears distinct echoes from his earlier dialogues written in Italian but also refers to Copernicanism. This article shows how that image does not merely work as an embodiment of its author as the prophet of the new cosmology, but rather reflects the (paradoxical) conditions under which Bruno thought that we are able to comprehend the incomprehensible infinite materiality of a sentient universe.

> Io curo poco il Copernico, disse il Nolano,
> e poco mi curo, che voi o altri l'intendano;...
>
> GIORDANO BRUNO, *Cena delle Ceneri*

Beyond all doubt, Giordano Bruno (1548–1600) was an important Copernican. Between 1583 and 1585, about four decades after the publication of the *De revolutionibus*, the philosopher from Nola, a runaway friar and ordained priest from the Convento San Domenico Maggiore in Naples, published in London a series of dialogues written in Italian which discuss and develop important aspects the new cosmology.[2] For the next couple of centuries these texts, intended for

---

1 I wish to thank Konrad Eisenbichler and Miguel Ángel Granada for reading this manuscript and for their many invaluable suggestions and corrections—all errors of course being mine.

2 *La cena delle Ceneri, Della causa principio ed uno, De l'infinto, universo e mondi.* For a fine intellectual biography, see Rowland I., *Giordano Bruno. Philosopher/Heretic* (New York: 2008); Ciliberto M., *Giordano Bruno* (Rome: 1990) 7–28 for Bruno's early years. For a concise recent summary on Bruno research, see Saiber A., *Giordano Bruno and the Geometry of Language*

 | DOI 10.1163/9789004281127_011

an audience of educated English and French courtiers, were printed only once in a small press run by John Charlewood; yet they were to secure Bruno's fame as a philosopher for centuries to come.[3] His position and achievements as a forerunner to the so-called scientific revolution (as well as his many misconceptions) have been amply discussed.[4] This article looks instead at the specific literary strategies which Bruno employed to situate Copernicus vis-à-vis what he called *nolana filosofia*. In what follows I shall contextualise one brief passage from Bruno's Latin poem *De minimo* (1591), a text that was published late in what turned out to be an involuntary short career as a philosopher. Bruno was arrested in 1592 by the Venetian Inquisition and burnt at the stake eight years later as a heretic in Rome.

Let me first briefly outline a few well-known facts about Bruno as a cosmologist. In an environment hostile to his ideas, *il Nolano* (as he liked to call himself, thus also trying to forge a new identity for himself as an author)[5] not only defended the heliocentric system, but went much further than what he considered to be the limited mathematical and geometric calculations of Copernicus (whom he nevertheless regarded an enormously gifted astronomer).[6]

The *Nolano* developed an intellectually highly concise theory according to which the cosmos consists of countless solar systems that form an infinitely large material universe.[7] Thus, from the very beginning of his Copernican literary career Bruno

> had been repeating that there were elements in the new cosmology that were fundamentally erroneous and misleading. He centred his criticism

---

(Aldershot – Burlington: 2005) 9–12. For a concise introduction to Bruno's philosophy, see Blum P.R., *Giordano Bruno* (Munich: 1999) 38–96 and *passim*. For Bruno's cosmology, see Michel P.H., *The Cosmology of Giordano Bruno* (London: 1973). For an excellent work on Bruno's science see Gatti H., *Giordano Bruno and Renaissance Science* (Ithaca: 1999).

3 His 'most important literary coup' as Blum, *Giordano Bruno* 39 says.

4 The most recent and comprehensive account is Tessicini D., *I dintorni dell'infinito. Giordano Bruno e l'astronomia del cinquecento* (Pisa – Rome: 2007).

5 On this aspect see now Hufnagel H., *Ein Stück von jeder Wissenschaft. Gattungshybridisierung, Argumentation und Erkenntnis in Giordano Brunos italienischen Dialogen* (Stuttgart: 2009) 69–84 with many references.

6 See for instance the often quoted praise in Bruno, *Cena* 27–28.

7 Gatti, *Bruno and Science* 47: 'Bruno's infinitism was, perhaps, his major contribution to the cosmological discussion of his time, and it has remained at the centre of critical attention from his day to ours.' See ibidem 99–101 for a concise and instructive summary of the contemporary context and the subsequent reception of this idea by scholars of science and followers of Frances Yates's Hermetic reading of Bruno's cosmology.

> on Copernicus's mathematical methodology and his lack of physical reasoning, because he thought that Copernicus was confusing mathematical concepts and physical realities.[8]

With his realist approach, Bruno underestimates the heuristic potential of mathematics that was to become the decisive factor for the development of modern science.[9] Yet, as Hilary Gatti in her magisterial study of Bruno's notion of science has emphasised, Bruno was not so much a scientist in the mould of Kepler, but a philosopher of science *ante litteram* who devoted enormous energies to account for the far-reaching implications as well as the difficulties of the new infinitist cosmology.[10] In the *Cena delle Ceneri*, the first and most controversial of the so-called London Dialogues, the *Nolano* claims to have liberated the human mind, which had been caught for centuries in the fetters of a finite, geocentric, and anthropocentric cosmology. In what has an utterly preposterous

8 Gatti, *Bruno and Science* 83.

9 In this respect Bruno follows the Peripatetic tradition; for the *locus classicus* see Aristotle, *Phys.* 193b 31–194a 11. For Bruno's position, "De innumerabilibus, immenso et infigurabili", in idem, *Opera latine conscripta*, ed. F. Tocco, 3 vols. in 8 parts (Naples – Florence: 1879–1891) vol. I/1 (lib 1–3), 170: 'Mathematice enim circularis motus non est in materia, quaecunque et qualiscunque sit, immo neque ullam formam vere in materia esse Platonici dixerunt (et non omnino male), neque hominem verum, neque verum equum. Verum punctum veramque lineam geometra nusquam suam inveniet, quinimo (si sapit) neque esse credat, nisi haec aliter, quam vulgo definita sunt, ille definiat. Quod ad motus ergo attinet eorum nulla, vel quae propius accedat, mensura reperitur.' The following passage aptly summarises Bruno's mature position on the nature of planetary movements in a complex series of spirals (rather than circles) that cannot be described adequately by geometrical figures, Bruno, *De immenso* 173: 'En ubi dii illi physici metaphysicique motores, quorum interitu, vultus tum naturae tum sophiae tum virtutis, omnis redit. Et hic motor, nullis geometriae regulis, nullis circulis eundem rebus vultum restituit, sed semper innovaturus in gyrum spiralem et hunc, quem nullo etiam geometrico canone possis satis assequi, compellit ire. Quid ergo opus est circulos circulis, orbes orbibus apponere, quasi veritatem aliquando geometricam adsequuturus, et non tuam ineptiam naturae, sed naturae ordinem ad tuas symmetrias aptaturus aliquando?'

10 Gatti, *Bruno and Science* 51; on this topic, see also Bönker-Vallon A., "Introduction", Bruno Giordano, *De l'infinito universo e mondi*, trans. and ed. A. Bönker-Vallon (Hamburg: 2007) lxx, xc–xcii; Leinkauf T., "Introduction", in Bruno Giordano, *De la causa, principio ed uno*, ed. T. Leinkauf (Hamburg: 2007) liv–lv and lxxiv–lxxxi with extensive bibliography; Védrine H., *La conception de la nature chez Giordano Bruno* (Paris: 1967) 281–288 and *passim*.

ring of messianism,[11] Bruno employs a set of complex literary strategies in order to point to himself (or to his alter ego, the Nolan) as the prophet of a dawning new age,[12] of a universe which is alive and sentient throughout and in which all celestial bodies do not move on fixed, crystalline and therefore mathematically predictable spheres, but are guided instead by their intrinsic souls.[13] Bruno did not only consider himself a prophet; as a philosopher he was well aware of the difficulties inherent in conveying theses messages, which for most of his contemporaries were not easy to accept. One of the many serious difficulties in his message lies in his claim to realist physics in an infinite cosmos. For Bruno's contemporaries the most fundamental problem consisted in making plausible what goes against all sensory experience and reason: the daily rotation of the earth around its axis and its annual motion. Vision cannot prove it because one cannot see that the earth is moving. Similarly, it is difficult to imagine an infinite universe in other than metaphorical or highly abstract terms.[14] In order to pave the way for his set of theorems, Bruno had to erode the validity of vision and to find new ways to bring physical (as opposed to mathematical) evidence to support his claims. Indeed, large parts of the *Cena delle Ceneri* are devoted to this task, that is, to replace the evidence generated by visual proof with an awareness that, in order to reflect the immensity of the one infinite material universe, any discourse must needs be fragmented, distorted like an uneven mirror and dependent on time.[15] In this mannerist

11 See for instance Bruno Giordano, "Cena delle Ceneri", in idem, *Dialoghi italiani, nuovamente ristampati con note da Giovanni Gentile*, 3rd edn., ed. G. Aquilecchia, 2. vols. (Florence: 1985) vol. 1, 35.: 'Con ciò un solo, benché solo, può e potrà vencere, ed al fine arà vinto, e trionfarà contra l'ignoranza generale; e non è dubio se la cosa de' determinarsi, non co' la moltitudine di ciechi e sordi testimoni, di convizii e di parole vane, ma co' la forza di regolato sentimento, il qual bisogna che conchiuda al fine; perché, in fatto, tutti gli orbi non vagliono per uno che vede, e tutti i stolti non possono servire per un savio.'

12 Cfr. Ingegno A., *Cosmologia e filosofia nel pensiero di Giordano Bruno* (Florence: 1978) 26–53.

13 For a succinct account see Gatti, *Bruno and Science* 83–85.

14 See for instance the description of the countless solar systems as an endless wood in which some trees will appear to form a centre, which in truth are relative to other trees. Bruno, *De immenso* (III, 4) 149.

15 Bruno thus neglects the rules set up by Counter Reformation theologians who prohibited ambiguity in all literary production. See Cox V., *The Renaissance Dialogue. Literary Dialogue in its Social and Political Contexts. Castiglione to Galileio* (Cambridge: 1992) chap. 1 for a discussion of the different genres of the dialogue and their assessment by Renaissance censors. See for instance Bruno, *Cena* 28–29 where Teofilo says of Copernicus that he had put together 'quelli abietti e rugginosi fragmenti ch'ha possuto haver le mani da la antiquità, le ha ripoliti, accozzati e risaldati in tanto [...] ch'ha resa la Causa, già

mode of reasoning Bruno developed an intricate set of modes to present his *nolana filosofia*: the *Cena delle Ceneri* thus unfolds in the context of a fictitious conversation that again becomes the topic of another conversation.[16] Bruno's ideas thus appear in the reflections of two different conversations; the author of the *Cena* becomes the object of a doubly indirect speech: Teofilo, the Nolan's mouthpiece, reports on a conversation that had taken place during an Ash-Wednesday supper. This series of literary *mises en abyme* is used to articulate the difficulties in communicating the new science; it is well epitomised in Teofilo's comment: 'And now what shall I say of the Nolan? Perhaps it is not appropriate for me to praise him, since he is so close to me as I am to myself.'[17] In another much-quoted passage, Bruno says that in the process of writing a text he is like a painter who cannot step back from the canvas because he is and remains part of the canvas.[18] Here Bruno adopts a strategy that is characteristic of his philosophy in general: the use of a particular literary form that evokes images.[19] The metaphorical character of such verbal statements functions as

---

ridicola, abietta e vilipresa, onorata, preggiata, piú verisimile che la contraria [...].' See also, with similar cosmological undertones, Bruno, *Cabala* 835–836.

16 See for instance the beginning of the first Dialogue of Bruno, *Cena* 19–24. For a discussion of the issue of Brunos mannerist literary stile, with many references to further literature, see Nuzzo E., "Le figure metaforiche nel linguaggio filosofico di Giordano Bruno", in Meroi F. (ed.), *La mente di Giordano Bruno* (Florence: 2004) 13–60, here 19–21.

17 Bruno, *Supper* 87; Bruno, *Cena* 29–30: 'Or che dirrò io del Nolano? Forse, per essermi tanto prossimo, quanto io medesmo a me stesso, non mi converrà lodarlo?' As P.R. Blum has argued, truth is therefore where the Nolan stands; *idem*, "Brunos Brunianismus", in Heipke K. – Neuser W. (eds.), *Die Frankfurter Schriften Giordano Brunos und ihre Voraussetzungen* (Weinheim: 1992) 201–210 and Gatti, *Bruno and science* 49. On this passage see also the interesting remarks in Granada, *La reivindicación* 107–131, who deals extensively with this issue, and especially with Bruno's implicit criticisms of Christian as well as Peripatetic doctrines inherent in this claim.

18 Bruno, *Cena* 16: 'Se nel ritrare vi par che i colori non rispondano perfettamente al vivo, e gli delineamenti non vi parranno al tutto proprii, sappiate ch'il difetto è provenuto da questo, che il pittore non ha possuto essaminar il ritratto con que' spacii e distanze, che soglion prendere i maestri de l'arte; perché, oltre che la tavola, o il campo era troppo vicino al volto e gli occhi, non si possea retirar un minimo passo a dietro, o discostar da l'uno e l'altro canto, senza timor di far quel salto, che feo il figlio del famoso defensor di Troia.' Cfr. also the remarks in Gatti, *Bruno and Science* 49.

19 Bruno Giordano, "Cantus circeus", in idem, *Opera latine conscripta*, ed. F. Tocco, 3 vols. in 8 parts (Naples – Florence: 1879–1891) vol. II/1, 220: 'Et ideo sicut in scriptura extrinseca atque pictura quae serviunt oculis extrinsecis duo requiruntur: ratio videlicet formae atque figurae characterum et imaginum, et materia atque subiectum in quo formae illae et imagines possint subsistere, manere et perdurare. Ita etiam in scriptura

a representation of the difference between signifier and signified as well as of the materiality of language, with its concomitant associative and/or evocative power. Moreover, this form of *discorso* allows Bruno to mythologise his own person, as well as to use images of finite embodiments that function as, albeit inadequate, representations of the infinite cosmos. The effort to determine the relationship between finite and infinite thus constitutes an important aspect of Bruno's philosophy.

In his Latin poems, published more than half a decade after the so-called *dialoghi italiani*, Bruno developed several interrelated and amazingly concise philosophical notions to address this difficult issue; probably the most important among them was his theory of the *minima* and monads.[20] When considering the individual exemplar of a species we may meaningfully speak of size, of different qualities, but always and only in relationship to other, no less individual exemplars. All entities are made up of one single infinitely large substance (matter) from which the different forms emerge and to which they, after a certain amount of time, return. Every individual being thus composed, it partakes in the universal substance: everything is in everything (*quod omnia in omnibus*).[21] Metaphorically, the monads are like fragments of a shattered mirror that reflects the universe in its entirety. From the perspective of any intellect, knowledge about the entire universe is and must remain, therefore, finite and paradoxical; according to Bruno true insight comes about not by looking to the infinite skies, but rather by a form of radical introspection.[22] Under

intrinseca atque pictura, quae serviunt oculis intrinsecis, duo sunt necessaria.' See also, ibidem 235. In order to describe these complex transactions, Ciliberto has coined the phrase 'ut pictura philosophia,' Ciliberto M., "Introduction", in Giordano Bruno, *Le ombre delle idee* (Milan: 1997) 17; see also Bruno Giordano, "Explicatio triginta sigillorum", in idem, *Opera latine conscripta*, ed. F. Tocco, 3 vols. in 8 parts (Naples – Florence: 1879–1891) vol. II/2, 133–134; Ordine N., *Giordano Bruno und die Philosophie des Esels* (Munich: 1999) 214 with footnote 38 for references; Ordine N., *La soglia dell'ombra. Letteratura, filosofi a e pittura in Giordano Bruno* (Venice: 2003) 67–70. On the central role of the imagination for Bruno's doctrine of an infinite universe, see De Bernart L., *Immaginazione e scienza in Giordano Bruno. L'infinito nelle forme del' esperienza* (Pisa: 1986).

20 For what is probably the best introduction to this topic, see Gatti, *Bruno and Science* 128–142; see also Schettino E., "The necessity of the *Minima* in the Nolan Philosophy", in Gatti H. (ed.), *Giordano Bruno. Philosopher of the Renaissance* (Aldershot: 2002) 299–325.

21 On the doctrine 'quod omnia in omnibus' see for instance Bruno, Imaginum 94–95; and in general, Leinkauf T., Mundus combinatus. *Studien zur Struktur der barocken Universalwissenschaft am Beispiel Athanasius Kirchers SJ* (Berlin: 1993) 83–91.

22 In a literally spectacular sense Bruno links the opposing view, according to which we have to know ever more celestial spheres outside ourselves, to a form of Narcissism that fails to acknowledge that truth is inside us rather than outside; cfr. Bruno, *De immenso* 125. It is

such conditions very few, very ingenious individual human beings may attain a more or less concise image of truth by turning to their physical and mental interiors; for it is there that they will find an embodied, sentient mirror of the infinite universe. Although this form of consciousness is subject to fortuna[23] and vicissitude,[24] it remains the only means to obtain (a fragmented) notion of the universe. Bruno thus seeks to develop an internal culture that entails what Ferdinand Fellmann has aptly characterised as 'intensives Denken'; a form of total cognitive and somatic immersion of the individual; this habit of mind is designed to result in a particular mental and physical disposition. In that process, the constant presence of polyvalent artefacts, namely image/word combinations, statues or seals, as Bruno calls them, plays a significant role because such imaginary or material objects may aid us in focusing the motions of our bodies and minds.[25] Bruno was a master in the ancient art of memory, and the construction and location of such statues in the memory is one of the preconditions for the practice of this art.[26]—These admittedly very sketchy remarks

---

interesting to note that Bruno here adapts and inverts the Neoplatonic myth of Narcissus, which has the function to represent a figuration of the soul oblivious of its divine homeland; see for instance Kodera S., *Disreputable Bodies. Magic, Gender, and Medicine in Renaissance Natural Philosophy* (Toronto: 2010) 65–69.

23 For quotes on the role of fortune, see Bruno Giordano, "Spaccio della bestia trionfante", in idem, *Dialoghi italiani, nuovamente ristampati con note da Giovanni Gentile*, 3rd edn., ed. Giovanni Aquilecchia, 2 vols. (Florence: 1985) vol. 2, 569–589. Bruno Giordano, "Candelaio. Comedia", in idem, *Opere Italiane*, ed. E. Canone (Paris, G. Giuliano: 1582; reprint Florence: 1999) vol. 1, fol. 127v–129r [286–289]. In Bruno's system, Fortuna secures the unity of all things by treating everyone equally.

24 On Bruno's doctrine of 'vicissitudo' see Granada M.A., *La reivindicacion de la filosofia en Giordano Bruno* (Barcelona: 2005) 245–258; Severini E.M., "Vicissitudine e tempo nel pensiero di Giordano Bruno", in Meroi F. (ed.), *La mente di Giordano Bruno* (Florence: 2004) 225–258; Ordine, *Soglia dell'ombra* 59–61.

25 Cfr. Gatti, *Bruno and Science* 185–187, 192–193 and 197–200.

26 On Bruno's art of memory, see Yates F.A., *The Art of Memory* (London: 1969); Sturlese R., "Il *De imaginum signorum et idearum compositione* di Giordano Bruno ed il significato filosofico dell' arte della memoria", *Giornale critico della filosofia italiana* 69 (1990) 297–306; Sturlese R., "Per un'interpretazione del *De umbris idearum* di Giordano Bruno", *Annali della Scuola Normale Superiore di Pisa* 3, 22 (1992) 942–967; Bolzoni L., *The Gallery of Memory. Literary and Iconographic Models in the Age of the Printing Press* (Toronto: 2001); Clucas S., "Giordano Bruno's *De imaginum, signorum et idearum compositione.* Art, Magic and Mnemotechnics", *Physis. Rivista internazionale di storia della scienza* 38 (2001) 75–98; Clucas S., "*Simulacra et Signacula.* Memory, Magic and Metaphysics in Brunian Mnemonics", in Gatti H. (ed.), *Giordano Bruno. Philosopher of the Renaissance* (Aldershot – Burlington: 2002) 251–272.

of Brunian infinitist metaphysics have immediate consequences for the relationship Bruno constructs between himself and Copernicus.

As I shall show in the following example, a close reading reveals that Bruno creates such images in the context of a literary strategy which is designed to work as an embodied simulation of the monadic condition of individuals; in other words: these images seek to trigger images of past, present and future aspects of a monad we may call *Nolanus*.

> Heic ego te appello, veneranda praedite mente,
> Ingenium cuius, obscuri infamia secli
> Non tetigit, et vox non est suppressa strepenti
> Murmure stultorum, generose Copernice, cuius
> Pulsarunt nostram teneros monumenta per annos
> Mentem, cum sensu ac ratione aliena putarem
> Quae manibus nunc attrecto, teneoque reperta[27]

This passage is from one of Bruno's last works: published 1591, the *De immenso* outlines the Nolan's ideas about the infinite universe in an impressive Latin poem inspired by Lucretius' *De rerum natura*.[28] What at first might appear to the modern reader as a conventional and somehow bombastic praise of Copernicus turns out to be a precise characterisation of how Bruno constructed the relationship between his own person and the astronomer and, by extension, between the heliocentric cosmology and what Bruno termed his *nolana filosofia*. Let us look more closely at the exact wording of these first six lines of chapter 9, book 3 of the *De immense*, which is entitled *De lumine Nicolai Copernici.* With distinct echoes from the *Cena*,[29] Bruno praises Copernicus's immaculate character and his genial voice which are to be distinguished from the inarticulate mutterings of the ignorant crowd in a disgraceful epoch. The New Astronomy is here set in a historic context; more precisely, Bruno says it was (re)discovered (*repertus*) in a dark age of ignorance. This is by no means

27 'Here I appeal to you, noble Copernicus, adorable gifted mind, whose character was left untouched by the dark infamous times and whose voice was not subdued by the noisy grunts of the foolish, whose literary works shook our mind in younger years, when I believed to be contrary to sense and reason, what I am now gripping fast with my hands and am holding as proven /rediscovered.'—Bruno, *Immenso* 188–189.

28 For an introduction to this text, see Gatti, *Bruno and Science* 79–85; Granada M.A., *Giordano Bruno, universo infinito, union con Dios, perfeccion del hombre* (Barcelona: 2002) chap. 2; Blum, *Giordano Bruno* 135–142; Tessicini, *Dintorni dell'infinito* 68–74.

29 Bruno, *Cena* 28.

irrelevant, because in Bruno's infinite universe everything is in a constant process of change, hence time is the decisive factor for any object trying to assume a definite and unique shape.[30]

According to Bruno's metaphysics, matter is infinite and eternal; the individual and transient forms perpetually emerge from and return into this universal substance. As a consequence, time also structures the relationship between Bruno and Copernicus and their respective works: Bruno tells us that for years during his youth he had been an opponent of the heliocentric doctrine, but then gradually became an adherent to it.[31] He says that the doctrine had been 'hitting his mind for years'.[32] Obviously we may read this as a metaphorical description of his irritation, his "headaches" with a doctrine, which, back, then, apparently conformed neither to sense nor to reason, [*cum sensu ac ratione aliena putarem*]. Upon closer scrutiny, however, it seems strange that Bruno should choose this particular metaphor; for, according to his metaphysics, the mental and the physical aspects of the universe do not interact because they are incapable of touching each other: the mind cannot be "pushed around" so to speak.[33] We must, therefore, suppose that the figure of speech is either inappropriately gross (which is highly unlikely in a key passage in one of the Nolan's own *monumenta*). More likely, Bruno deliberately chose the paradoxical expression in order to refer to a specific and intertwined set of ideas that he now, whilst writing *De immenso*, had come to regard as obsolete. In fact, the idea that the lower and material parts of the soul are moved from outside and by higher powers has a long tradition in the Neoplatonic theory of prophecy, that is, in the transmission of important messages from the celestial realm to human souls. In his *Theologia platonica*, Marsilio Ficino explains that some human beings are able to foretell the future because the *idola* or signs emitted by the stars push or hit (*pulsant*) our mind whilst we are asleep and thus more susceptible to celestial messages.[34] The *idola* are thus inducing amazing

---

30 Bruno, *De immenso* 169–170: 'Linea non unquam poterit punctumque resumi, | Sumptum quod fuerat, verum paris atque propinqui | Consimili speciemque modo referentis, ut illa | Praesenti haud tanto differt in tempore, quanto | Obtusi ad normam sensus appareat esse | Quicquam vel multum: est porro distantia tanta, | Quantam non solum rota temporis efficit eius, | Verum etiam illorum quae circum, atque undique currunt.' On the decisive role of time see also for instance Bruno, *Spaccio* 592.

31 On a tentative schedule for Bruno's gradual conversion to Copernicanism cfr. Gatti H., "Copernico", *Bruniana e Campanelliana* 14 (2008) 511–520, here 514–516.

32 'pulsarunt nostram ... mentem ... per annos'.

33 Cfr. Bruno, *De immenso* 505–506.

34 Cfr. Ficino Marsilio, *Theologia Platonica. Platonic theology*, ed. and trans. M.J.B. Allen et al, 6 vols. (Cambridge Mass. – London: 2001–2006) vol 4, 144: '... idola animarum superiorum

divinatory powers into us that arrive through these benevolent shocks or strikes of the celestial *idola.* This influence from above reaches us in very mechanical ways; the *pulsatio* of the cosmic images thus accounts for the phenomenon of action at a distance and is vital for the Neoplatonic theory of magic. Ficino's ideas were well known to Bruno; they had developed in the context of a geocentric astrology that explained how all material beings on earth are guided by the influence of the stars (which are moved again by external, angelic forces); the only (possible) exception being the higher parts of the human soul because it is endowed with free will.[35] For Bruno this finite cosmology is, of course, dated; the idea that knowledge comes from above, from some heavenly or supercelestial realm, has to be discarded because in the infinite universe there is no such thing as above or below: knowledge is to be found instead inside everything, including inside us.[36] I therefore read the line in the passage quoted above not only as a playful and ironic reference to Ficino's theories on prophesy, but also as an indication that Bruno wished his reader to be aware of the fact that he perceived his past involvement with this tradition as a total immersion in the geocentric astrological *Weltbild* that included all its concomitant ramifications into physics, metaphysics and moral philosophy.[37] Bruno

---

pulsant semper idolum nostrum; hunc nos influxum haud agnoscimus...' ibidem, 148: 'Vis illa supernorum idolorum pluvias inductura ad pluvias praefigurat caelorum rotas,.... Vis haec suscitat eas imagines quae in vacante phantasia rationeque sunt ad pituitam pertinentes, ut subito cogitemus flumina, imbres, [...] atque similia.'

35 For the entire context, cfr. Ficino Marsilio, "Disputatio contra iudicium astrologorum", in Kristeller P.O. (ed.), *Supplementum ficinianum,* 2 vols. (Florence: 1937) vol. 2, *passim* and 74: 'Corpus non agit directe nisi in corpus et virtutem corporalem. Intellectus autem et voluntas sunt virtutes immateriales, quod patet quia abstrahunt omnia et abstracta diligunt. Ergo celum non agit in ista, dico directe. Agit autem directe in corpus, reflexe in sensum quia corporalis, in intellectum vero neutro modo. Sed potest intellectus et voluntas sensui inclinatio assentiri.'

36 Gatti, *Bruno and Science* 211; for two very amusing passages that testify to Bruno's negative attitude to Ficino's concept of prophecy and sideral inspiration in general, see Bruno Giordano, *Il Candelaio. Chandelier,* trans. Y. Hersant, ed. G. Aquilecchia – G. Bàrberi-Squarotti (Paris: 1993) 9: 'A chi inviarrò quel che dal sirio influsso celeste, in questi più cuocenti giorni, et ore più lambiccanti, che dicon caniculari, mi han fatto piovere nel cervello le stelle fisse, le vaghe lucciole del firmamento mi han crivellato sopra, il decano de dudici segni m'ha balestrato in capo, e ne l'orecchie interne m'han soffiato i sette lumi erranti?' ibidem 51: 'Quanto ben dimostrano che essi [scil. i pedanti] son quelli soli a quai Saturno ha pisciato il giudizio in testa, le nove damigelle di Pallade un cornucopia di vocaboli gli han scarcato tra la pia e dura matre [...].'

37 See for instance the chapter heading in Bruno Giordano, "De triplici minimo et mensura", in idem, *Opera latine conscripta,* ed. Tocco F., 3 vols. in 8 parts (Naples – Florence:

thus recounts the story that in his younger years, the Copernican theory really 'shook his mind', because back then his entire intellectual life was imbued with the old ways of thinking and a corresponding regimen of the body he now regards as erroneous and obsolete. It is tempting to surmise that it was precisely the conventual discipline to which the young Fra Giordano had been subjected for so many years that allowed for the Nolan's constant awareness of the somatic resonances these doctrines had for a daily practice, as well as of the material potential of words to enhance such intellectual habits.[38] In the *Cena* similar ideas about the psycho-physical immersion of every individual being into a system of thought are reflected in the framework of a particular literary strategy. Here the Nolan is reported to have first offended the pedant Torquato, thereby causing a common uproar, upon which the Nolan patiently explains to this Aristotelian pedant that:

> I am instead as much a friend to you as I am to myself. [Because] I could hate you on these grounds no more than I could hate myself when I was younger, more childish, less wise and less discreet. This is the reason why instead of getting angry with you, I pity you and pray to God that, as He gave me this knowledge, so (if it does not please him to make you capable of seeing) at least He make you aware that you are blind. [...] I mean that, even though I have never been so barbarous, uncouth and boorish in discussion and disputation, I was once as ignorant as you. In this way, while I consider your present my past, and you consider my past as your present, I shall love you and you will not hate me.[39]

1879–1891) vol. I/3, 232: 'Conclusio, ut ex virtute consuetudinis credendi falsis sensus etiam ipse perturbatur.'

38 Clucas, "Imaginum compositione" 93 accordingly says that Bruno's mnemonics deals with images, signs and that the 'art does not consist simply in a rational understanding, but in a practical *experience* of the structures of universal reality, the operator enacts or performs his belief in the unity of the cosmos.' On the contemplative aspects of Bruno's art of memory, see Clucas S., "*Amorem, artem, magiam, mathesim*. Brunian Images and the Domestication of the Soul", *Zeitsprünge* 3 (1999) 19–22. On Bruno's monastic experience, see Miele M., "L'organizzazione degli studi dei Domenicani di Napoli al tempo di Giordano Bruno", in Canone E. (ed.) *Giordano Bruno. Gli anni napoletani e la 'peregrinati' Europea* (Cassino: 1992) 29–50.

39 Bruno Giordano, *The Ash Wednesday Supper*, ed. E.A. Gosselin – L.S. Lerner (Toronto: 2001) 187–188; Bruno, *Cena* 135f.: '[...] vi son cossí amico come di me stesso. [...] Tanto dunque io posso odiar voi per questa caggione, quanto me medesmo, quando ero piú giovane, piú putto, men saggio e men discreto. Cossí, in loco ch'io mi devrei adirar con voi, vi compatisco, e priego Idio che, come ha donato a me questa cognizione, cossí (se non gli

Bruno uses here a significant chiastic literary structure[40] that is perfectly apt to convey a sense of the physical entanglement of all beings in the infinite universe. Torquato becomes thus tied into a cats-cradle of statements, which are designed to realign and level incompatible positions. *Sub specie infinitatis*, even Torquato,—this Peripatetic beast—inhabits the infinite universe; therefore he is part of the same universal substance as the Nolan, both out of necessity to change their physical forms, and, in tandem with them, their mental dispositions. This inner logic in Bruno's reasoning, with the prominent role it gives to time and change,[41] is of course unable to eclipse what is on the surface of this text, namely Teofilo's grossly condescending remarks; for they highlight the ignorance of a particular schoolmaster who can be nothing else but one of those many asses who repeat what any self-appointed 'Mercury' will teach them as ultimate truth.[42] Torquato is therefore blind, but only in a metaphorical way, because he is too stupid to understand that the earth moves around the sun. This is one of the many instances in which Bruno adapts a Platonic metaphysics of light to his own purpose.

In this context it is instructive to note that the chapter heading for *De immenso* III, 9 proposes to speak of the *lumen* of Copernicus, a word that has a wide range of different concrete and metaphorical meanings: in our context it is carefully chosen to evoke a set of divergent associations. Of course, and quite conventionally, *lumen* refers to Copernicus's ingenuity, to his intellect. Yet, the word also denotes eyesight: in this case it has an interesting textual reference to a quote from the *Cena*: 'the Nolan replied that in judging and determining he saw neither through the eyes of Copernicus nor those of Ptolemy, but through

---

piace di farvi capaci del vedere) almeno vi faccia posser credere che sete ciechi. [...] Voglio che, quantunque mai son stato, conversando e disputando, cossí salvatico, malcreato ed incivile, son stato però un tempo ignorante come voi. Cossí, avendo io riguardo al stato vostro presente conforme al mio passato, e voi al stato mio passato conforme al vostro presente, io vi amarò e voi non m'odiarete.'

40 On a related and fascinating application of this trope in a contemporary dissident medical discourse, see Bidwell-Steiner M., *Große Welt-kleine Welt-verkehrte Welt. Die philogyne Naturphilosophie der Renaissance-Denkerin Oliva Sabuco de Nantes y Barrera* (Innsbruck – Vienna: 2009) 93–94, 124; and in a more general perspective the study by Pechriggl A., *Chiasmen. Antike Philosophie von Platon zu Sappho—von Sappho zu uns* (Bielefeld: 2006).

41 On this question see also Bruno Giordano, *De gli heroici furori. Des fureurs héroïques. Introduction and notes by M.A. Granada, transl. by P.H. Michel = Oeuvres complètes 7*, ed. by Y. Hersant – N. Ordine (Paris:1999) 43–45 *with notes* and the discussion of the function of the famous 'ruota della metamorfosi' in Brunos and the waters above and below the firmament, in ibid., Miguel A. Ganada, *Introduction, passim*, esp. XC–CII.

42 On the topic of the false divine messengers, see Bruno, *Cena* 32.

his own eyes.'[43] By using eyesight as a physical metaphor for cognition, Bruno emphasises the extent to which he had adopted Copernicus's views to his own "vision" of the infinite universe. *Lumen* is also a lamp or torch; again this is evocative of the passages in which the Filoteo of the *Cena* describes the labyrinthine ways in which the Nolan and his party wound through the London night to reach the house where the dinner (*cena*) was held, and, later in the dialogue, the Nolan's lament that he had not even been given a lamp for his way home, and that his heliocentric doctrine was not to be discussed in daylight, but only in the dim flickering of candles or torches (*lumina*). Bruno's adversaries, Nundinio and Torquato, are described as asses among the blind who regard these pedants as endowed with eyesight and who 'employ bladders as lanterns'.[44] If we translate *lumen* as torch or candle, then the chapter heading suggests that Copernicus had brought light back amongst the ignorant, like a timid flickering candle that had been obscured by centuries of darkness. The word *lumen* also denotes a "star"; in which case we have to take into account that in contemporary technical as well as metaphysical discourses, *lumen* means light as it becomes visible in a medium, such as in air or in a mirror; this word has to be distinguished from *lux*, which designates the state of light as it is extant in its source of light, such as the sun.[45] In this case, "Copernicus's star" would be a celestial object that emits only reflected light, that is, a second class light. This celestial body is therefore not to be confused with the sun, which seems to be the metaphorical description Bruno wants to reserve for his own person.[46] To put it more bluntly: already in the chapter heading Bruno insinuates that his star shines brighter than Copernicus's ever did. In doing so, Bruno suggests that it was he who put Copernicus's *monumenta* in their proper

43 Bruno, *Supper* 85; Bruno, *Cena* 27: 'Al che rispose il Nolano, che lui non vedea per gli occhi di Copernico, né di Ptolomeo, ma per i proprii.'

44 Bruno, *Supper* 193; Bruno, *Cena* 142: 'vessiche per lanterne'.

45 For the distinction see for instance Ficino Marsilio, *Commentaire sur le Banquet de Platon. De amore*, ed. M. Raymond (Paris: 1956) 234.

46 On Bruno's distinction between reflected and emitted light, see for instance, Bruno, *De immenso* 182. 'Hinc quoque subiecti lucis discrimen apertum est. | Scilicet aequoreo lucem de corpore fundunt | Tellus, Luna, Venus, Mavors, reliquique planetae; | Maior at haec inter rutilo sol fulget ab igne. | … Utque Dei templum, lux ardens, araque coeli | Militia in tanta, in populo celeberrima tanto, | De statione sua spectare errantia circum.' For a discussion of Bruno's self-fashioning as a prophet see Ingegno, *Cosmologia e Filosofia* 26–43, who says that Copernicus becomes a 'divino segno' for Bruno who tries to establish a complementary relationship between himself and the astronomer (ibidem 41); Copernicus is the last exponent of an occidental cosmology that is superseded by Bruno's new philosophy (ibidem 53).

place, that is, into "his own hands". Again, this may be read as a figure of speech that reminds us of the fact that Bruno had not unconditionally accepted the Copernican heliocentric doctrine. He had claimed to go further than the astronomer: rather than stopping at mathematical or geometric calculations, the Nolan had "embraced" Copernicanism in the literal sense by applying the theory to astrophysics. The last line of the cited passage is ample proof for this. It refers us to a world in which knowledge has to be grasped with one's hands and has to be held tightly, actions that radically differ from the contemplative Platonic gazing at unchangeable eternal forms. From the perspective of the history of ideas this positive assessment of touch, the most physical of all senses (which was intimately connected to sexuality), entails a decidedly anti-Platonic and, with it, anti-Christian position. According to these traditions the sense of sight stood at the top of the hierarchy of the senses (not least because vision was sensitive to potentially noumenal light and hence served as a medium for the vision of Platonic forms).[47]

For Bruno, some kind of hands-on knowledge has to be applied not only to the planet earth and to the human realm, but rather to the entire, infinite universe; for the stars in the innumerable skies are not different from the terrestrial sphere, just as human beings are integral parts of an infinite universe that is alive and sentient throughout.[48] Again, it is fascinating to observe how Bruno applies this central tenet to his encomium of Copernicus: for in the citation above he is not saying "Copernicanism" or "doctrine", but is rather talking of *monumenta* which the astronomer left behind, that is, of his books, his literary works which the Nolan now holds fast in his hand and won't let go of again. Keeping in mind Bruno's Neapolitan background,[49] we may imagine that this tight grip will prevent a theft of the *De revolutionibus*. What might be read as a (quite Brunian) joke opens the field for another set of serious considerations on the ways in which true knowledge is transmitted to and between human beings.[50] In the *Cabala del cavallo Pegaseo*, published six years earlier, Bruno

---

47 For a characteristic Christian Neoplatonic position on the hierarchy of the senses, see Ficino, *De amore* 179–181.

48 Bruno Giordano, *De immenso*, 237; Ciliberto M., *Umbra profonda. Studi su Giordano Bruno* (Rome: 1999) chap. 2, esp. 279–281.

49 For an example of a pedant who is robbed in the streets of Naples, see Bruno, *Candelaio* 196–205.

50 Cfr. the remarks in Bruno, *Cena* 46–48 on the false prophets, the stupidity of the multitude which is prone to be seduced by false prophets because it is so difficult to independently appropriate, recognise or discover the truth; see also the following footnotes.

had discussed his most radical solution to this problem: Promethean theft.[51] Here, true knowledge is acquired not through learning, but rather by theft. Contrary to Aristotle and Ficino, Bruno maintains that human beings are the cleverest animals because they have hands that make them capable of theft.[52] Even though their intellect is the same as that of any other living being, it is the hand as the organ of organs that distinguishes humans from other living beings. The message in these lines from the *De immenso* is then that Bruno has stolen a book in which Copernicus (all too timidly) outlines a heliocentric doctrine; the Nolan now applies this knowledge to the one and only place where it is appropriate: to the physical world, as he says a few lines later in the *De immenso,* pointing to himself as a divine prophet.[53]

What emerges from this is a *monumentum* we may call *Iordanus*, the philosopher-prophet of a new age, who is holding a copy of the *De revolutionibus.*

51 Bruno, *Cabala* 878–879: 'Che fero? [...] hanno inceppate le cinque dita in un'unghia, perché non potessero come l'Adamo stender le mani ad apprendere il frutto vietato dall'arbore della scienza, per cui venessero ad esser privi de frutti de l'arbore della vita, o come Prometeo (che è metafora di medesimo proposito) stender le mani a suffurar il fuoco di Giove per accendere il lume nella potenza razionale. Cossì li nostri divi asini privi del proprio sentimento et affetto, vegnono ad intendere non altrimente che come gli vien soffiato a l'orecchie dalle revelazioni o de gli dèi o de' vicarii loro.' On this topic, see Papi F., *Antropologia e civiltà nel pensiero di Giordano Bruno* (Florence: 1968) 237–247; Jorn O., "Corporeità in Bruno, senso e figura", *Bruniana e Campanelliana* 8 (2002) 159–180; Kodera S., "Introduction", in Bruno Giordano, *Cabala del Cavallo Pegaseo/Die Kabala des pegaseischen Pferdes*, ed. S. Kodera (Hamburg: 2008) LXXV–XC.

52 See Aristotle *Anim.* 431b 20 ff.; Ficino, *Theologia Platonica,* vol. 6, 183; for Bruno's postion, cfr. Bruno Giordano, "Cabala del cavallo pegaseo", in idem, *Dialoghi italiani, nuovamente ristampati con note da Giovanni Gentile, 3rd edn.,* ed. G. Aquilecchia, 2 vols., (Florence: 1985) vol. 2, 887: '[...] che sarrebe se posto che l'uomo avesse al doppio d'ingegno che non have, e l'intelletto agente gli splendesse tanto più chiaro che non gli splende, e con tutto ciò le mani gli venesser transformate in forma de doi piedi, rimanendogli tutto l'altro nel suo ordinario intiero: dimmi dove potrebbe impune esser la conversazion de gli uomini, come potrebero instituirsi e durar le fameglie et unioni di costoro parimente o più che de cavalli, cervii, porci, senza esserno devorati da innumerabili specie de bestie per essere in tal maniera suggetti a maggiore e più certa ruina? e per conseguenza dove sarrebono le instituzioni de dottrine, le invenzioni de discipline, le congregazioni de cittadini, le strutture de gli edificii, et altre cose assai che significano la grandezza et eccellenza umana, e fanno l'uomo trionfator veramente invitto sopra l'altre specie? Tutto questo, se oculatamente guardi, si referisce non tanto principalmente al dettato de l'ingegno, quanto a quello della mano, organo de gli organi.'

53 Bruno, *Immenso* 189: 'Coeperunt veri fontes, pulcherrimaque illa | Emicuit rerum species, (nam me Deus altus | Vertentis secli melioris non mediocrem | Destinat, haud veluti media de plebe, ministrum).'

This *monumentum* (the word literally means: reminder) could be one of the thousands of memory statues which Bruno—a master of this art—was capable of storing in his mind. This image recalls Raphael's *School of Athens,* the famous fresco, at its centre, Plato. He is holding a copy of the *Timaeus*, though in a rather absentmindedly way (and therefore certainly not streetwise for an adept of the *School of Nola*).[54]

Bruno's emphasis on the particular materiality of the text of the *De revolutionisbus* is reflected in a key passage from *Cena delle Ceneri* and the Nolan's heated disputation with the doctors Torquato and Nundino. Here, the material of the book chases away the asinine pedants and their erroneous visualisations of a geocentric universe; the book becomes decisive in a debate on the orbits of earth and moon.[55] The Nolan's Aristotelian adversary pens down a set of eight concentric circles and uses the upper half of them to illustrate the Ptolemaic cosmos, with the earth at the centre. The lower part of the diagram depicts what Torquato considers to be an important aspect of the Copernican heliocentric system; the Pedant claims that the moon is carried around the earth and that both move around the sun.[56] Smitho (who, as the dialogue unfolds, gradually becomes an adherent to Bruno's infinitism)[57] accuses Torquato of having just looked at the illustration and of having failed to read and/or to understand the text according to which the moon and the earth are revolving on the same epicycle around the sun. Therefore, Theofilo's presentation of the disposition of the earth and the moon had been correct. The Nolano then explains to the dinner party that his Aristotelian opponent simply had mistakenly assumed that the point left by the foot of the compass signified the earth and says that the text according to Teofilo *clearly* states 'that the earth and the moon were as if contained in the same epicycle'.[58] I do not wish to further discuss the validity of this claim or its lack of evidence, which are remembered as Bruno's Copernican Mistake.[59] I would rather like to focus on

54 There is, of course, another well-known passage in *Causa* 224, which reflects on the materiality of the text: when Filoteo hands the following dialogues of the *Della causa* to the interlocutors of the first dialogue, he also appropriates the famous Augustinian phrase—'tolle, lege!' ("prendete, leggete!")—for the *nolana filosofia.*

55 Bruno, *Cena* 139–141; Gatti, *Bruno and Science* 61 says that, with the exception of Frances Yates and her followers, this episode has been 'surprisingly ignored'.

56 Bruno, *Supper* 192; Bruno, *Cena* 140–141.

57 See the interesting remarks in Hufnagel, *Wissenschaft* 90–93.

58 Bruno, *Supper* 192–193; Bruno, *Cena* 141: 'la terra e la luna essere contenute come da medesmo epiciclo.'

59 I refer to the superb discussion of the problem in Gatti, *Bruno and Science* 60–68, who also explains the thermodynamic reasons that probably led Bruno to adapt this model; on

Bruno's strategy in communicating the *nolana filosofia* in this particular episode and relate it to the invocation of Copernicus in the *De immenso*. In the *Cena*, the dispute between the Nolan and the Pedant is settled because their hosts, "diligent noblemen", have the book brought to the dinner table where the party has gathered. The wording of the text makes it clear that the diagram had been misleading, for the reasons stated above. Bruno here not only conveys the idea that vision is deceptive, but also invokes the particular materiality of the drawing and, by extension, of the printed word. Before the party gathered around the supper-table may begin to read or to recite Copernicus's words, a copy of the *De revolutionibus* has to be displayed before their eyes. In order to truly digest the materiality of the words[60] one has to hold the text in one's hands; which is exactly what the party does. In this passage, Bruno highlights the fact that written words (let alone when they are written in a language that is not one's own) are assimilated in more indirect ways than the visual evidence created by a diagram, which in this particular case was drawn from memory. Reading is a more complex process of assimilation than merely looking. In Bruno's terminology, words have to be digested; accordingly, Teofilo recounts that the guests at the Supper 'chewed' the words and not in the original language, but in English (*mastigando in lor lingua*). At this point, the two defeated pedants take their leave and we have to imagine them as members of a dinner party who have broken one of their teeth, as Bruno writes in the preface to the following Italian Dialogue, the *Della causa, principio ed uno.*[61]

---

the topic, see also Tessicini, *Dintorni dell'infinito* 17–22; Granada M.A., "L'héliocentrisme de Giordano Bruno entre 1584 et 1591: la disposition des planètes inférieures et les mouvements de la terre", *Bruniana e Campanelliana* 16 (2010) 31–37.

60 On the topic of malleable words which are changing like every other natural being, see Gatti, *Bruno and Science* 203–208; Ordine, *Philosophie des Esels* chap. 14.

61 This of course recalls a famous passage in Bruno Giordano "Della causa, principio ed uno", in idem, *Dialoghi italiani, nuovamente ristampati con note da Giovanni Gentile*, 3rd edn., ed. G. Aquilecchia, 2 vols. (Florence: 1985) vol. 1, 197–199 which describes the heated debate of the Ash Wednesday Supper in the terms of a metabolism: 'Non vi maravigliate, fratello, perché questa non fu altro ch'una cena, dove gli cervelli vegnono governati dagli affetti, quali gli vegnon porgiuti dall'efficacia di sapori e fumi de le bevande e cibi. Qual dunque può essere la cena materiale e corporale, tale conseguentemente succede la verbale e spirituale [...].' (ibidem 197). See also for instance the remarks on digestion as fundamental form of intellectual contemplation in Bruno Giordano, "Lampas triginta statuarum", in idem, Bruno Giordano, *Opera latine conscripta*, ed. Tocco F., 3 vols. in 8 parts (Naples Florence: 1879–1891) vol. III, 3: 'Animae veritatem esse cibum, utpote quae in eius substantiam, veluti proprium nutrimentum, transmutabilis est, [esse] constat. Perfectio et finis huius nutrimenti est lumen contemplationis, quo animus noster oculis intelligentiae primum quidem solem primae veritatis, deinde ea, quae circa ipsum sunt,

What the reader of the *Cena* is supposed to understand by this episode is that the Nolan has won a battle against the pedants; that he has beaten those guardians of a wrongheaded philosophy with their own weapons, that is, with words. Bruno thus creates awareness that words have a specific material character and that in order to be assimilated they have to be digested and transformed. The acquisition of knowledge is not invoked here in biblical metaphors, as defloration and/or as rape: acts that are (at least from the daft patriarchal and therefore sexist perspective in which such acts are usually recounted) directed *outwardly*; instead, in Bruno the act of becoming wise implies the experience of a metabolism, a form of appropriation which, of necessity, has to happen *inside* ourselves.[62] It is therefore only logical that in the rest of this chapter of the *De immenso*, Bruno cites and paraphrases large parts from the preface of the *De revolutionibus*. He wishes his reader to visualise him as though he were a prophetic memory statue reading out an important text, namely passages from Copernicus's dedicatory letter to Pope Paul III and from the preface of the *De revolutionibus*.[63]

We may conclude that Bruno tells his reader that he had stolen Copernicus's words and *monumenta*, that he had acquired them by means of an active appropriation in order to use them in a different context: his own. The Nolan, therefore, did not convert to the new doctrine as Saul of Tarsus did when he fell, blinded, on his way to Damascus or through a sudden and passive inspiration of a prophetic text. For Bruno, the passive reception of truths, a blind faith in what comes from above, is the distinctive mark of asses, not of Promethean philosophers. For the latter we are aware that truth resides inside all living and material beings who, like the innumerable stars of the infinite universe, are moved by an intrinsic and sentient force, a universal soul. These few verses from the *De immenso* are thus a good example of the many ways in which

---

valet intueri. Inventio particularium est veluti prima cibi apprehensio, collatio ipsorum in sensibus internis est veluti digestio quaedam, intellectus perfecta informatio est tandem cognitionum perfectionis nostrae in praesenti statu, ad quem, veluti ad animae virilitatem et perfectam consistentiam, omnes, qui natura scire desiderant, promoveri concupiscunt. Hoc habitu imbuti et istiusmodi corroborati alimento, in ea intelligimur esse dispositione, qua per artem et scientiam in operibus intelligentiae progredimur.'

62 In order to convey this idea Bruno adapted the myth of Actaeon in his dialogue *De gli eroici furori*: experiencing nature unveiled (Diana) means the annihilation of the individual (Actaeon, who is devoured by his own hunting dogs); the literature on Bruno's Actaeon is vast, but see Fellmann F., "Introduction", in Bruno Giordano, *Von den heroischen Leidenschaften*, trans. and ed. C. Bacmeister (Hamburg: 1989) XVIII–XL.

63 Bruno, *Immenso* 189–197. In the following chapter of the *De immenso* (III, 10) Bruno will criticise Copernicus.

Bruno elaborates his thoughts; his terminology is carefully chosen and evocative of a kaleidoscopic sequence of topics central to the *nolana filosofia*: that is, infinitist physics, the role of time and—last but not least—the embodiment of thought in different persons and by means of different organic features at different historical moments. As so often in his work, Bruno manages to create a tight literary space. We have shown how the few lines we have examined are capable of initiating an intensive speculative series of associations, of echoes from other texts. Bruno called such literary structures *idola* or statues; he intended them to be used as inner fields for invention, as spaces in which very different speculative movements may take place.[64] When we perform this act of appropriating their different meanings, such artefacts work like simulators for the monadic structure Bruno had postulated for the entire infinite universe. As we have seen, the passage from the *De immenso* distinctly echoes and with amazing succinctness summarises central ideas Bruno had elaborated in his earlier Italian texts in a great variety of different literary stiles and in another language. Other than in the *Cena* and in the *Della causa*, however, we do not find a report about the Nolan who is in a more or less antagonistic dialogue with Aristotelian (and other) pedants who fail to acknowledge the truth of the infinite universe and his prophet. In the *De immenso*, Bruno presents instead an image of himself to his readers (as well as to himself) that functions as a memory image for his relationship with Copernicus: the Nolano, the talking statue of a prophet, holding the *De revolutionibus* in his hands with a firm grip, reading out a few pages of a book he claims to have thoroughly appropriated.

Selective Bibliography

Bidwell-Steiner M., *Große Welt-kleine Welt-verkehrte Welt. Die philogyne Naturphilosophie der Renaissance-Denkerin Oliva Sabuco de Nantes y Barrera* (Innsbruck – Vienna: 2009).

Blum P.R., *Giordano Bruno* (Munich: 1999).

64 Bruno, *Lampas* 5–6 describes this form of invention and says that the emerging forms represent the special disposition of their operator, that is in the particular historic setting and chance. '[...] per formationem, quemadmodum est statuae (idoli) confectio ex cera eadem variis pressa modulis, ut ex eadem quidem nunc hominem, nunc equum, nunc avem, nunc vas etc. possimus informare. [...hoc inventionis species] potius naturae operanti vel casui vel fortunae assimilatur, cui proportionatur inventio, quae in locis verbalibus positivis explicatur.'

Blum P.R., "Brunos Brunianismus", in Heipke K. – Neuser W. (eds.), *Die Frankfurter Schriften Giordano Brunos und ihre Voraussetzungen* (Weinheim: 1992) 201–210.

Bolzoni L., *The gallery of memory: literary and iconographic models in the age of the printing press* (Toronto: 2001).

Bönker-Vallon A., "Introduction", in Bruno Giordano, *De l'infinito universo e mondi/ Über das Unendliche, das Universum und die Welten*, trans. and ed. (Hamburg: 2007).

Bruno Giordano, "Candelaio. Comedia", in idem, *Opere Italiane*, ed. E. Canone (Paris, G. Giuliano: 1582; reprint Florence: 1999).

Bruno Giordano, "Cantus circeus", in idem, *Opera latine conscripta*, ed. F. Tocco, 3 vols. in 8 parts (Naples – Florence: 1879–1891) vol. II/1, 179–257.

Bruno Giordano, "Cena delle Ceneri", in idem, *Dialoghi italiani, nuovamente ristampati con note da Giovanni Gentile*, 3rd edn., ed. G. Aquilecchia, 2 vols. (Florence: 1985) vol. 1.

Bruno Giordano, "De compositione imaginum", in idem, *Opera latine conscripta*, ed. F. Tocco, 3 vols. in 8 parts (Naples – Florence: 1879–1891) vol. II/3, 87–322.

Bruno Giordano, "De innumerabilibus, immenso et infigurabili", in idem, *Opera latine conscripta*, ed. F. Tocco, 3 vols. in 8 parts (Naples – Florence: 1879–1891) vol. I/1 (lib 1–3).

Bruno Giordano, "De triplici minimo et mensura", in idem, *Opera latine conscripta*, ed. F. Tocco, 3 vols. in 8 parts (Naples – Florence: 1879–1891) vol. I/3, 121–361.

Bruno Giordano, "Della causa, principio ed uno", in idem, *Dialoghi italiani, nuovamente ristampati con note da Giovanni Gentile*, 3rd edn., ed. G. Aquilecchia, 2 vols., (Florence: 1985) vol. 1.

Bruno Giordano, "Explicatio triginta sigillorum", in idem, *Opera latine conscripta*, ed. F. Tocco, 3 vols. in 8 parts (Naples – Florence: 1879–1891) vol. II/2, 121–160.

Bruno Giordano, *Il Candelaio. Chandelier*, trans. Y. Hersant, ed. and comm. G. Aquilecchia – G. Bàrberi-Squarotti (Paris: 1993).

Bruno Giordano, "Lampas triginta statuarum", in idem, *Opera latine conscripta*, ed. F. Tocco, 3 vols. in 8 parts (Naples – Florence: 1879–1891) vol. III, 1–258.

Bruno Giordano, *De gli heroici furori. Des fureurs héroïques*, trans. P.H. Michel, ed. and comm. M.A. Granada (Paris: 1999).

Bruno Giordano, "Spaccio della bestia trionfante", in idem, *Dialoghi italiani, nuovamente ristampati con note da Giovanni Gentile*, 3rd edn., ed. G. Aquilecchia, 2 vols. (Florence: 1985) vol. 2.

Bruno Giordano, *The Ash Wednesday supper*, trans. and ed. E.A. Gosselin – L.S. Lerner (Toronto: 2001).

Bruno Giordano, "Cabala del cavallo pegaseo", in idem, *Dialoghi italiani, nuovamente ristampati con note da Giovanni Gentile*, 3rd edn., ed. G. Aquilecchia, 2 vols. (Florence: 1985) vol. 2.

Ciliberto M., "Introduction", in Bruno Giordano, *De umbris idearum/Le ombre delle idee* (Milan: 1997).

Ciliberto M., *Umbra profonda. Studi su Giordano Bruno* (Rome: 1999).

Ciliberto M., *Giordano Bruno* (Rome: 1990).

Clucas S., "*Amorem, artem, magiam, mathesim*. Brunian Images and the Domestication of the Soul", *Zeitsprünge* 3 (1999) 5–24.

Clucas S., "Giordano Bruno's *De imaginum, signorum et idearum compositione*. Art, Magic and Mnemotechnics", *Physis. Rivista internazionale di storia della scienza* 38 (2001) 75–98.

Clucas S., "*Simulacra et Signacula*. Memory, Magic and Metaphysics in Brunian Mnemonics", in Gatti H. (ed.), *Giordano Bruno. Philosopher of the Renaissance* (Aldershot – Burlington: 2002).

Cox V., *The Renaissance Dialogue. Literary Dialogue in its Social and Political Contexts. Castiglione to Galileio* (Cambridge: 1992).

De Bernart L., *Immaginazione e scienza in Giordano Bruno. L'infinito nelle forme del' esperienza* (Pisa: 1986).

Fellmann F., "Introduction", in Bruno Giordano, *De gli heroici furori/Von den heroischen Leidenschaften*, trans. and ed. C. Bacmeister (Hamburg: 1989) VII–XL.

Ficino Marsilio, *Commentaire sur le Banquet de Platon. De amore*, ed. R. Marcel (Paris: 1956).

Ficino Marsilio, "Disputatio contra iudicium astrologorum", in *Supplementum ficinianum*, ed. P.O. Kristeller (Florence: 1937) vol. 2, 11–76.

Ficino Marsilio, *Theologia Platonica. Platonic theology*, ed. and trans. M.J.B. Allen et al., 6 vols. (Cambridge Mass. – London: 2001–2006).

Gatti H., "Copernico", *Bruniana e Campaneliana* 14 (2008) 511–520.

Gatti H., *Giordano Bruno and Renaissance science* (Ithaca: 1999).

Granada M.A., "L'héliocentrisme de Giordano Bruno entre 1584 et 1591: la disposition des planètes inférieures et les mouvements de la terre", *Bruniana e Campanelliana* 16 (2010) 31–50.

Granada M.A., *Giordano Bruno, universo infinito, unión con Dios, perfección del hombre* (Barcelona: 2002).

Granada M.A., *La reivindicación de la filosofía en Giordano Bruno* (Barcelona: 2005).

Hufnagel H., *Ein Stück von jeder Wissenschaft. Gattungshybridisierung, Argumentation und Erkenntnis in Giordano Brunos italienischen Dialogen* (Stuttgart: 2009).

Ingegno A., *Cosmologia e Filosofia nel pensiero di Giordano Bruno* (Florence: 1978).

Jorn O., "Corporeità in Bruno, senso e figura", *Bruniana e Campanelliana* 8 (2002) 159–180.

Kodera S., *Disreputable Bodies. Magic, Gender, and Medicine in Renaissance Natural Philosophy* (Toronto: 2010).

Kodera S., "Introduction", in Bruno Giordano, *Cabala del Cavallo Pegaseo/Die Kabala des pegaseischen Pferdes*, ed. S. Kodera (Hamburg: 2008) LXXV–XC.

Leinkauf T., "Introduction", in Bruno Giordano, *De la causa, principio ed uno/Von der Ursache dem Prinzip und dem Einen*, ed. T. Leinkauf (Hamburg: 2007).

Leinkauf T., *Mundus combinatus. Studien zur Struktur der barocken Universalwissenschaft am Beispiel Athanasius Kirchers SJ* (Berlin: 2009).

Miele M., "L'organizzazione degli studi dei Domenicani di Napoli al tempo di Giordano Bruno", in Canone E. (ed.), *Giordano Bruno. Gli anni napoletani e la 'peregrinatio' Europea* (Cassino: 1992) 29–50.

Nuzzo E., "Le figure metaforiche nel linguaggio filosofico di Giordano Bruno", in Meroi F. (ed.), *La mente di Giordano Bruno* (Florence: 2004) 13–60.

Ordine N., *Giordano Bruno und die Philosophie des Esels* (Munich: 1999).

Ordine N., *La soglia dell'ombra. Letteratura, filosofi a e pittura in Giordano Bruno* (Venice: 2003).

Papi F., *Antropologia e civiltà nel pensiero di Giordano Bruno* (Florence: 1968).

Pechriggl A., *Chiasmen. Antike Philosophie von Platon zu Sappho—von Sappho zu uns* (Bielefeld: 2006).

Rowland I., *Giordano Bruno. Philosopher/Heretic* (New York: 2008).

Saiber A., *Giordano Bruno and the Geometry of Language* (Aldershot – Burlington: 2005).

Schettino E., "The necessity of the *Minima* in the Nolan Philosophy", in Gatti H. (ed.), *Giordano Bruno. Philosopher of the Renaissance* (Aldershot – Burlington: 2002) 299–325.

Severini E.M., "Vicissitudine e tempo nel pensiero di Giordano Bruno", in: Meroi F. (ed.), *La mente di Giordano Bruno* (Florence: 2004) 225–258.

Sturlese R., "Il *De imaginum signorum et idearum compositione* di Giordano Bruno ed il significato filosofico dell' arte della memoria", *Giornale critico della filosofia italiana* 69 (1990) 297–306.

Sturlese R., "Per un'interpretazione del *De umbris idearum* di Giordano Bruno", *Annali della Scuola Normale Superiore di Pisa* 3, 22 (1992) 942–967.

Tessicini D., *I dintorni dell'infinito. Giordano Bruno e l'astronomia del cinquecento* (Pisa – Rome: 2007).

Tocco F., *Le opere latine di Giordano Bruno esposte e confrontate con le italiane* (Florence: 1889).

Védrine H., *La concéption de la nature chez Giordano Bruno* (Paris: 1967).

Yates F.A., *The Art of Memory* (London: 1969).

CHAPTER 10

"Copernicus Found a Treasure the True Value of Which He Did Not Know at All". The *Life of Copernicus* by Pierre Gassendi

Claus Zittel

1 Introduction[1]

'Of course everyone speaking of Copernicus has scribbled off from Gassendi [...]', Georg Christoph Lichtenberg remarks in a letter that heralds his own life of Copernicus. He hurries to add: 'and this is exactly what I will do also.'[2] Lichtenberg's unvarnished confession hits one fundamental point: Gassendi's relevance as a critical mediator of Copernicus's teachings cannot be assessed based on the quotas of citations but has to be deduced by the text itself. In view of the innumerable hagiographic clichés that stick tenaciously to Copernicus's name and establish his fame, looking at Gassendi's early biography provides the opportunity to see through some of the formative principles of the subsequent legends.

In 1654, 21 years after Galilei's conviction, Pierre Gassendi (1592–1655) published a collection of astronomers' lives, which also contains a vita of Copernicus.[3] In that year, Gassendi is at the peak of his fame and, at the same time, the end of his life. He worked coevally with Hobbes and Descartes and was regarded as their equal in the 17th century. Today, he is at best remembered

1 A German version of this article, focussed more on a functional analysis of the vitas, has been published in Enenkel K. – Zittel C. (eds.), *Die* Vita *als Vermittlerin von Wissenschaft und Werk* (Berlin: 2013) 123–156.

2 'Freilich haben alle, die von Kopernikus reden, den *Gassendi* ausgeschmiert, und das werde ich denn auch tun.' Lichtenberg Georg Christoph, *Schriften und Briefe*, ed. W. Promies (Munich: 1968–1992) vol. IV, 950. From a letter by Lichtenberg to Jeremias David Reuß on the occasion of Lichtenberg's *Copernicus biography* (ibid. vol. III, 138–188).

3 Gassendi Pierre: *Tychonis Brahei, Equitis Dani, Astronomorum Coryphaei, Vita* [...] *Accessit Nicolai Copernici; Georgii Peuerbachii & Joannis Regiomontani, Astronomorum celebrum, Vita* (Paris, Vidua Mathurin Dupuis: 1654); 2nd edn. (The Hague, Adriani Vlacq: 1654); 3rd edn. (The Hague, Adriani Vlacq: 1655). The 1654 edition only with the Copernicus frontispiece, the 3rd (most of the time erroneously regarded as the 2nd) edition now also has the Brahe frontispiece.

 DOI 10.1163/9789004281127_012

as the author of the fifth objection against Descartes' *Meditationes*, by some also as the refounder of atomism and epicureism. Gassendi's work, however, has also an essential share in the transformation and reformulation of natural philosophy in the 17th century, which is usually ascribed to Descartes.[4] In intellectual history, Gassendi is among those revealing figures, who appear like dwarfs to us today but were giants to their contemporaries. What is little known today is that Gassendi was also the leading astronomer of France, who had—among other things—been the first to watch the transit of Mercury, observed the moon and published numerous astronomical works.[5] For his merits in astronomy, a moon crater was named after him.

Gassendi's lives of famous astronomers is therefore the rare stroke of luck in the field of scientists' vitas that were not published by disciples or friends but come from the pen of a colleague of the highest repute in the same discipline. At the early age of sixteen, Gassendi became professor of rhetoric, a few years later of philosophy; in 1645, he was appointed professor of mathematics at the famous *College Royal*, which helped him to gain a strong international influence. He had a lot of interests and talents in fields as diverse as medicine, astronomy, physics, the study of antiquity, literature and languages. His mentor, Peiresc,[6] had introduced him to the circles of the humanist scholars, through his educational journeys and his network of letters he was in touch with the leading scholars of Europe. He was friends with Hobbes and Mersenne, exchanged letters with Schickhardt and Hevelius, was a priest and humanist scholar, provost and experimental scientist, materialist and sceptical natural philosopher. Above all, however, he also possessed a sense for historical research, which predestined him as an author of vitas. The vita of scientists as a genre was virtually reinvented by him: with his biographies he set new standards of the critical evaluation and explication of past scientific achievements.

4 Cf. Tack R., *Untersuchung zum Philosophie- und Wissenschaftsbegriff bei Pierre Gassendi* (Meisenheim: 1974); Detel W., *Scientia Rerum Natura Occultarum, Methodologische Studien zur Physik Pierre Gassendis* (Berlin: 1978).

5 Regarding Gassendi's astronomical activities cf. Humbert P., *L'oeuvre astronomique de Gassendi* (Paris: 1937); Zittel C., "Die Lunatiker von Aix-en-Provence: Peiresc-Mellan-Gassendi", in Feist U. – Rath M. (ed.): *Et in imagine ego. Facetten von Bildakt und Verkörperung* (Berlin: 2012) 276–300.

6 Regarding Peiresc's eminent impact cf. Miller P., *Peiresc's Europe: Learning and Virtue in the Seventeenth Century* (New Haven – London: 2000).

2 *Vitas* in the Historiography of Philosophy and Science

In the more recent history of science, scientists' biographies from previous centuries do not play a noteworthy role, the points of reference rather being the scientific protagonists' texts and experimental practices. They are completely ignored in the history of philosophy, which really ought to consider the same scientists. This is not surprising in that those biographies bear the odium of spreading idealised heroic stories, in some cases, e.g. Newton's, in undisguised ignorance also of known facts.[7] In addition, historiographers of philosophy mostly take the view that biographical circumstances do not matter, only the theory counts, liberated from of its historical ballast.

In art history, the genre of the vita has a significantly higher status than in other disciplines. Artists' *lives* are not only regarded as an indispensable source of information but also as a vehicle for the transport of theory. They were foundational to the field and inform it to this day.[8] In the scientific historiography regarding the early modern period, however, a genre that is essential for the formation, legitimisation, distribution and regulation of science is almost totally ignored. Even the testimony of the editors of the *Biographica* in the new complete Copernicus edition is symptomatic of this when they pass a judgement on Gassendi's Copernicus vita, edited by them: one had to add hardly anything to Franz Hipler's view from 1873, that is to say that Gassendi had to be given credit '[...] for diligently collecting the material that had been published until then and for having intertwined it with love, taste and a rich knowledge of the science of astronomy to give a well-ordered and thankworthy life image.' His vita, however, did not yield '[...] any heretofore unknown insights with regard to the biographical facts.'[9] I will demonstrate in the following that a

7 Cf. Iliffe R. – Keynes M. – Higgitt R. (eds.), *Early Biographies of Isaac Newton, 1660–1885*, 2 vols. (London: 2006).

8 Cf. Kliemann J., "Giorgio Vasari: Kunstgeschichtliche Perspektiven", in Ganz P. – Gosebruch M. – Meier N. – Wuttke M. (eds.), *Kunst und Kunsttheorie 1400–1900* (Wiesbaden: 1991) 29–74; Rubin P., *Vasari: Art and History* (New Haven: 1995); Verstegen I., "Birth Dates, Death Dates, and the Beginnings of Modern Art History", *Storiografia: Rivista annuale di storia* 10 (2006) 1–19; Davis C. – Feser S. – Lemelsen K. – Nova A. (eds.), *Die Vite Vasaris. Entstehung – Topoi – Rezeption* (Venice: 2010); Ketelsen T., "Die Künstler-Viten Giorgio Vasaris als Wissensform und als Wissenschaftssteuerung", in Enenkel K. – Zittel C. (eds.), *Die* Vita *als Vermittlerin von Wissenschaft und Werk. Form und Funktionsanalystische Untersuchungen zu frühneuzeitlichen Biographien von Gelehrten, Wissenschaftlern und Künstlern* (Münster: 2012) 331–344.

9 '[...] das bis dahin publizierte Material fleißig gesammelt und mit Liebe, Geschmack und reicher Kenntnis der astronomischen Wissenschaft zu einem wohlgeordneten und dankenswerten Lebensbilde verwebt [...] aber hinsichtlich der biographischen Fakten keine bis

vita can be more than just a provider of historical facts and must therefore be assessed according to different criteria.

There were many more such vitas in early modern times than is commonly known. Especially the philosophers' lives fell victim to collective oblivion.[10] Most of them have been forgotten for the simple reason that the earlier Latin editions that often contain them are not being read any more. The narrative patterns of those vitas have remained unexplored accordingly so that little is known to what extent they follow—for example—classical biographies[11] such as e.g. Plutarch's, Diogenes Laërtius', the *Pythagoras* and *Plotin* vitas by Porphyry, Lucian's parodies of philosophers' lives,[12] hagiography or artists' vitas;[13] or if exactly the opposite is true.

The ten typologies of vitas as suggested by Thomas F. Mayer and D.R. Woolf in their introduction to their collection of essays, *The Rhetorics of Life-Writing*,[14] show the way. There still is, however, a substantial need for differentiation

---

dahin unbekannten Erkenntnisse.' Kirschner S. – Kühne A., "Die Rezeption von Copernicus im Spiegel seiner Biographien", in Seising R. – Folkerts M. – Hashagen U. (eds.), *Form, Zahl, Ordnung. Studien zur Wissenschaftsgeschichte* (Stuttgart: 2004) 467–479, here: 472.

10 John Aubrey's biography of Thomas Hobbes (1680) is famous; Fontenelle writes the vitas of Leibniz and Newton as eulogies. Le Bovier de Fontenelle Bernard, *The Life of Sir Isaac Newton; with an Account of his Writings.* (London: 1728); URL= <http://www.maths.tcd.ie/pub/HistMath/People/Newton/Fontenelle/Fonten.html>; "Lebens-Beschreibung des Herrn von Leibnitz verfaßt von dem Herrn Fontenelle", in Leibniz, *Monadologie*, ed. D. Till (Frankfurt a.M. – Leipzig: 1996). Gassendi's life was summed up in several vitas, e.g. by Sorbière. Cf. Taussig S. – Turner A.J., *Mémoire de Gassendi: vies et célébrations écrites avant 1700* (Turnhout: 2008).

11 Regarding the philosophers' biographies in antiquity cf. Friedrich L., *Die griechisch-römische Biographie nach ihrer litterarischen Form* (Leipzig: 1901); Momigliano A., *The Development of Greek Biography* (Cambridge Mass. – London: 1993); Schirren T., *Philosophos Bios. Die antike Philosophenbiographie als symbolische Form. Studien zur Vita Apollonii des Philostrat* (Heidelberg: 2005). For the time before 1700, Karl Enenkel has recently examined the biographical and autobiographical literature; for the 17th century, a comparative study is still missing. Cf. Enenkel K., *Die Erfindung des Menschen. Die Autobiographik des frühneuzeitlichen Humanismus von Petrarca bis Lipsius* (Berlin: 2008).

12 Manetti Gianozzo, *Vite di Socratis e Senecae*, ed. A. de Petris (Florence: 1979).

13 Regarding the structure and the narrative patterns especially of Vasari's vitas, research has been increasing lately. Cf. on this especially Blum G., "Kunstgeschichte als 'große Erzählung' und Bildsystem. Zur Geschichtstheologie von Vasaris Vite (1550)", in Ganz D. – Thürlemann F. (eds.), *Das Bild im Plural. Mehrteilige Bildformen zwischen Mittelalter und Gegenwart* (München: 2010) 271–288. Verstegen I., "Vasari's progressive (but non-historicist) Renaissance", *Journal of Art Historiography* 5 (2011) 1–19.

14 Mayer T.F. – Woolf D.R., *The Rhetorics of Life-Writing in Early Modern Europe. Forms of Biography from Cassandra Fedele to Louis XIV* (Ann Arbor: 1995). On this, critically, cf. the introduction to Enenkel – Zittel (2013).

because it is exactly Gassendi's vitas that do not fall into line with any of the patterns presented there. These biographies differ perceptibly from their classical and humanist precursors not only by eschewing fictional and apologetic elements but also—as will subsequently be demonstrated—their strict reference to research, their doxographical orientation, their critical processing of the historical tradition and their independence of judgement.

Scientists' lives, however, were not only prefixed to work editions or presented united in collections of vitas; some—as for example Descartes' vita by Baillet,[15] Gassendi's wonderful vita of Peiresc (1641),[16] which was immediately translated into English, or his voluminous biography of Epicurus,[17] which had been reconstructed on the basis of classical sources—were edited singly. All this testifies to the fact that such scientists' vitas are by no means a parasitical genre that only accompanies someone else's scientific work but that they are independent research contributions with a scientific claim to themselves. Not least Gassendi's astronomers' lives attest how powerful such vitas could be; they were edited for the second time the very same year, for the third in the year after.

3 Gassendi's *Lives of Astronomers*

As is easily perceivable on the title page, Gassendi's astronomers' lives on the contrary were published together, as a collection of four vitas, namely Brahe's, Copernicus', Peurbach's and Regiomontanus'. Brahe's life is indeed highlighted with giant letters as the main piece, the other vitas rank as add-ons below.

15 Baillet Adrien, *La vie de monsieur Des-Cartes*, 2 vols. (Paris, Daniel Horthemels: 1691).

16 Gassendi Pierre, *Viri illustris Nicolai Claudii Fabricii de Peiresc, senatoris aquisextiensis vita* (Paris, Sébastien Cramoisy: 1641. More recent editions: *Vie de Tycho Brahé, Copernic, Peurbach et Regiomontanus*, trans. J. Peyroux (Bordeaux: 1996); *Vie de l'illustre Nicolas-Claude Fabri de Peiresc, conseiller au Parlement d'Aix*, trans. R. Lassalle (Paris: 1992); *Vita Epicuri, Peireskii, Tychonis Brahei, Copernici, Peurbachii, & Regiomontanio. Contenu dans Opera omnia*, vol. v. (Lyon, Laurent Anisson – Jean-Baptiste Devenet: 1658); *Vie de Copernic par Gassendi*, trans. M. Thirion (Paris: 1973).

17 Gassendi Pierre: *De Vita et Moribus Epicuri* (Lugduni, Guillelmi Barbier: 1647); *De vita et moribus Epicuri: libri octo* (The Hague, Adrianum Vlacq: 1656); *Vie et moeurs d'Epicure* (french-latin), ed. S. Taussig (Paris: 2006). On this: Taussig S., "*La vie d'Épicure* par Gassendi", in Paganini G. – Tortarolo E. (eds.), *Der Garten und die Moderne, Epikureische Moral und Politik vom Humanismus bis zur Aufklärung* (Stuttgart Bad Canstatt: 2004) 139–161; idem, "Epicurus Redivius By Pierre Gassendi: Biography Opposed to the Love of Self", in *Intellectual News* 11 (2003–2004) 19–27.

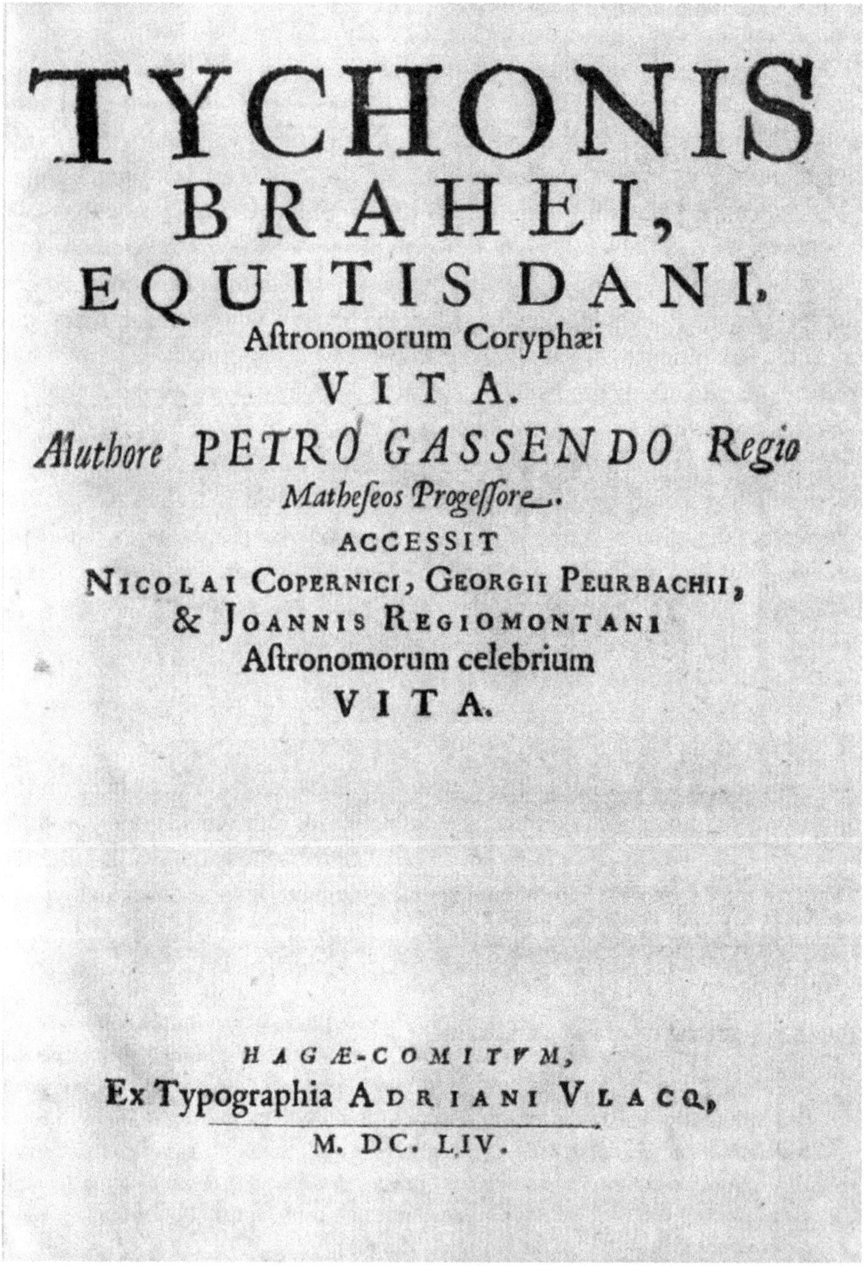

TYCHONIS
BRAHEI,
EQUITIS DANI.
Aſtronomorum Coryphæi
VITA.
Authore PETRO GASSENDO *Regio*
Matheſeos Profeſſore.
ACCESSIT
NICOLAI COPERNICI, GEORGII PEURBACHII,
& JOANNIS REGIOMONTANI
Aſtronomorum celebrium
VITA.

HAGÆ-COMITUM,
Ex Typographia ADRIANI VLACQ,
M. DC. LIV.

FIGURE 10.1 Gassendi Pierre, *TYCHONIS BRAHEI, EQUITIS DANI, Astronomorum Coryphaei, Vita.* First edition: Paris, Vidua Mathurin Dupuis: 1654.

Nevertheless, all four vitas would have to be examined in detail and questioned as to their mutual relationship. Here, I can but hint to some differences and will then concentrate on the Copernicus vita, which is relevant for the present volume.

Generally speaking, we have here four lives of astronomers who stood for differing world systems. Written at a time when, since Galilei's conviction, those theories had been fiercely debated, these lives at the same time present a political issue, depending on the respective astronomical theory and how it is exposed. Accordingly, these vitas to a lesser degree deal with the lives and doctrines of the scientific personalities they pay tribute to, but rather present a critical processing of their theories for the discussions that were then topical in science. To be more precise: These lives of astronomers are a cryptic apology for Galilei. They examine the sources critically, convey knowledge and at the same time enact a skilful camouflage under the guise of which also delicate material can be presented and evaluations insinuated.

In the lives of astronomers, Gassendi goes back to the cosmographical expositions from his *Institutio astronomica*[18] of the systems of Ptolemy, Copernicus and Brahe and embeds them in longer biographies. A historic-genetical exposition therefore replaces the purely systematic one. The choice of a different format of exposing the material does therefore not mean that the explanation of the world systems in a new dress will be addressed more plausibly but the mode of explaining these systems changes.

4 The Brahe Vita

The Brahe vita is clearly more comprehensive than the other three astronomers' lives; it was completed in 1654 as the last vita and it brings 'new facts' because—as was the case with his great Peiresc vita—Gassendi could rely on unpublished material, letters, documents and reports from contemporary witnesses. It is regarded as the first comprehensive vita of a scientist in early modern times but unjustly so because as early as 1642 Gassendi had published his big *Vita of Peiresc* who also had been an eminent scientist and astronomer. The lives of Brahe and Copernicus contain diagrams that were taken from Gassendi's *Institutio Astromica* (1647); both are furnished with frontispieces by Jacob van Meurs, but not so the last double vita.

18 Gassendi Pierre: *Institutio astronomica Iuxta Hypotheseis tam Veterum, quam Copernici, et Tychonins, Dictata à Petro Gassendo Regio Mathesos Professore. Eiusdem Oratio Inauguralis iteratò edita* (Paris, Louis de Heuqueville: 1647).

FIGURE 10.2 Frontispiece to Gassendi Pierre, *TYCHONIS BRAHEI, EQUITIS DANI, Astronomorum Coryphaei, Vita*, 3rd edition (The Hague, Adriani Vlacq: 1655). Copper engraving.

Single important turning points in Brahe's life are not described purely biographically but starting from the situation of scientific problems; thus, the readers can reconstruct the path of thoughts, e.g. when in Book 4 on the basis of the discussion with Rothmann about his cannonball experiments (1590), Brahe's objections are explained that subsequently lead him to his compromise world system.[19] The vita ends in a comprehensive depiction of Brahe's last days, apparently in order to put a light to the unclear circumstances of his death. It is noteworthy that Gassendi is the only one to highlight the achievements and skills of Brahe's sister Sophie whose collaboration in Brahe's observatory is otherwise neglected by the historiography of science.

In the 1655 print, which appeared as the second, corrected edition but really is the third, the six books of the Brahe vita span 221 pages.[20] Together with the preface, an appendix of forty pages that presents an oratorio, panegyric poems and further material, and the detailed subject index, the vita comprises roughly 350 pages. The vita of Copernicus in the same print covers around forty pages; it is significantly shorter. Both pre-Copernicans, Georg Peurbach (1421–1463) and his pupil Regiomontanus (Johannes Müller, 1436–1476) are dealt with in a double vita of forty pages because they collaborated closely and still propagated the Ptolemaic system in their teaching books. Here, their significance as forerunners of the Copernican system and their improvements of astronomical instruments is emphasised. Gassendi especially commends the exactness of Regiomontanus' observations. Thus, Gassendi's collection of lives presents advocates of three different world systems in unequal elaborateness. One further striking difference between the vitas is that the more comprehensive ones, regarding Brahe and Peiresc, are arranged chronologically, visibly marked in the margin with continuous years beside the text so that a second outer frame is constituted which at a glance reveals exactly when the events described happened and in which years a lot of things came to pass. The shorter vitas obviously do not require such guidance.

19 Cf. Gassendi, *Tychonis Brahei, Equitis Dani, Astronomorum Coryphaei, Vita* (Paris, Vidua Mathurin Dupuis: 1654) 111–112.

20 In the 1654 print, the vitas together have 496 pages. The Brahe vita stretches over 384 pages, out of which 54 pages are taken up by the Praefatio, the actual vita of Brahe is 304 pages long, followed by an index. The Copernicus vita has only 50 pages, starting with a new pagination. The vitas of Peurbach and Regiomontanus are continued with the same page count until page 110, followed by a joint index of twelve pages. The 1655 edition, however, presents all the vitas' texts with a consolidated continuous page count, thus emphasising the unity of the collection more strongly.

With a reference to the life of Brahe, Gassendi is sometimes alleged to have been a supporter of the Tychonian system,[21] but in the respective vita there is not a single piece of evidence to this effect. The Brahe vita could easily have been published on its own, the addition of further lives can therefore be seen as a corrective.

The introduction is preceded on sixty pages by a history of astronomy, which is doubly remarkable; on the one hand because we are here looking at an early form of the historiography of science, on the other hand because for Gassendi, as opposed to tradition, astronomy is a historical discipline that should collect observation data from earlier centuries also with the help of philology in order to gain an ever-widening basis of reliable hypotheses about the stars' orbits. Gassendi first presents the Ptolemaic, then the Copernican world system, albeit not as the climax of the development but with the comment, it would also be advocated by those, 'qui sequi aut Ptolemaicam non sustinent, aut Copernicaeam non audent.'[22] Repeatedly, critical clues are interspersed, e.g. by quoting objections of Kepler's. Above all, however, Gassendi points out that it was neither Copernicus nor Galilei but in fact Kepler, who described the planetary orbits as ellipses and consequently ascribes to him the real astronomical innovations. Additionally, the life of Brahe contains detailed explanations of Brahe's experiments, his measured data and theories.[23] Brahe is praised as the best and most exact observer whose data represent a degree of exactitude that sets the standards for centuries to come.

This vita is far from amounting to nothing more than a mere biography; it is a research report, explains backgrounds, discusses the hypotheses, examines alternatives and arrives at a well-balanced overall image of Brahe's historical achievements.

---

21 E.g. Lange F.A., *Geschichte des Materialismus* (Iserlohn 1866, reprint Frankfurt a.M. 1974) vol. I, 242; Fisher S., "Pierre Gassendi", *The Stanford Encyclopedia of Philosophy* (Winter 2009 Edition), Zalta E.N. (ed.), URL = <http://plato.stanford.edu/archives/win2009/entries/gassendi/>.

22 Gassendi, *Tychonis Brahei, Equitis Dani, Astronomorum Coryphaei, Vita* [...] *Accessit Nicolai Copernici, Georgii Peuerbachii & Joannis Regiomontani, Astronomorum celebrium,* 3rd edn, book V (The Hague, Adriani Vlacq: 1655) 383.

23 Kirschner S. – Kühne A., "Die Rezeption von Copernicus im Spiegel seiner Biographien", in Seising R. – Folkerts M. – Hashagen U. (eds.), *Form, Zahl, Ordnung. Studien zur Wissenschaftsgeschichte* (Stuttgart: 2004) 467–479.

5 The Life of Copernicus

Irrespective of its brevity, this also applies to the Copernicus vita[24] on the basis of which I want to present some of the various functions that a vita by Gassendi has to fulfil.

Gassendi was—like his friend and patron Peiresc—an early factionist and defender of Galilei. Initially, Gassendi openly avowed himself to Galilei, out of consideration for his clerical office as well as for his person, formulating cautiously and indirectly after the latter had renounced his theory in the wake of the inquisition trial. Gassendi's style changes, it becomes more refined, tactically adroit, ambiguous and ironic, constantly leaving it in abeyance if there is irony or not. Sanctimoniously, Gassendi asks for example, 'if it might be permitted on this occasion to reprint the eulogy'[25] which (the Jesuit) Clavius had delivered on Copernicus, generally with delight quoting positive appraisals coming from the mouths of renowned scholars of the catholic church from the time before the indexing. Also, he does not omit the church's appreciation of Copernicus's calendar reform and declares innocently that it 'may well be worth'[26] including the enthusiastic letter by cardinal Schönberg about the Copernican system. When he mentions criticism at all, he puts it into the mouths of Protestants like Osiander or the Huguenot Ramus.

While Descartes e.g. cannot imagine the truths that the natural light of reason reveals as contradicting the divine truth, things are easier for the sceptic Gassendi because to him the natural light of reason only opens preliminary insights. As a result, the position that Gassendi can officially take in the contemporaneous struggles about faith and knowledge is—outlined in a simplified way—that religious dogmas and tenets of knowledge belong into different areas. Exactly because he would believe in the immobility of the earth, he did not need any justification for it and as long as he was certain in his belief he could hypothetically toy with all kinds of possibilities on the level of knowledge. This attitude was not without risk, for this dalliance might produce results that disproved the position of faith. Indeed, Gassendi had earlier on with his experiments with bodies dropping from a ship's mast refuted a central

24 Below, the *vita* is referred to the new edition in Kühne A. – Kirschner S. (eds.), *Biographia Copernicana. Die Copernicus-Biographien des 16. bis 18. Jahrhunderts* (Berlin: 2004) 74–106. This volume includes also a German translation: "Das Leben des Kopernikus, Domherren im Ermland, des berühmten Astronomen", ibid. 107–149.

25 'In hac vero occasione licebit Elogium heic adscribere [...].' Gassendi, *Biographia Copernicana* 88.

26 '[...] esse operae-pretium potest.' Ibid. 88.

proposition of Aristotle's and demonstrated a further argument in favour of the earth's motion.

Gassendi had openly expounded his official position in innumerable letters; on the other hand, his scientific contribution in support of Copernicanism was well known. He could presuppose this when he started drawing up his *Life of Copernicus.*

For an opening, Gassendi skilfully describes the reason for the vita's composition. It was the fruit of a serene innocent request and at the same time it expressed a common wish; consequently, it was exactly the opposite of a partial pamphlet: Gassendi's friend, the poet Jean Chapelain, paying a visit, had cast a glance on the proofs of Gassendi's Tycho Brahe biography and thereby caught a glimpse of some of Tycho's verses about Copernicus, which amused him greatly. On this occasion he asked Gassendi to supplement Brahe's life with a Copernicus vita, 'although for his name had become uncommonly renowned, hardly anyone far and wide knew what kind of a man he had really been.'[27] Moreover, the books that gave account of him were rare and at best known to but a few. Copernicus was famous in the 17th century but hardly anyone had closer knowledge about him and reliable information was scarce.[28] Gassendi declined the request at first but since the printing of the Brahe vita was delayed 'in the face of a considerable shortage of professionals in the printing trade' he made use of the time.[29]

---

27 'scilicet, cùm illlius nomen celebre adeò euaserit, quis tamen, aut qualis vir fuerit, longè, latéque ignorari.' Ibid. 74.

28 There were earlier vitas of Copernicus, e.g. a compilation of sources by Melchior Adam, and short vitas of Peuerbach and Regiomontanus from Georg Tannstetter's pen; a critical revision of the whole material did not exist up to then. Cf. Adam Melchior, *Vitae* (Frankfurt a.M. – Heidelberg: 1615–1620) – all vols. online: URL= <http://www.uni-mannheim.de/mateo/camenaref/adam.html>. On this: Seidel R., "Melchior Adams 'Vitae' (1615–1620) und die Tradition frühneuzeitlicher Gelehrtenbiographik: Fortschritte und Grenzen eines wissenschaftlichen Paradigmas um 1600", in: Kozielek G. (ed.), *Oberschlesische Dichter und Gelehrte vom Humanismus bis zum Barock* (Bielefeld: 2000) 179–204; Werle D., "Melchior Adams Gelehrtenbiographien und ihr Bezug zur Enzyklopädistik", in Schierbaum M. (ed.), *Enzyklopädistik 1550–1650: Typen und Transformationen von Wissensspeichern und Medialisierungen des Wissens* (Berlin et al., 2009) 105–125. Tannstetter Georg, "Viri mathematici", in: idem (ed.) *Tabulae eclypsium magistri Georgii Peurbachii; Tabula primi mobilis Ioannis de Monteregio* (Vienna: 1514) aa3v–aa6v. Transl. in Graf-Stuhlhofer F., *Humanismus zwischen Hof und Universität. Georg Tannstetter (Collimitius) und sein wissenschaftliches Umfeld im Wien des frühen 16. Jahrhunderts* (Vienna: 1996) 156–171.

29 Gassendi, *Leben des Kopernikus* 107.

This opening, which distances itself from all apologetics, contradicts the common expectation that through a life of Copernicus a scientific hero was to be enthroned, a founding legend of the new astronomy to be written and a central theme of canonisation was concurred. As became clear very quickly, Gassendi's Copernicus image differs largely from the glorifying hagiographies of the 19th century and the fictional eulogies of earlier and later biographies of scientists.[30] Astonishingly, it rather basically resembles contemporary reconstructions of the history of science. There are multiple reasons for this. On the one hand, the *Vita* is directed at specialists, it was written in Latin and larded with astronomical and mathematical details. Although written much better than many eulogies on scientists today, Gassendi's vitas were not introductory but went far beyond mere biographies in that they presented research in the historical context of the discipline as well as the effective history of the work. The vitas circulated in the hands of the experts, and immediately after the publication of the lives of astronomers, Gassendi sent two copies to Hevelius and exchanged observation data in an accompanying letter. The lives consequently also served the purpose to forge strategic scientific alliances.

The Copernicus image of later times was designed and cemented rather by laymen who did not read beyond the first pages of *De Revolutionibus* and took it for granted that Copernicus had honoured all the claims he had made there. Not so Gassendi; he was in a position to comprehend the mathematical and empirical deliberations and to evaluate them in the light of the subsequent developments in astronomy. His evaluation is matter-of-fact throughout. Nowhere does he claim that Copernicus's system had simplified astronomy, nowhere a better agreement of his theory with the observation data is mentioned; on the contrary, in the historical passages, much is added and silently corrected in the knowledge how little Copernicus's data were exact compared to those of Brahe and Kepler.

This raises a couple of interesting questions not only regarding today's Copernicus research but also in view of what the life of a scientist of such calibre was able to achieve. In order to assess the vita's function better, it will be necessary to first explain its structure and then elaborate on the philosophical and historical scientific context.

30 Cf. on this Roling B., "John Aubrey und die Philosophenviten des 17. Jahrhunderts", in Enenkel K. – Zittel C. (eds.), *Die Vita als Vermittlerin von Wissenschaft und Werk. Form- und Funktionsanalytische Untersuchungen zu frühneuzeitlichen Biographien von Gelehrten, Wissenschaftlern, Schriftstellern und Künstlern* (Münster: 2013) 207–236.

The Copernicus vita can be subdivided into thirteen segments, which—according to the dedicatory text to Jean Chapelain, the dearest friend—follow this structure:

First of all (1), the occasion which led to the composition of this vita (107–108) is described, then (2) Copernicus's biography and his studies. Copernicus is presented as scientist, artist, physician, cartographer and economist, his travels, important friends and astronomical observations are portrayed. Having looked around in the world, he finds time for his studies by retreating into seclusion. Copernicus, that is the bottom line of the sketch of his character, was talented for the vita activa and made for the vita contemplativa.[31] Then the impetus for research is named (pp. 108–111). Gassendi uses the depiction of the universal attainments of Copernicus (3), who had studied all books of astronomy, to elaborate on the history of astronomy and, against its backdrop, to introduce the readers to the Copernican world of thought (111–114). After this outline of the basic ideas, Gassendi goes into detail. An explanation of the Copernican hypotheses follows here (4) and of his instruments as well as a disclosure of his observation data (pp. 114–116). The following sections (5–7) are dedicated to *De Revolutionibus*, Copernicus's chief work: First of all, Gassendi gives a general depiction of the main thesis from book I on the basis of the observations disclosed under (3) (116–120), then he expounds books II–VI (pp. 120–123) and eventually the posthumous extensions (123–124). Subsequently (8), the appraisals by Copernicus's contemporaries and successors of his achievements are quoted (124–136), then (9) the publication history and its circumstances are reported (130–136), the Copernicus portraits (10) are described (136–138), his character (11) is sketched out (138–141) and eventually (12) his death and posthumous reputation are described (141–143). The vita concludes (13) with an account of the fate of the Copernican work (143–149).

---

31 Hans Blumenberg in his standard work, *Die Genesis der kopernikanischen Welt* (Frankfurt a.M.: 1975) 310–312, under the chapter heading "Der Theoretiker als Täter" ('the theorist as culprit') has conjured up the image of Copernicus the perpetrator, who makes the sky stand still and the earth move. For this purpose, Blumenberg quotes John Donne and discusses this formula in the context of heretical notions as a blasphemous act. Gassendi's Copernicus vita, which Blumenberg passingly mentions, does not support this line because on the one hand, numerous panegyric poems are quoted, which emphasise that the heliocentric theory had been introduced contra naturae iura, i.e. against the evidence of sensual perception, and thereby commended Copernicus's perpetration explicitly. On the other hand, these eulogies are not found in the scientific section of the vita but in the character sketch. Thus Gassendi distinguishes between different levels of impact, which become blurred in Blumenberg's generous perpetrator profile.

It can be seen from this first stage of going through the vita, that Gassendi does not only explain what position Copernicus took, and when; but he describes the circumstances and conditions in which he arrived at his theses, what kind of special education and knowledge he brought with him, what observations he made and what alternative theories and systems he had at hand in order to integrate his observation data, and why he then decided in favour of his system. Gassendi's sense of historical consideration must be emphasised; almost always there is an introductory historical outline of the most important opinions on a problem. The vita's narration herein corresponds to his understanding of astronomy; for Gassendi, astronomy is not (as it used to be in antiquity) a static science examining the immutable course of the stars but a science based on experience that requires as many observations as possible in order to formulate on their basis hypotheses that future generations can supplement and interpret in the light of better theories – astronomy thus is continuously progressing. Astronomy is also a historical and philological science supposed to collect with the help of philology the records and observations from past centuries to widen the basis of comparison.

Today, one of the focal questions in the discussion around *De Revolutionibus*, sc. with what consequences Osiander through his preface diminished Copernicus's tenet to a mere hypothesis,[32] is therefore not a relevant question to Gassendi. When Gassendi refers to astronomical hypotheses as merely probable assumptions, this only corresponds to what in his view applies to scientific assumptions generally. What is more, he clearly sees that Osiander's preface did not correspond to Copernicus's purposes; it had been Osiander's intention

> to protect Copernicus – although he did not regard the movement of the earth merely as a hypothesis but as a doctrine that matches the truth – from those who would take offence at this claim, by weakening it to the effect as if Copernicus had not advocated the said motion as a doctrine but had assumed it simply as a hypothesis.[33]

It had been Kepler who realised that Osiander was the author of the preface; as late as in the 19th century, however, there were erudite historians of astronomy

32 On the discussions of the Copernican system cf., Westman R.S., *The Copernican Question: Prognostication, Skepticism, and Celestial Order* (Berkeley: 2011).

33 'Copernicus Motum Terrae habuisset, non solùm pro Hypothesi, sed pro vero etiam placito; ipse tamen ad rem, ob illos, qui heinc offenderentur, leniendam, excusatum eum faceret, quasi talem Motum non pro dogmate, sed pro Hypothesi mera assumpsisset;' Gassendi, *Biographia Copernicana* 96.

who thought that Copernicus had written it himself. For Gassendi, however, Copernicus's intention is not an issue; he is only interested in the question if the theory formulated by him can make a claim to be more than a hypothesis. Slyly, Gassendi has a Protestant soften the Copernican tenet. The subtext, consequently, runs like this: the Protestants weakened Copernicus's doctrine, the catholic canon believed in it as a truth.

By delineating Copernicus's path of thinking, Gassendi attempts to make it comprehensible how Copernicus came about to believe in his doctrine as a truth. In so doing, it suits him that he can fashion the short vita's setup more freely and fashion the single sections in their central parts according to his own conception of science.

In contrast to Gassendi's Epicurus vita, *The Life Of Copernicus* does not contain an apologetic part that would answer to accusations and slander. The exposition of diverse valuations pretends to be documentary, Gassendi's own point of view is veiled; appraisal is always quoted as coming from others, especially from high authorities of the church. Accordingly, one has to be cautious quoting Gassendi—and not only him—, whishing to distil no matter what doctrines from his writings; whoever knows to read correctly, however, realises clearly his favouring of the Copernican system. I will touch on the few reservations later.[34]

---

34 The terse commentary on the Copernicus *Biographica* in the Collected Works claims far too certainly that Gassendi had been a Copernican. Cf. Kirschner – Kühne, *Biographia Coperniana* 75. Generally, it is hard to identify Gassendi's exact position in almost all his writings, for ample parts are presentations and scholarly and polemical discussions where he hardly ever speaks in his own name. Conceivably, his preferences lie with a moderate scepticism according to which, by means of the observation of nature, the mind can reach plausible assumptions about the truths hidden by nature. Gassendi's physics foregoes the cognition of the essentialities, takes its orientation on the visible appearances and, by taking its starting point from there, reaches plausible assumptions on the causes of the phenomena whereat the whole setting of alternatively possible views is presented. These are excellent premises for the composition of a reflected vita of a scientist. Also Gassendi's gigantic work, *Syntagma philosophicum*, published posthumously in the *opera omnia*, does not yield unambiguous evidence. In section 2, book 3, Gassendi discusses the astronomical hypotheses, explicitly taking a stand in favour of the Tychonian system. At the end and most comprehensively, the Copernican hypothesis is presented and a lot of space is dedicated especially to its further development by Galilei and Kepler, whereby one can gather a certain sympathy but certainly not a vote. Gassendi emphasises the importance of the further development of astronomy by Copernicus, additionally stressing the strong arguments that speak in its favour and naming the weaknesses of the objections. Moreover, he underlines he advantage of simplicity and inner harmony, however maintaining his vote for Brahe in this text. Clearly, Bruno's theses about the

Interestingly, Gassendi does not acknowledge Copernicus as a revolutionary but places him in a line with Ptolemy. When Gassendi, instead of speaking of ruptures and revolutions, elaborates the continuity from the ancient world systems up the Copernican one, he does so not only out of caution towards the Inquisition but also for epistemic reasons. Gassendi could elegantly realise his proposition to honour a scientist on the index, whose positions he considered dated for scientific reasons. Vis-à-vis the church he protected himself additionally by consequently treating the Copernican system as a hypothesis and through the documentary attitude; scientifically, he adds corrections and presents more recent developments in astronomy. These, however, needed to be introduced with more caution. In 1616, *De Revolutionibus* was put on the index with the cue *Donec corrigatur* ('until it be corrected'); such a formula was omitted in the case of the indexing of Galileo's *Dialogo* in 1633. Gassendi, however, did not shy away from addressing Galileo's trial about the heliocentric doctrine:

> It would be appropriate in this place to name some eminent persons from the crowd of its adherents but on the one hand these are sufficiently known... and on the other hand they wish to rather not be named so as not to give the impression that they had but little respect for the decree by which the holy congregation of cardinals is said to have disapproved this doctrine for the Inquisition twenty years ago in the case of Galileo [...] about which Riccioli recently [...] has composed a defence statement.[35]

Gassendi was a follower of central tenets of Galileo's; but it was not equivalent to be an adherent of Galileo and of Copernicus, and with regard to Galilei the formulations had to be more cautious. It was a means to this end to smuggle in veiled hints to the images that accompanied the vitas.

What is remarkable here in the first place is that Gassendi at the very beginning of his biography emphasises Copernicus's proximity to painting and here also reports of the legend of his self-portrait:

---

universe's infinity and the multitude of worlds are rejected, which, however, are preserved through their rejection (similarly also in the Brahe vita).

35 'Memorandi heic aliqui praecipui ex eius Sectatoribus viderentur, sed partim ij sunt abundè perspecti [...] partim nominari detrectant, ne esse videantur parùm reuerentes decreti, quo sacra Cardinalium Inquisitorum congregatio illam improbasse ante annos viginti in Galilaeo dicitur; & de quo [...] Ioannes Baptista Ricciolus Tractatum, Apologiae instar [...]'. Gassendi, *Biographia Copernicana* 107f.

> Even though he cared much for all sections of mathematics, he nevertheless occupied himself especially with the theory of perspective and on this occasion learned how to paint, in which field he progressed at least so far that it is said he had also painted himself excellently with the aid of his mirror image. The decision to deal with painting at all, however, comes from his intention to undertake an educational journey, mainly to Italy, and his aim to elaborate everything on this journey that seemed noteworthy to him not only in broad strokes but rather, as best he could, graphically exactly.[36]

This obvious styling of the scientist as an artist and the question that is raised with it, i.e. if there are interdependences between the lives of artists and scientists, would require a separate examination. For not only with regard to Galilei, as Horst Bredekamp[37] has recently repeated indefatigably, but also for the biographies of his precursors it was an important factor that scientist be artistically active. Gassendi emphasises this also for Tycho Brahe, extensively quoting from his Copernicus poem. Gassendi himself enjoyed the fame of being the best philosopher among the littérateurs and the best littérateur among the philosophers.[38] In any case, this is the backdrop against which Gassendi turns to the portraits of Copernicus. Accordingly, he frames them with a characterisation of Copernicus after the humanist ideal of the *uomo universale*: Copernicus was good, philanthropic, erudite, 'exceedingly versed in all branches of science', from numismatics and astronomy to medicine (a second Asclepius), worldly wise and philologically well grounded in the ancient languages.[39]

Let us now take a glance at the frontispiece:

36 'Cùm parteis verò omneis Matheseos curaret, tum Perspectiuae speciatim incubuit, eiusque occasione Picturam tum addidicit, tum eò vsque calluit, vt perhibeatur etiam se ad speculum eximiè pinxisse. Consilium autem pingendi ex eo cepit, quòd peregrinationem, ac potissimùm in Italiam cogitans, in animo haberet, non modò adumbrare, sed graphicè etiam, quantum posset, exprimere quicquid occurreret obseruatu dignum.', ibid. 75.

37 Cf. Bredekamp H., *Galilei der Künstler – Die Zeichnung, der Mond, die Sonne* (Berlin: 2007).

38 'le meilleur philosophe des litterateurs, et le meilleur litterateur des philosophes.' Edward Gibbon, *Essai sur l'étude de la littérature* (Dublin, Guillaume Hallhead: 1777) 13.

39 Gassendi, *Biographia Copernicana* 138f.

FIGURE 10.3 Frontispiece to Gassendi's *Nicolai Copernici Varmiensis Canonici, Astronomie illustris Vita*. Copper engraving.[40]

---

40 Subscriptio: 'Non docet instabiles Copernicus ætheris orbes, Sed terræ instabiles arguit ille vices.' The epigram and the subscriptions to the images are taken from Boissard: 'Nicolaus Copernicus from Thorn in Prissia, mathematician, born in the year 1473, died 1543. Not the erratic circles of the ether does Copernicus teach, but exposes the volatile changes of the earth.'

When touching upon Copernicus's physiognomy, Gassendi explains his frontispiece as follows:

> Now that there would be something to say about his face and his build, there is nothing left to us but to represent his countenance in the same way as it can be found depicted in Boissard among the copperplate images of famous men. I have only had added to the painting a fur-trimmed cloak of the fashion as our Bulliardus saw him, according to his report, in the image which can be seen in Strasbourg, namely in the church on the famous mechanical clock, the image which the splendid Bernegger, who had a portrait procured from Prussia thirty years ago, wanted to have reproduced on the title page of the Galilean dialogues.[41]

What Gassendi refers to, is the frontispiece of the Latin edition of Galileo's *Dialogue On The Two Main World Systems* of 1635; contrary to Della Bella's famous engraving for the indexed Italian first edition of the *Dialogo*,[42] the frontispiece does not show a bearded old man but a young beardless Copernicus with a fur cloak. It also puts an orrery in his hand, which differs from the precursor Boissard[43] and which Gassendi does not mention.[44] The earlier Copernicus portrait was probably unknown to Della Bella. His image is the first representation of Copernicus in Italy and the very first to show him with an orrery (and not with a lily of the valley).

Copernicus there wears a long cloak with a fur collar and ostentatious fastening cords, galloons and buttons—a prestigious East European attire with

41 'Nunc, cùm de eius facie, corporisque habitu superesset quidpiam dicendum; nihil tamen licet iam nobis aliud, quàm exhibere ipsius effigiem, cuiusmodi expressa habetur inter virorum illustrium, quae apud Boissardum sunt caelatae, icones. Curaui solùm ipsi appingi diploidem pelliceam, qualem se obseruasse Bullialdus noster memorauit in ea effigie, quae Argentorati visitur, qua Ecclesiae parte celebre horologium machinale exstat, qualemque eximius Berneggerus, qui ante annos plus minus triginta aduehi ex Prussia effigiem procurauerat, repraesentari voluit in fronte Galileanorum Dialogorum [...].' Gassendi, *Biographia Copernicana* 97.

42 Regarding della Bella's portrayal of Copernicus cf. Zittel C., "Zeichenkunst und Wissenschaft. Stefano della Bellas Frontispize zu Werken Galileos", in Albrecht A. – Cordibella G. – Remmert V. (eds.) *Tintenfass und Teleskop. Galileo Galilei im Kontext von Wissenschaft und Kunst* (forthcoming Berlin: 2014).

43 Boissard Jean Jacques, *Icones quinquaginta virorum illustrium*, part 3 (Frankfurt a.M., n.p., 1598).

44 Regarding the Copernicus images cf., Metze G., *Die Entwicklung der Copernicus-Porträts vom 16. Jahrhundert bis zum 18. Jahrhundert* (Munich: 2004).

FIGURE 10.4 Tobias Stimmer, *Copernicus* (Strasbourg cathedral). Painted wood.

FIGURE 10.5 *Copernicus*, from: Boissard Jean Jacques, *Icones quinquaginta virorum illustrium*, part 3 (Frankfurt a.M., n.p.: 1598) 314. Engraving.

a clerical headgear which not without reason is missing in other portraits of Copernicus. Through the biretta, Della Bella makes Copernicus a cleric, which might be interpreted as a conscious camouflage.

Mathias Bernegger (1582–1640), from Strasbourg, was a friend of Kepler's and among the most decided and influential Copernicans. For the frontispiece, engraved by van Heyden, Bernegger procured a template that orients itself on Tobias Stimmer's Copernicus image on the Strasbourg clock, which in itself supposedly goes back to the self-portrait mentioned by Gassendi. Bernegger

FIGURE 10.6 Jacob van der Heyden's Frontispiece to Galileo, *Systema Cosmicum* (1635).[45]

conflated this template with the iconography of the engraving by Della Bella from which he borrowed the whole composition, the fur cloak and the simple orrery, which, however, now is held pointing upwards. The fact that Gassendi

45 Galilei Galileo, *Systema Cosmicum. Dialogus de systemate mundi*, ed. M. Bernegger (Strasbourg, David Hautt: 1635).

commissioned the engraver Jacob van Meurs to equip Boissard's image especially with a fur (adding a report by Bulliardus as an explanation) and appends the orrery in the same position, doubly alludes to the later Copernicanism under the banner of Galilei. The frontispiece for Gassendi's Copernicus vita is therefore simultaneously a visible and veiled apology of Galilei.

The pictorial allusion within the text is a little more concealed. Comparing the depiction of the Copernican world system from Copernicus's *De Revolutionibus* with the depiction that supposedly represents his world system in Gassendi, astonishing differences become visible.

The print of the first edition shows nine circles, presumably an effect of the wrongly placed headings. In the manuscript, there are eight circles and seven headings that obviously mark the space between the lines and not the lines themselves, i.e. not planetary orbits but spheres. The circles or rings in the diagram in Copernicus do not, as might be expected, represent the orbits on which the planets migrate through the open space around the earth[46] but rotating planetary spheres. The ancient notion that planets do not move independently but travel with spheres that circulate around the earth is simply reassumed here in a model with a new reference point, and now the planets are carried around the earth on rotating spheres. Hence the title 'On The Orbits Of The Celestial Spheres', which emphasises the traditionality of the venture. The spherical theory is improved, not dismissed, the reference system is simply changed and such a change of the reference point does not affect, as Otto Neugebaur formulated as early as in 1957,[47] the model's structure which is related to antiquity more than to subsequent planetary models.

Gassendi, too, does not ascribe to Copernicus—as many have done later on—that according to him the planets revolve steadily on circular orbits, but it is the skies that do so. He clearly sees that the orientation merely is not towards the earth any more but towards the sun. Gassendi simply replaces the diagram, disfigured by the woodcut artist from *De Revolutionibus*, with a diagram from Galilei's *Dialogo*;

46 Cf. on this Cohen, *Revolutionen* 168–170.

47 Neugebaur O., *The exact sciences in Antiquity* (Providence: 1957), on this, Cohen, *Revolutionen* 190–193.

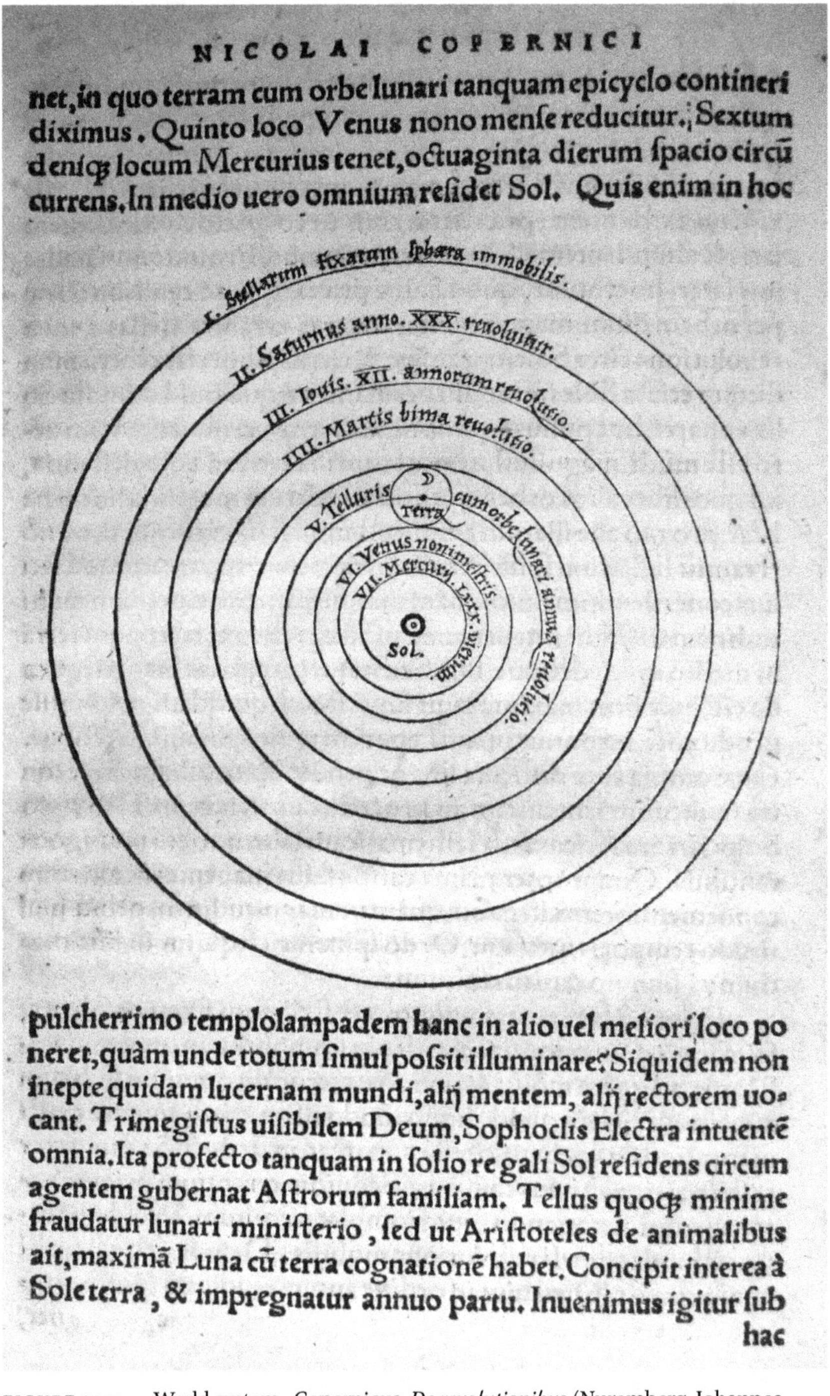

NICOLAI COPERNICI

net, in quo terram cum orbe lunari tanquam epicyclo contineri diximus. Quinto loco Venus nono menſe reducitur. Sextum denique locum Mercurius tenet, octuaginta dierum ſpacio circũ currens. In medio uero omnium reſidet Sol. Quis enim in hoc

pulcherrimo templo lampadem hanc in alio uel meliori loco poneret, quàm unde totum ſimul poſsit illuminare? Siquidem non inepte quidam lucernam mundi, alij mentem, alij rectorem uocant. Trimegiſtus uiſibilem Deum, Sophoclis Electra intuentẽ omnia. Ita profecto tanquam in ſolio regali Sol reſidens circum agentem gubernat Aſtrorum familiam. Tellus quoque minime fraudatur lunari miniſterio, ſed ut Ariſtoteles de animalibus ait, maximã Luna cũ terra cognationẽ habet. Concipit interea à Sole terra, & impregnatur annuo partu. Inuenimus igitur ſub hac

FIGURE 10.7 World system, Copernicus, *De revolutionibus* (Nuremberg, Johannes Petreius: 1543), fol. 9v. Woodcut.

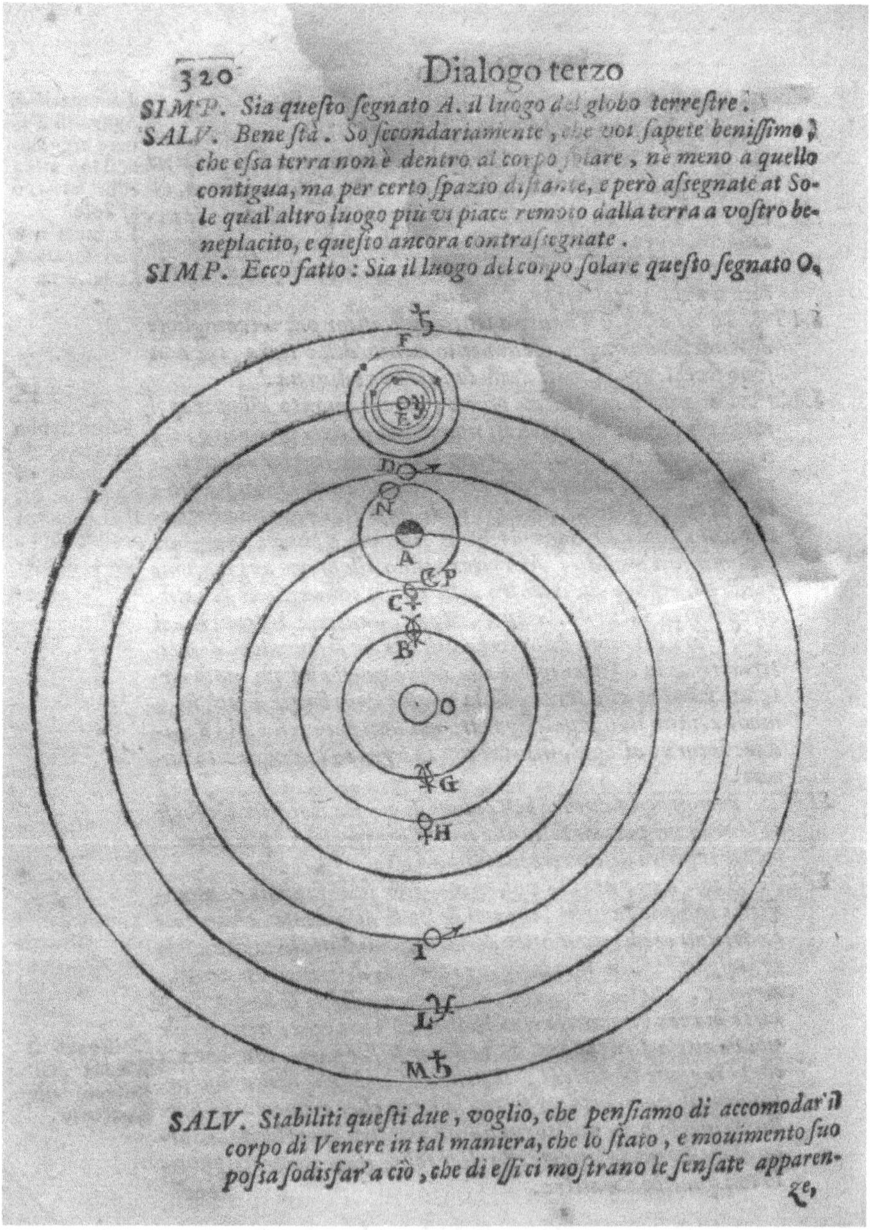
320 Dialogo terzo

SIMP. Sia queſto ſegnato A. il luogo del globo terreſtre.

SALV. Bene ſtà. So ſecondariamente, che voi ſapete beniſſimo, che eſſa terra non è dentro al corpo ſolare, nè meno a quello contigua, ma per certo ſpazio diſtante, e però aſſegnate al Sole qual'altro luogo più vi piace remoto dalla terra a voſtro beneplacito, e queſto ancora contraſſegnate.

SIMP. Ecco fatto: Sia il luogo del corpo ſolare queſto ſegnato O.

SALV. Stabiliti queſti due, voglio, che penſiamo di accomodar'il corpo di Venere in tal maniera, che lo ſtato, e mouimento ſuo poſſa ſodisfar'a ciò, che di eſſi ci moſtrano le ſenſate apparenze,

FIGURE 10.8 World system, Galilei, *Dialogo sopra i due massimi sistemi del mondo Tolemaico e Copernicano* (Florence: 1632) 320. *Woodcut.*

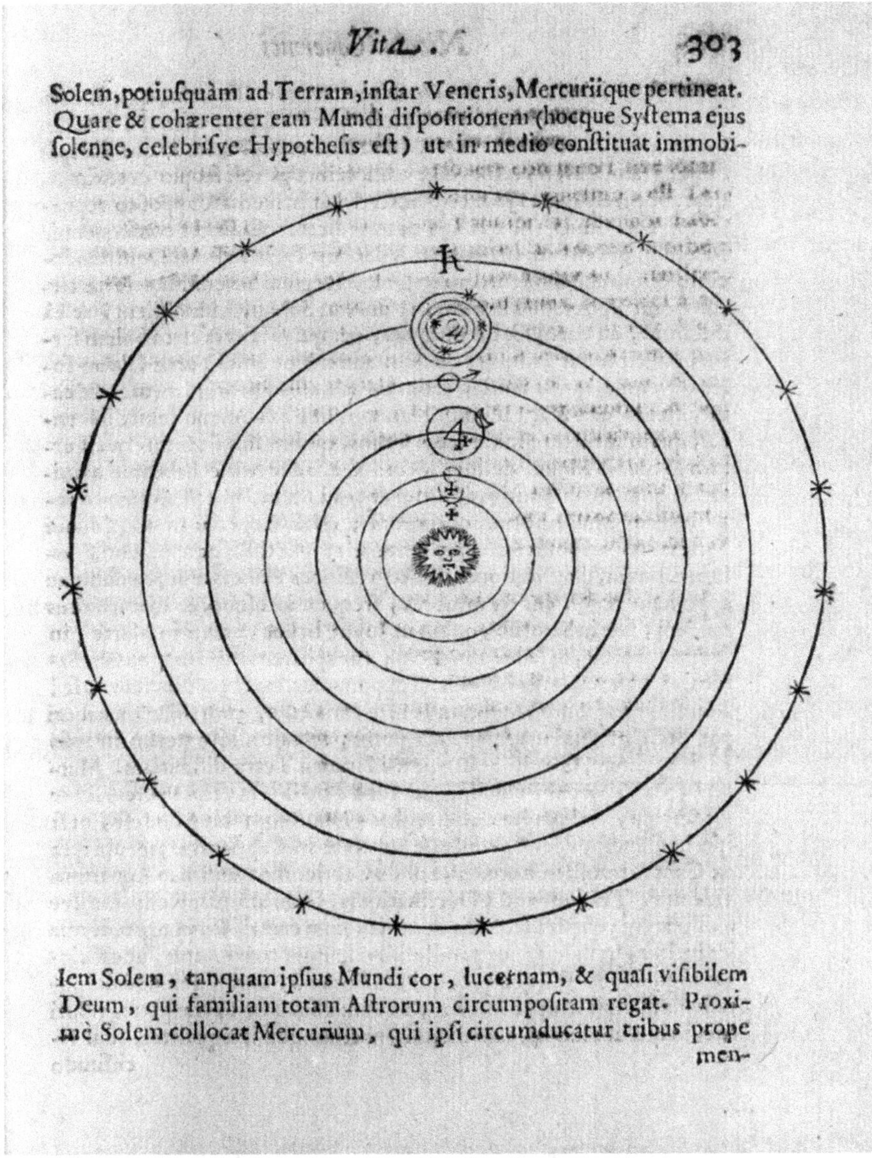
Vita. 303

Solem, potiuſquàm ad Terram, inſtar Veneris, Mercuriique pertineat. Quare & cohærenter eam Mundi diſpoſitionem (hocque Syſtema ejus ſolenne, celebriſve Hypotheſis eſt) ut in medio conſtituat immobi-

lem Solem, tanquam ipſius Mundi cor, lucernam, & quaſi viſibilem Deum, qui familiam totam Aſtrorum circumpoſitam regat. Proximè Solem collocat Mercurium, qui ipſi circumducatur tribus prope men-

FIGURE 10.9 World system, Gassendi, *Nicolai Copernici Varmiensis Canonici, Astronomi illustris Vita.* Woodcut.

he embellishes it with a sphere of fixed stars and passes it off as Copernicus's diagram where only the minor correction is made that now the moons of Jupiter are represented. In fact, he does not show spheres any more but the circular orbits of the planets in their correct number: a pictorial crypto-apology for Galilei.

Panofsky has elaborated the aesthetic privileging of the circle over the ellipsis[48] with a view of Galilei as the conflict between classicist and mannerist ideals of style. Ewa Chojecka[49] has attempted to transfer this approach to the case of Copernicus and detected a similar connection between aesthetic and scientific notions there. She claims that Copernicus's visual culture forced him to favour the circle. One could additionally assert Platonic concepts to which Copernicus adhered; consequently, primarily philosophical and not astronomical main ideas would have guided him towards his world image. The circular movement would not be an empirical ascertainment but an ideal handed down from antiquity, which was further affirmed by the aesthetic ideal of perfection in the Renaissance. Gassendi shared Chojecka's view and he quotes from Copernicus a corresponding ideal of harmony, which incited him to favour the circular orbit:

> In the first division, however, he outlined a general notion or description of the world, namely according to his own hypothesis whereby the motion [...] is ascribed to the earth. That is to say that there he taught firstly, that the world was shaped like a sphere and that the earth was spherical as well and together with the water formed a globe; further, that the movements of the sky are even, eternal and either circular or composed of circular movements. [...] Thus, he conceived his famous hypothesis, placing the sun immovably in the centre (heliostatic, rather than heliocentric), quasi as the heart of the world itself, as its light and, as it were, visible god who reigns the entire flock of stars grouped around her.[50]

---

48 Cf. Panofsky E., *Ästhetische Prägung: Galileo Galilei als Kritiker der Bildkünste*, ed. H. Bredekamp (Berlin: 2013).

49 Chojecka E., "Die Kunsttheorie der Renaissance und das wissenschaftliche Werk des Kopernikus", *Zeitschrift für Kunstgeschichte* 35 (1972) 257–281.

50 '[...] at Priore exhibuit generalem Mundi ideam, siue descriptionem, iuxta propriam Hypothesin, in qua motus Terrae [...] tribuitur. Scilicet, vbi primùm docuit esse Mundum sphaericum, Terram etiam sphaericam esse, & in vnum globum cum aqua coïre; ac insuper motum caelestem esse aequalem, perptetuum, & aut circularem, aut ex circularibus compositum; [...] Quare & cohaerenter [...] eius solemne, celebrisve Hypothesis [...] vt in medio constituat immobilem Solem, tanquam ipsius Mundi cor,

The privileging of the circle, however, was exactly the problem of the Copernican theory. It was clear as daylight for Gassendi that the ultimate elaboration of the heliocentric theory only came from Brahe and above all Kepler and that the imagined circular orbits were a downright metaphysical burden that also impeded Galilei's system. Clinging on to the notion of a harmonious cosmos was especially fatal because it led to all kinds of theoretical contortions:

> When he initially realised that the astronomers had been forced to defy current notions in order to preserve the even movement on the celestial circular orbits by leading and measuring them not around the actual centre but around another (on the so-called equant circle); that the same were not able to gather together the main point, that is the shape and harmonious disposition of the world, elsewhere from their tangled pile of hypotheses; that thus they fared like a painter who might optimally represent hands, the head, feet and the rest of the body parts, each on its own but did not paint them conjunctly in the correct relation to a whole body and brushed together from them rather a monster than a human being.[51]

As a result of this, another reason becomes clear for Gassendi's strong emphasis on the importance of painting for Copernicus and his training as a painter because in hindsight this becomes discernible as co-causal for his fixation on an old style of thinking. But even Galilei adhered to the circular orbit. The adoption of the circular diagram in the context of the critical analysis of the circular orbits' consequences also points to the problematic Copernican heritage when it comes to Galilei. Gassendi knew exactly that Galilei had also been trained as an artist and informed by Renaissance ideals.

Clinging to an old style of thinking, which marred empirical observation through its dictate of harmony, had to be a thorn in the side of an empirical researcher like Gassendi. Gassendi, himself famous for the exactness of

lucernam, & quasi visibilem Deum, qui familiam totam Astrorum circumpositam regat.' Gassendi, *Biographia Copernicana* 81–82.

51 'Principio, cùm caetera inter, animaduerteret cogi Astronomos aduersus communem notionem peccare, dum, vt aequabilem motionem in caelestibus circuitionibus tuerentur, non ipsam penes centrum proprium, sed penes alienum (circuli nimirùm Aequantis vocati) desumebant, ac metiebantur: Neque posse aliunde ipsos rem praecipuam, hoc est, Mundi formam, dispositionemque concinnam ex farragine illa sua hypotheseon colligere; sed iis perinde euenire, ac si quis Pictor manus, caput, pedes, membra caetera, optimè illa quidem, sed non vnius corporis comparatione depicta adunaret, sicque ex illis monstrum potiùs, quàm hominem compingeret [...].' Ibid. 77.

his observations in the transit of Mercury, dedicated himself in detail to the respective achievements of Copernicus's. His critical method also consists in the examination of the plausibility of Copernicus's own specifications, with knowledge of the astronomical practices. Gassendi emphasises that Copernicus, 'true to the paragon Ptolemy', built instruments and made observations; here, he names a weak point of *De Revolutionibus*, i.e., that Copernicus relied upon the data of Ptolemy all too often. In this context, it is virtually a specimen of revealing historiography when Gassendi analyses the instruments mentioned by Copernicus. So he says for example that even if Copernicus describes the design of the quadrant, it did not follow that he had actually used this instrument and the same was true of the armillary sphere. These instruments would indeed allow for much more accurate observations than those made by Copernicus; anyway, it had only been authenticated that Copernicus used Ptolemaic measuring sticks and the Jacob's staff. Thus Gassendi suggests to his readers that Copernicus had only utilised the simple device of the Biltmore stick. Galilei's art of observation is now indirectly played off against Copernicus's specifications of the planets' and the moon's periods; this was '[...] a truly admirable consideration on which he would have congratulated himself if he had already been able to see what we have since then discovered with the telescope, namely that four similar moons revolve around Jupiter.'[52]

The representation of Copernicus's inaccuracy is then dressed in a praise of the candour with which he confessed to not having any exact observation material: Rheticus is quoted amply who reports that Copernicus had regarded his examinations as 'consciously moderate and not exaggerated.'[53] 'Therefore and not out of indolence or aversion to work he avoided on purpose the accuracy that goes into the smallest detail.'[54] Free of the 'downrightly anxious precision' with which others carried out the celestial observation,[55] Copernicus had admitted that

> If I succeeded to determine the truth precisely to the sixth of a degree which makes 10 minutes [of the arc], I would no less rejoice than

52 'Mira sanè excogitatio, de qua sibi ipse fuisset gratulatus, si perspexisset, vti deinceps telescopio deteximus, quatuor pareis Lunas Ioui circumferri [...].', ibid. 83.

53 '[...] exquisitiones mediocreis, non nimias esse voluit.' Quotation of Rhetikus, ibid. 100.

54 'Itaque consultò, non inertia, aut taedio defatigationis eas comminutiones vitauit [...]', ibid. 100.

55 'vt planè scrupulosè'. Ibid. 100.

> Pythagoras allegedly did after he had found the theorem about the orthogonal triangle.[56]

What reads like a cheerful anecdote, must have been hair-raising to the knowledgeable readers in Gassendi's time. For Brahe and Kepler, a deviation of 10 arc minutes was unacceptable and a reason to exclude such an imprecise theory as being out of the question. Tycho's determination of the planets' positions was accurate to one arc minute. Gassendi's own Mercury observations were geared to Brahe's premises of exactness. On a second glance one notices that only Brahe and Kepler are praised openly; Gassendi extols 'Brahe's superhuman and heroic endeavour' as well as his 'excellent observations'. He and Kepler with his Rudolphine Tables, which he drew up 'with nearly unbelievable toil [...]', have only accomplished wherefore Copernicus had prompted Rheticus with 'numerous admonitions, remarks and instructions.'[57]

After all, scientific innovations became possible only after Copernicus, but he had at least stimulated them: Gilbert's magnet philosophy, Galilei's explanation of the tides, that planets move on elliptic orbits, and Tycho's subversion of his world system. Gassendi's conclusion, which one can still fully subscribe to, reads: 'Therefore it seems rather advisable if we seize on a word by Kepler who said, Copernicus had found a treasure the true value of which he did not know at all.'[58]

6 Conclusion

In order to assess Gassendi's scientific attitude towards this, we first of all need to liberate us from today's clichéd image of a Copernican revolution. Many 18th to 21st-century historians proclaimed with great naturalness the Copernican revolution as a fact, and at least since Thomas Kuhn we have been trained to understand such revolutions as a change of paradigm. More recent studies in the history of science, however, have unveiled the idea of a 16th-century Copernican revolution as a figment coming from 18th-century

---

56 'Ego vero [...] si ad sextanteis, quae sunt scrupula decem, veritatem adducere potero, non minùs exsultabo animis, quàm ratione Normae reperta Pythagoram accepimus.', ibid. 100.

57 'Tycho Brahe ingenti, ac verè heroico ausu [...]' – '[...] exquisitas obseruationes [...]' – '[...] incredibili labore [...]' – '[...] multa monens, subijciens, praecipiens [...]', ibid. 105.

58 'Quare praestat potius, occasione eius, quod Kepplerus ait, eum fuisse thesaurum inventum a Copernico, cuius ipsemet pretium ignorarit.' Gassendi (1654) 50.

historiographers.[59] It is worth the while, therefore, to locate the Copernicus vita in the light of this debate.

Especially I. Bernard Cohen in his standard work on *Revolutions in Science* suggests four test criteria,[60] by which one could examine if scientific innovations were revolutions. Firstly, one should ask if a scientist referred to his work as a revolution himself or if his contemporaries acknowledged it as revolutionary; secondly, if the documentary evaluation of the later scientific development of the field attested a revolution; thirdly, how later historians and, fourthly, how present-day experts related to it. Cohen shows that the presumptive Copernican revolution does not conform to any of the four principles.

Ptolemy was well aware that with the assumption of an even circular motion, the planetary movements could not be represented adequately, which is why he assumed an equation point, regarded from which an uneven movement appears even on a circular arc. An increased precision resulted from this auxiliary construction but Copernicus wanted to eliminate it; in order to do so, he had to introduce a complicated system of epicycles and epicycles on epicycles, i.e. of circles on circles, which in no way equalled a simplification of the presumptive Ptolemaic chaos. The calculations themselves became neither simpler nor more exact. As Owen Gingerich could show, his results were not numerically better than the Ptolemaic ones, but even slightly more complicated; as Gingerich also assumes, no one calculated planetary positions by means of the Copernican epicycle model towards the end of the 16th century.[61]

Therefore, the Copernican model of even circular movements did not improve the astronomical observations and was moreover dismissed by Kepler who with his ellipses ultimately reintroduced a sort of equation point. As Cohen summarises:

> Had it not fallen into the hands of Tycho Brahe and Kepler, the Copernican system would have contributed to a perpetuation of the Ptolemaic system in a somewhat more complicated shape.[62]

59 Cohen I.B., *Revolutionen in der Naturwissenschaft* (Frankfurt a.M., 1994) 86.

60 Cf. ibid., 82f.

61 Ibid. 167–195, see Cohen's comments on Gingerich, here: 185.

62 'Wäre es nicht Tycho Brahe und Kepler in die Hände gefallen, hätte das Kopernikanische System zur Verewigung des Ptolemäischen Systems in einer etwas komplizierten Form beigetragen.' Here, Cohen approvingly quotes Otto Neugebaur. Cf. Cohen, *Revolutionen* 193.

It is remarkable that Gassendi largely anticipated Cohen's historical method of validation and its results, for his historical approach permits him to outline how the Copernican system was valued by the contemporaries, how it had affected astronomy since then, how one stated one's position towards it from a topical point of view. Precisely as subsequently Cohen, Gassendi reaches the view that it had not been Copernicus's objective to revolutionise the ancient sciences but to improve the old system, and that it fell only to Kepler and Galilei to devise a new system, sufficiently viable for a new dynamic astronomy because it analysed the powers that are responsible for the planets' movements.

Certainly, Gassendi's lives of the astronomers are special cases, and yet they set the standards for the whole genre.[63] In Gassendi's vitas, the historical scientific appraisal and systematic subsumption of Brahe's and Copernicus's achievements against the backdrops of the history of astronomy and contemporaneous philosophical and epistemological discussions reach a critical-historical level of reflection that is not always reached even today. His vitas may therefore be regarded as exemplary of what a vita was able to achieve if it came from an author who combined a historical sense and a commanding expertise in every detail with literary mastership.

English translation: Wolfgang Neuber.

Selective Bibliography

Adam Melchior, *Vitae* (Frankfurt a.M. – Heidelberg: 1615–1620) – all vols. online: URL= <http://www.uni-mannheim.de/mateo/camenaref/adam.html>.

Baillet Adrien, *La vie de monsieur Des-Cartes*, 2 vols. (Paris, Daniel Horthemels: 1691).

Blum G., "Kunstgeschichte als 'große Erzählung' und Bildsystem. Zur Geschichtstheologie von Vasaris Vite (1550)", in Ganz D. – Thürlemann F. (eds.), *Das Bild im Plural. Mehrteilige Bildformen zwischen Mittelalter und Gegenwart* (Munich: 2010) 271–288.

Blumenberg H., *Die Genesis der kopernikanischen Welt* (Frankfurt a.M.: 1975).

Boissard Jean Jacques, *Icones quinquaginta virorum illustrium, part 3 (Frankfurt a.M., n.p.: 1598)*.

Bredekamp H., *Galilei der Künstler—Die Zeichnung, der Mond, die Sonne* (Berlin: 2007).

Chojecka E., "Die Kunsttheorie der Renaissance und das wissenschaftliche Werk des Kopernikus", *Zeitschrift für Kunstgeschichte* 35 (1972) 257–281.

63 Cf. the vita of Barrow by Abraham Hill, quoted in Roling (2013) 79, where it says: 'Another Camerarius, or Gassendus would make another Life of Melanchthon or Piereskius.'

Cohen I.B., *Revolutionen in der Naturwissenschaft* (Frankfurt a.M.: 1994).

Davis C. – Feser S. – Lemelsen K. – Nova A. (eds.), *Die Vite Vasaris. Entstehung – Topoi – Rezeption* (Venedig: 2010).

Detel W., *Scientia Rerum Natura Occultarum, Methodologische Studien zur Physik Pierre Gassendis* (Berlin: 1978).

Enenkel K., *Die Erfindung des Menschen. Die Autobiographik des frühneuzeitlichen Humanismus von Petrarca bis Lipsius* (Berlin: 2008).

Enenkel K. – Zittel C. (eds.), *Die Vita als Vermittlerin von Wissenschaft und Werk* (Berlin: 2013).

Fisher S., "Pierre Gassendi", *The Stanford Encyclopedia of Philosophy* (*Winter 2009 Edition*), Zalta E.N. (ed.): URL= <http://plato.stanford.edu/archives/win2009/entries/gassendi/>.

Friedrich L., *Die griechisch-römische Biographie nach ihrer litterarischen Form* (Leipzig: 1901).

Galilei Galileo, *Systema Cosmicum. Dialogus de systemate mundi*, ed. M. Bernegger (Strasbourg, David Hautt: 1635).

Gassendi Pierre: *Tychonis Brahei, Equitis Dani, Astronomorum Coryphaei, Vita* [...] *Accessit Nicolai Copernici, Georgii Peuerbachii & Joannis Regiomontani, Astronomorum celebrium, Vita* (Paris, Vidua Mathurin Dupuis: 1654); 2nd edn. (The Hague, Adriani Vlacq: 1654); 3rd edn. (The Hague, Adriani Vlacq: 1655).

Gassendi Pierre "Das Leben des Kopernikus, Domherren im Ermland, des berühmten Astronomen", in Kühne A. – Kirschner S. (eds.), *Biographia Copernicana. Die Copernicus-Biographien des 16. bis 18. Jahrhunderts* (Berlin: 2004) 107–149.

Gassendi Pierre, *De Vita et Moribus Epicuri* (Lugduni, Guillelmi Barbier: 1647).

Gassendi Pierre, *De vita et moribus Epicuri:* libri octo (The Hague, Adrianum Vlacq: 1656).

Gassendi Pierre, *Institutio astronomica Iuxta Hypotheseis tam Veterum, quam Copernici, et Tychonins, Dictata à Petro Gassendo Regio Mathesos Professore. Eiusdem Oratio Inauguralis iteratò edita* (Paris, Louis de Heuqueville: 1647).

Gassendi Pierre, *Vie de Copernic par Gassendi*, trans. M. Thirion (Paris: 1973).

Gassendi Pierre, *Vie de l'illustre Nicolas-Claude Fabri de Peiresc, conseiller au Parlement d'Aix*, trans. R. Lassalle (Paris: 1992).

Gassendi Pierre, *Vies de Tycho Brahé, Copernic, Peurbach et Regiomontanus*, trans. J. Peyroux (Bordeaux: 1996).

Gassendi Pierre, *Vie et moeurs d'Epicure* (french-latin), ed. S. Taussig (Paris: 2006).

Gassendi Pierre, *Viri illustris Nicolai Claudii Fabricii de Peiresc, senatoris aquisextiensis vita* (Paris, Sébastien Cramoisy: 1641).

Gassendi Pierre, *Vita Epicuri, Peireskii, Tychonis Brahei, Copernici, Peurbachii, & Regiomontanio. Contenu dans Opera omnia*, vol. v (Lyon, Laurent Anisson – Jean-Baptiste Devenet: 1658).

Gibbon Edward, *Essai sur l'étude de la littérature* (Dublin, Guillaume Hallhead: 1777).

Graf-Stuhlhofer F., *Humanismus zwischen Hof und Universität. Georg Tannstetter (Collimitius) und sein wissenschaftliches Umfeld im Wien des frühen 16. Jahrhunderts* (Vienna: 1996).

Humbert P., *L'oeuvre astronomique de Gassendi* (Paris: 1937).

Iliffe R. – Keynes M. – Higgitt R. (eds.), *Early Biographies of Isaac Newton, 1660–1885*, 2 vols. (London: 2006).

Kirschner S. – Kühne A., "Die Rezeption von Copernicus im Spiegel seiner Biographien", in Seising R. – Folkerts M. – Hashagen U. (eds.), *Form, Zahl, Ordnung. Studien zur Wissenschaftsgeschichte* (Stuttgart: 2004) 467–479.

Kliemann J., "Giorgio Vasari: Kunstgeschichtliche Perspektiven", in: Ganz P. – Gosebruch M. – Meier N. – Wuttke M. (eds.), *Kunst und Kunsttheorie 1400–1900* (Wiesbaden: 1991) 29–74.

Lange F.A., *Geschichte des Materialismus* (Iserlohn 1866, reprint Frankfurt a.M.: 1974) vol. I.

Le Bovier de Fontenelle Bernard, *The Life of Sir Isaac Newton; with an Account of his Writings* (London: 1728), URL= <http://www.maths.tcd.ie/pub/HistMath/People/Newton/Fontenelle/Fonten.html, 12/2013>.

Le Bovier de Fontenelle Bernard, "Lebens-Beschreibung des Herrn von Leibnitz verfaßt von dem Herrn Fontenelle", in Leibniz, *Monadologie*, ed. D. Till (Frankfurt a.M. – Leipzig: 1996).

Lichtenberg Georg Christoph, *Schriften und Briefe*, ed. W. Promies (Munich: 1968–1992) vol. III/IV.

Manetti Gianozzo, *Vite di Socratis e Senecae*, ed. A. de Petris (Florence: 1979).

Mayer T.F. – Woolf D.R., *The Rhetorics of Life-Writing in Early Modern Europe. Forms of Biography from Cassandra Fedele to Louis XIV* (Ann Arbor: 1995).

Metze G., *Die Entwicklung der Copernicus-Porträts vom 16. Jahrhundert bis zum 18. Jahrhundert* (Munich: 2004).

Miller P., *Peiresc's Europe: Learning and Virtue in the Seventeenth Century* (New Haven – London: 2000).

Momigliano A., *The Development of Greek Biography* (Cambridge Mass. – London: 1993).

Neugebaur O., *The exact sciences in Antiquity* (Providence: 1957).

Panofsky E., *Ästhetische Prägung: Galileo Galilei als Kritiker der Bildkünste*, ed. H. Bredekamp (Berlin: 2013).

Roling B., "John Aubrey und die Philosophenviten des 17. Jahrhunderts", in: Enenkel K. – Zittel C. (eds.), *Die Vita als Vermittlerin von Wissenschaft und Werk. Form- und Funktionsanalytische Untersuchungen zu frühneuzeitlichen Biographien von Gelehrten, Wissenschaftlern, Schriftstellern und Künstlern* (Münster: 2013) 207–236.

Rubin P., *Vasari: Art and History* (New Haven: 1995).

Schirren T., *Philosophos Bios. Die antike Philosophenbiographie als symbolische Form. Studien zur Vita Apollonii des Philostrat* (Heidelberg: 2005).

Seidel R., "Melchior Adams 'Vitae' (1615–1620) und die Tradition frühneuzeitlicher Gelehrtenbiographik: Fortschritte und Grenzen eines wissenschaftlichen Paradigmas um 1600", in: Koziełek G. (ed.), *Oberschlesische Dichter und Gelehrte vom Humanismus bis zum Barock* (Bielefeld: 2000) 179–204.

Tack R., *Untersuchung zum Philosophie- und Wissenschaftsbegriff bei Pierre Gassendi* (Meisenheim: 1974).

Tannstetter Georg, "Viri mathematici", in: idem (ed.) *Tabulae eclypsium magistri Georgii Peurbachii; Tabula primi mobilis Ioannis de Monteregio* (Vienna: 1514) aa3v–aa6v.

Taussig S., "*La vie d'Épicure* par Gassendi", in Paganini G. – Tortarolo E. (eds.), *Der Garten und die Moderne, Epikureische Moral und Politik vom Humanismus bis zur Aufklärung* (Stuttgart – Bad Canstatt: 2004) 139–161.

Taussig S., "Epicurus Redivius By Pierre Gassendi: Biography Opposed to the Love of Self", in *Intellectual News* 11 (2003–2004) 19–27.

Taussig S. – Turner A.J., *Mémoire de Gassendi: vies et célébrations écrites avant 1700* (Turnhout: 2008).

Verstegen I., "Birth Dates, Death Dates, and the Beginnings of Modern Art History", *Storiografia: Rivista annuale di storia* 10 (2006) 1–19.

Verstegen I., "Vasari's progressive (but non-historicist) Renaissance", *Journal of Art Historiography* 5 (2011) 1–19.

Werle D., "Melchior Adams Gelehrtenbiographien und ihr Bezug zur Enzyklopädistik", in Schierbaum M. (ed.), *Enzyklopädistik 1550–1650: Typen und Transformationen von Wissensspeichern und Medialisierungen des Wissens* (Berlin et al.: 2009) 105–125.

Westman R.S., *The Copernican Question: Prognostication, Skepticism, and Celestial Order* (Berkeley: 2011).

Zittel C., "Die Lunatiker von Aix-en-Provence: Peiresc-Mellan-Gassendi", in Feist U. – Rath M. (eds.), *Et in imagine ego. Facetten von Bildakt und Verkörperung* (Berlin: 2012) 276–300.

Zittel C., "Zeichenkunst und Wissenschaft. Stefano della Bellas Frontispize zu Werken Galileos", in Albrecht A. – Cordibella G. – Remmert V. (eds.), *Tintenfass und Teleskop. Galileo Galilei im Kontext von Wissenschaft und Kunst* (Berlin: 2014).

CHAPTER 11

Hero of the Bourgeois World. Copernicus and His Afterlife in German Literature

Wolfgang Neuber

Nicolaus Copernicus's literary vestiges amount to quite a number of texts in (German) literature; for better or for worse they are, however, much less well-remembered than his reception in the histories of science and philosophy. Here, when it comes to the tradition of the German-speaking countries, Kant's, Nietzsche's and Freud's comments on the man and his achievement are among the most famous ones. Kant's interest in Copernicus is primarily motivated by his attempt to bring about a Copernican revolution in philosophy, to prioritise epistemology over sensual evidence; in his *Critique of Pure Reason* he says:

> Failing of satisfactory progress in explaining the movements of the heavenly bodies on the supposition that they all revolved around the spectator, he [sc. Copernicus] tried whether he might not have better success if he made the spectator to revolve and the stars to remain at rest. A similar experiment can be tried in metaphysics [...].[1]

Nietzsche's position is very much in the same vein as Kant's. In *Beyond Good and Evil* (1886), he uses Copernicus as a metaphor for himself and celebrates the man for having vanquished sensual evidence:

> Thanks, for the time being, to that Pole Boscovich, who, together with the Pole Copernicus, was the greatest and most victorious opponent of evidence. For while Copernicus persuaded us to believe, against all senses, that the earth does *not* stand still, Boscovich taught us to disavow the belief in the last that was established about the earth.[2]

1 Kant Immanuel, *Critique of Pure Reason*, trans. N. Kemp-Smith (London: 1950) 51 (B XVIf.). Kant's Critique was first published in 1781, the quote is from the preface to the second edition, 1787. Cf. Russell H.N., "Copernicus' Role in Kant's Revolution", *Journal of the History of Ideas* 20 (1959) 274–281; cf. also Steffen Schneider's essay in this volume.

2 'Dank vorerst jenem Polen Boscovich, der, mitsammt dem Polen Kopernicus, bisher der grösste und siegreichste Gegner des Augenscheins war. Während nämlich Kopernicus uns

 | DOI 10.1163/9789004281127_013

It has been noted that this praise comes as a double-edged sword.[3] Copernicus seems to have vanquished the senses but on second sight this statement is undermined by the phrases 'vorerst' ('for the time being') and 'überredet hat zu glauben' ('persuaded to believe'). It is not so much Copernicus's astronomical discovery that Nietzsche glorifies but his opposition to predominant opinions and his abundant rhetorical skills that eventually changed people's views.[4] The mighty individual conquers widely shared delusions through his powers of speech. It is quite in keeping with this observation that Nietzsche's interest in Copernicus has been interpreted as rooted in his concept of the 'Übermensch' ('superman') as an over-compensation of the Copernican wound.[5]

Around 1917, Sigmund Freud was fascinated by this concept; he refers to it in two articles from that time. In his lectures towards an introduction to psychoanalysis, he gives a rather brief summary:

> In the course of time, mankind in its naïve self-love has had to suffer two big wounds from science. The first, when it learnt that our earth is not the centre of the universe but a minute particle of a world system that is unimaginable in its size. For us, it is tied to the name Copernicus, although Alexandrian science had proclaimed something similar before.[6]

überredet hat zu glauben, wider alle Sinne, dass die Erde *nicht* feststeht, lehrte Boscovich dem Glauben an das Letzte, was von der Erde "feststand", abschwören [...].' Nietzsche F., *Jenseits von Gut und Böse*, in idem, *Sämtliche Werke, Kritische Studienausgabe in 15 Bänden* (KSA), ed. G. Colli – M. Montinari. (Munich – New York: 1980) vol. v, § 12, 26. In fact, Nietzsche is mistaken about 'Boscovich's' nationality; Rugjer Josip Bošković (1711–1787) was actually born in Ragusa (Dalmatia). Cf. Zittel C., "Boscovich", in Ottmann H. (ed.), *Nietzsche-Handbuch* (Stuttgart: 2000) 404–409; cf. also Zittel's essay in this volume.

3 Cf. Günther F.F., "Die kopernikanische Wende als anthropologische Denkfigur bei Nietzsche und Elias", in Günther F. – Holzer A. – Müller E. (eds.), *Zur Genealogie des Zivilisationsprozesses. Friedrich Nietzsche und Norbert Elias* (Berlin: 2010) 175–191, here 186f.

4 Cf. Schröder R., "War die copernicanische Reform der Astronomie ein Weltbildwandel?", in Markschies C. – Zachhuber J. (eds.), *Die Welt als Bild. Interdisziplinäre Beiträge zur Visualität von Weltbildern*, Arbeiten zur Kirchengeschichte 107 (Berlin: 2008) 91–112; cf. on Copernicus's rhetorical strategies also the essay by Jonathan Schüz in this volume.

5 Cf. Kluwe S., "Trauma und Triumph: Die kopernikanische Wende in Dichtung und Philosophie. Stationen der Rezeptionsgeschichte von Gryphius bis Nietzsche", in: Brodersen K. – Gebhardt H. (eds.), *Weltbilder*, Heidelberger Jahrbücher 47 (Berlin – Heidelberg – New York: 2003) 179–220, here 214.

6 'Zwei große Kränkungen ihrer naiven Eigenliebe hat die Menschheit im Laufe der Zeiten von der Wissenschaft erdulden müssen. Die erste, als sie erfuhr, daß unsere Erde nicht der Mittelpunkt des Weltalls ist, sondern ein winziges Teilchen eines in seiner Größe kaum vorstellbaren Weltsystems. Sie knüpft sich für uns an den Namen Kopernikus, obwohl schon

Very much at the same time, an article on a difficulty of psychoanalysis is more detailed, introducing the term 'narcissistic' (illusion):

> For us, the destruction of this narcissistic illusion is tied to the name and the works of Nic. Copernicus in the sixteenth century. Long before him, the Pythagoreans had questioned the preferred position of the earth, and Aristarchos of Samos had vocalised in the third century BC that the earth was much smaller than the sun and moved around that heavenly body. The great discovery of Copernicus's, too, had been made before him. But when it found common recognition, human self-love had experienced its first, the *cosmological*, wound.[7]

If Copernicus had inflicted the first narcissistic wound on mankind, the second one came from Darwin and the theory of evolution. The third wound eventually would come through Freud himself:

> The third and most sore wound, however, human obsession with grandeur shall experience through today's psychological research, which wants to demonstrate to the self that it is not even master in its own house, but depends on what unconsciously goes on in its psychic life.[8]

die alexandrinische Wissenschaft ähnliches verkündet hatte.' Freud S., *Vorlesungen zur Einführung in die Psychoanalyse* [1916–17], chapter 18: "Die Fixierung an das Trauma, das Unbewußte", in idem, *Gesammelte Schriften* (Leipzig – Vienna – Zurich: 1924) vol. VII, 282–295, here 294.

7 'Die Zerstörung dieser narzißtischen Illusion knüpft sich für uns an den Namen und das Werk des Nik. Kopernikus im sechzehnten Jahrhundert. Lange vor ihm hatten die Pythagoräer an der bevorzugten Stellung der Erde gezweifelt, und Aristarch von Samos hatte im dritten vorchristlichen Jahrhundert ausgesprochen, daß die Erde viel kleiner sei als die Sonne und sich um diesen Himmelskörper bewege. Auch die große Entdeckung des Kopernikus war also schon vor ihm gemacht worden. Als sie aber allgemeine Anerkennung fand, hatte die menschliche Eigenliebe ihre erste, die *kosmologische*, Kränkung erfahren.' Freud S., "Eine Schwierigkeit der Psychoanalyse", *Imago. Zeitschrift für Anwendung der Psychoanalyse auf die Geisteswissenschaften* 5 (1917), 1–7, here 4.

8 'Die dritte und empfindlichste Kränkung aber soll die menschliche Größensucht durch die heutige psychologische Forschung erfahren, welche dem Ich nachweisen will, daß es nicht einmal Herr ist im eigenen Hause, sondern auf kärgliche Nachrichten angewiesen bleibt von dem, was unbewußt in seinem Seelenleben vorgeht.' Freud, *Vorlesungen* 295.

'Arguably, the third wound, which is of a psychological nature, hits most sorely'; it is caused by the 'claim that the *self be not master in its own house*.'[9] These passages from Freud emphasise the threefold narcissistic wound inflicted upon mankind by the loss of its central position in the cosmos, aligning Freud's achievement with Copernicus's and Darwin's. The text's phrasing that it is 'psychoanalysis' which brings about this wound, not Freud, only thinly camouflages his expression of self-pride, in itself quite a narcissistic trait. Similar to Kant's and Nietzsche's claim to a part of the Copernican inheritance, Freud wants himself to be seen as the ultimate destroyer of human illusions.

•••

The focus of this essay, however, is not on philosophy or psychoanalysis but on literature. It is well known that before the acclaim that Copernicus found in the eyes of later days, a rather negative appraisal prevailed; earlier centuries would have responded critically or even negatively to Copernicus prompted by a theological reasoning rather than driven by historico-philosophical considerations. Heinrich Anshelm von Ziegler und Kliphausen is the last of Copernicus's biographers to appropriate such a religious argumentation. Not without a smirk, Ziegler opens his biography of Copernicus in the *Täglicher Schau=Platz der Zeit* (1695, *Diurnal Theatre of Time*) calling Copernicus '[t]he so-called Light & Eye of Mathematics and Astronomy'. Ziegler then continues:

> He established the miraculous conclusion that the globe be turning and on the contrary the sun with his host of stars stand still: which hard opinion, however, has been refuted by many scholars felicitously and especially from the Holy Scripture.[10]

With such a derisive and taunting remarks, Ziegler is put at the historical end of a line of biographers of Copernicus; he may well be the last significant

9 'Am empfindlichsten trifft wohl die dritte Kränkung, die psychologischer Natur ist'; sie ist verursacht durch die 'Behauptung [...], daß das *Ich nicht Herr sei in seinem eigenen Haus*.' Freud, "Eine Schwierigkeit" 7; the emphasis is Freud's.

10 'DAS so genannte Lumen & Oculus Matheseos & Astronomiae, Nicolaus Copernicus'; 'Er befestigte den Wunder=Schluß / daß sich die Erd=Kugel umdrehe / und hingegen die Sonne mit ihrem Sternen=Heere stille stehe: welche harte Meynung aber bereits von vielen Gelehrten treflich / und bevoraus aus Heil. Schrifft / widerleget worden.' Ziegler und Kliphausen Heinrich Anshelm von, *Täglicher Schau-Platz der Zeit* [...] (Leipzig, Gleditsch: 1695) 43.

intellectual figure to represent an authoritarian Biblicism in arguing against heliocentrism.[11]

By contrast, physico-theology in the early 18th century made its peace with the Copernican world system and adopted a position that compensates the loss of evidence of the planetary movements with the evidence of the immeasurable size of the cosmos. In alignment with Giordano Bruno, this facilitates to infer the infinity of the Creator (to whom corresponds the infinity of his creation). In the subsequent centuries—up to Nietzsche and Freud—a philosophical, rhetorical and literary adoration of Copernicus prevails, more often than not driven by a reverence based upon norms of Protestant bourgeois morals or the ideology of individualism respectively; a reverence that can be found in texts of fiction until the present day. In this paper, it is impossible to give but exemplary evidence for the time after 1750.

Daniel Heinse's novel *Ardinghello und die glückseligen Inseln* (1785, *Ardinghello and the Islands of the Fortunate*) is set in a fictitious past from which arises the prophecy of a coming heliocentric world model. A footnote explains: 'The system of the Prussian Copernicus was adopted in the Papal State the latest and Galileo had hardly been born at that time. One can consider the following a prophecy towards him.'[12]

In Conrad Ferdinand Meyer's *Huttens letzte Tage* (1871, *Hutten's Last Days*) the eponymous hero, a contemporary of Copernicus, reacts stoically when confronted with the miraculous new world system; he is utterly disinterested in cosmic order:

> First I shall serve my time on earth
> And if I am then not disburdened from service
>
> and new fief is distributed on the stars—
> Well then! I mean to stand my ground.[13]

---

11 On earlier references cf. also Kühlmann W., "Neuzeitliche Wissenschaft in der Lyrik des 17. Jahrhunderts. Die Kopernikus-Gedichte des Andreas Gryphius und Caspar Barlaeus im Argumentationszusammenhang des frühbarocken Modernismus", *Jahrbuch der Deutschen Schillergesellschaft* 23 (1979) 124–153.

12 'Das System des Preußen Kopernikus wurde am spätesten im Kirchenstaate angenommen, und Galilei war zu dieser Zeit kaum geboren. Man kann das Folgende für eine Prophezeiung auf ihn halten.' Heinse Daniel, *Ardinghello* (Leipzig: 1961) 252, footnote 24.

13 'Erst dien' ich aus auf Erden meine Zeit | Und bin ich dannzumal nicht dienstbefreit, Verteilt man auf den Sternen neues Lehn – | Wohlan! Ich denke meinen Mann zu stehn.' Meyer C.F. *Huttens letzte Tage* (Leipzig: 1913) 123.

Hutten's disinterest in cosmologic conditions does not, however, diminish Copernicus's achievement. On the contrary, the bigger this achievement is from a 19th-century readership's point of view the more audacious and philosophically sovereign Hutten's agnostic attitude appears—an attitude that confirms his secularist grounding and determination.

Adolf Prowe's festive drama "Copernicus"[14] (1874) on the occasion of the astronomer's 400th birthday makes every attempt to fashion Copernicus as the heroic individual that gives his life to profess a truth not acknowledged by his contemporaries. The text even goes so far as to align Copernicus's sufferings with those of Jesus Christ; on his deathbed, Copernicus says:

> It is accomplished—
> As you commanded, Lord, I have professed it:
> The sun stands still, the globe is moving![15]

It is hard to miss the text's allusion to Christ's last words on the cross, 'it is finished' ('es ist vollbracht'); the parallel is reinforced through Copernicus's statement that his spirit had been driven by the search for the ultimate truth, a search that requires self-sacrifice:

> It has cost my life—upwards
> now my restless spirit strives.[16]

On a somewhat different note, Arthur Schnitzler credits the Copernican turn especially with the phenomena of amazement and vertigo. He links this up with the new psychological intuition that man be not conscious of himself. In *Die Frage an das Schicksal* [*A Question of Fate*] (1889), from Schnitzler's *Anatol* cycle, Anatol hypnotises Max' mistress Cora. He makes her enact a couple of hypnotic suggestions (being a dancer, being bereaved, being a queen), which Max finds uncanny. Anatol responds to this remark:

> ANATOL I cannot find this … Not any more uncanny than life itself. Not any more uncanny than much that was only discovered in the course of centuries. How, do you think, our ancestors must have felt when they

14 Prowe A., *Copernicus. Ein dramatisches Gedicht. Festspiel zur vierten Säcularfeier der Geburt von Nicolaus Copernicus aufgeführt am Stadtheater zu Thorn* (Berlin: 1874).

15 'Vollendet ist es— | Wie Du befahlst, Herr, hab' ich es verkündigt: | *Die Sonne steht, der Erdenball bewegt sich*!' Ibid. 110. The emphasis is in the original.

16 'Mein Leben hat's gekostet – auf | Strebt jetzt mein ruheloser Geist.' Ibid. 111.

> suddenly heard the earth was spinning around? They must all have become vertiginous!
> MAX Yes . . . but that pertained to all![17]

Here the greater narcissistic wound—to employ Freud's phrasing—comes with the loss of the individual centre, the loss of the resting-within-himself, the loss of self-assuredness; it surpasses even the loss of the cosmological centre which was felt by the whole of mankind.

In Meyer and Prowe's texts, the preoccupation lies with Copernicus the heroic individual whereas Schnitzler rather accentuates the reception of the Copernican system. There are intermediate positions, however. Philipp Vandenberg's *Der Fluch des Kopernikus. Ein Renaissance-Roman* (1996, *The Curse of Copernicus. A Renaissance novel*) may serve as one of the more trivial examples. Here, a crime story set in 16th-century Europe ties on to 'the lost heritage of the great astronomer Nicolaus Copernicus which will change the world, shake the foundations of the Church and put the times out of joint'.[18] These words, taken from the blurb, are not much sillier or more commonplace than the novel itself, they probably do not even mark an all-time low in the reception of the Copernican revolution; they show, however, what the collective memory of even the greatest of theories may boil down to.

To sum it up, modern fiction especially appropriates those images of the Copernican revolution which over time have become common property and whose real or simply alleged effects serve as arguments or motivations of action. In most cases, however, Copernicus the heroic man becomes the protagonist. Helmuth M. Böttcher, an author who bestowed on the world such texts as *Butz, der Ameis* [*Butz the Ant*] (Leipzig – Zurich: 1922), *Der Affe Waldemar* [*The Monkey Waldemar*] (Rudolstadt: 1954) *Der Jünger, der den Dolch trug. Der Roman des Judas* [*The Disciple Who Carried the Dagger. The novel of Judas*] (Würzburg – Vienna: 1960) also wrote a book entitled *Umsturz des Himmels. Eine Novelle um Nikolaus Kopernikus* [*Subversion of Heaven. A novella around*

17 'ANATOL Das kann ich nicht finden . . . Nicht unheimlicher als das Leben selbst. Nicht unheimlicher als vieles, auf das man erst im Laufe der Jahrhunderte gekommen. Wie, glaubst du wohl, war unseren Voreltern zumute, als sie plötzlich hörten, die Erde drehe sich? Sie müssen alle schwindlig geworden sein! | MAX Ja . . . aber es bezog sich auf alle!' Schnitzler A.: *Anatol*, in idem, *Das dramatische Werk* vol. 1 (Frankfurt am Main: 1962) 30–41, here: 30.

18 Cf. Philipp Vandenberg, *Der Fluch des Kopernikus. Ein Renaissance-Roman* (Bergisch Gladbach: 1996), blurb: 'das verschollene Erbe des großen Astronomen Nikolaus Kopernikus, das die Welt verändern, die Kirche in ihren Grundfesten erschüttern und die Zeit aus ihren Fugen heben wird'.

Nicolaus Copernicus] (Berlin: 1962). In a narrative of 120 pages set in Bologna, 23-year-old Copernicus is affected by the omnipresent pathos of 'Renaissance' thought and the idea of 'humanism'; he eventually makes the vow to serve the truth of the 'spirit resting meekly before God in itself, responsible to itself and mankind'.[19]

By contrast, John Banville's *Doctor Copernicus. A Novel* (London: 1976), also focussing on the man Copernicus, is in a different league, possibly because thanks to its Irish origin there is no need to either mobilise the pathos of human Renaissance self-empowerment or to celebrate Copernicus as a Polish or German national hero respectively. This option, by the way, is explicitly refuted. Lukas, the Warmian Bishop and uncle of Copernicus, helps his nephew apodictically overcome his national qualms: 'You are not German, nephew, no, nor are you a Pole, nor even a Prussian. You are an Ermlander, simple. Remember it.'[20]

As a matter of fact, Banville's text touches on none of the 'famous' steps and situations of the run-of-the-mill hagiography of Copernicus: not the observation of the moon in Bologna, not his collaboration with Dominicus di Novara, not his appearance in Rome as the rising star of astronomy. Rather, the text shows a person who is not at all self-reliant, even a man who does not know that it might be desirable to assert himself ideologically as an individual. This makes *Doctor Copernicus* the most sophisticated textual example in my short sequence. As an Anglophone text, unfortunately, it lies beyond the boundaries of my investigation.

The criticism of Copernicus and his veneration alike are consistently engaged with the historico-philosophic consequences of the new cosmic system or with an individual construed as a lonesome hero of the intellect. Upon a first glance especially the first point may seem nothing short of self-evident, upon a second it is no longer. There are mathematic-astronomical deficits that remain after *De Revolutionibus Orbium Coelestium*[21] of 1543 and they cannot be ignored. Hans Blumenberg in his article "Kopernikus im Selbstverständnis der

19 'in Demut vor Gott in sich selber ruhenden, sich selbst und der Menschheit verantwortlichen Geistes'. Böttcher H.M., *Umsturz des Himmels. Eine Novelle um Nikolaus Kopernikus* (Berlin: 1962) 120.

20 Banville J., *Doctor Copernicus. A Novel* (London: 1976) 94.

21 Copernicus Nicolaus, *De revolvtionibvs orbium cœlestium, libri VI* (Nuremberg, Johann Petreius: 1543).

Neuzeit" (1964) dedicated a footnote[22] to these shortcomings and it is almost more important than the essay itself (as is the case with so many footnotes in that text). Blumenberg establishes especially three points:

1. [The] centre of the universe still remained "unoccupied" because Copernicus had to place the sun off-centre by 3 diameters, and the centres of the planets' orbits lay even more eccentrically—in the case of the orbit of Saturn even outside the orbit of Venus—, indeed the centre of the Earth's orbit in its turn had to conduct a circular movement so that Earth by itself was assigned 8 movements'.[23]
2. The Ptolemaic system was indeed more difficult to understand but easier to apply, 'because within it data complexes could be isolated [...], whereas in Copernicus the interweavement of orbit data of several planets—that is the real character of a system—is much more distinct'.[24]
3. Finally, regarding the predictability of planetary orbit data Copernicus's theory brought no gain whatsoever, 'indeed for the easily understandable reason that it is generally based upon the very same material of observations and upon still the same exactness of observation that had been available to Ptolemy'.[25]

In the reception of the Copernican model these facts find no attention at all. And yet, what Nietzsche and Freud positively formulated centuries later had not been seeded negatively in the first scepticism of the 16th and 17th centuries grounded on Biblicist polemics.

---

22 Blumenberg H., "Kopernikus im Selbstverständnis der Neuzeit", in *Akademie der Wissenschaften und der Literatur Mainz, Abhandlungen der geistes- und sozialwissenschaftlichen Klasse* (Wiesbaden: 1964) 337–368, here 341f.

23 '[Die] Mitte des Universums war auch weiterhin "unbesetzt", weil Kopernikus die Sonne um 3 Durchmesser hatte exzentrisch stellen müssen, und die Mittelpunkte der Planetenbahnen lagen noch exzentrischer (bei der Saturnbahn noch außerhalb der Venusbahn), ja der Mittelpunkt der Erdbahn mußte seinerseits eine Kreisbewegung ausführen, so daß der Erde allein schließlich 8 Bewegungen zukamen.' Ibid. 341f.

24 'weil in ihm Datenkomplexe isoliert werden konnten [...], während bei Kopernikus die Verflechtung der Bahndaten mehrerer Planeten—also der eigentliche Systemcharakter—viel ausgeprägter ist.' Ibid. 342.

25 'und zwar aus dem leicht einsichtigen Grunde, weil sie im ganzen auf eben demselben Material an Beobachtungen und noch auf derselben Beobachtungsgenauigkeit beruht, die schon Ptolemäus verfügbar waren.' Ibid. 342.

> At the same time it is an obvious historical fact that the *first* effect of the Copernican reform did not consist in a consciousness of humiliation and slight of man but in a consciousness of his belonging to the world of celestial bodies as a being that assigns himself a place in the world and the possession of the world. The interpretation of the new world system as a humiliation of man was essentially enabled by the fact that Copernicus's metaphysical and cosmological premises sank into oblivion while at the same time the increasing astronomical information made man invariably more aware of the proportions of the Earth's body to the universe and thus his excentricity.[26]

Finally, Blumenberg concludes with what makes up the substance of the Copernican theory:

> The greatness of the Copernican achievement lies not the destruction of an illusion and the replacement of appearance by truth; but it is the substantiation of this appearance, the detection of the mechanism of its formation—and consequently the opening of the way by which we can see through our projections into the world and revoke them. Since Copernicus, mankind has started learning to cope with the images produced by itself, to penetrate inexorably into the realm of their implicitness.[27]

26 'Dabei liegt als historische Tatsache klar zutage, daß die *erste* Wirkung der kopernikanischen Reform nicht in einem Bewußtsein der Erniedrigung und Kränkung des Menschen bestand, sondern im Bewußtsein seiner Zugehörigkeit zur Welt der Gestirne als sich selbst Weltort und Weltbesitz bestimmendes Wesen. Die Ausdeutung des neuen Weltsystems als einer Erniedrigung des Menschen wurde wesentlich dadurch möglich, daß die metaphysischen und kosmologischen Voraussetzungen des Kopernikus in Vergessenheit gerieten, während gleichzeitig die wachsenden astronomischen Kenntnisse die Proportion des Erdkörpers zum Weltall und damit die Exzentrizität des Menschen immer erschreckender ins Bewußtsein treten ließen.' Ibid. 347.

27 'Das Große der kopernikanischen Leistung ist nicht die Zerstörung einer Illusion und die Ersetzung des Scheins durch Wahrheit; sondern es ist die Begründung dieses Scheins, der Nachweis des Mechanismus seiner Entstehung—und damit die Erschließung des Weges, auf dem wir unsere Projektionen in die Welt hinein durchschauen und zurücknehmen können. Seit Kopernikus hat die Menschheit begonnen zu lernen, mit den von ihr selbst erzeugten Bildern fertig zu werden, unaufhaltsam in das Reich ihrer Selbstverständlichkeiten einzudringen.' Ibid. 354.

Just before this, Blumenberg quotes Copernicus's central sentence from *De Revolutionibus*: 'ne [...] quae telluris sint, attribuamus caelestibus'—'we shall not attribute to heaven what belongs to earth'.

•••

In the second part of my article I will try to explore if any aspect of Blumenberg's sharp-sighted observations had been accessible to earlier times. I will focus on two texts from the middle of the 18th century that follow historically after the extinct Biblicist polemics and the well-researched complex of physico-theology. Through the edition of *Copernicus-Biographien des 16. bis 18. Jahrhunderts. Texte und Übersetzungen* (2004, *Copernicus Biographies from the 16th to the 18th Centuries. Texts and translations*) both texts are within easy reach but they have been widely neglected in Copernicus scholarship because of its preoccupation with the history of science and philosophy. The first text is Johann Christoph Gottsched's "Commemorative Speech on the Immortally Merited Canon in Frauenburg Nicolaus Copernicus, As the Inventor of the True World Edifice"[28] from the year 1743; the second is Johann Gottfried Herder's "Something About Nicolaus Copernicus's Life, Towards His Image",[29] published anonymously in *Teutscher Merkur* in 1776.

Gottsched made his speech in May 1743 in his capacity as the rector of Leipzig University, in the presence of the princes Friedrich Christian and Xaver August, the sons of the Polish King August III (as Elector of Saxony: Friedrich August II). The date of the celebration and the political authority of the participants certainly ensured that Copernicus would be eulogised in Leipzig.

In its exordium, the speech argues that it is the duty of posterity to keep alive the memory of great men. Being assured of such adoration and posthumous fame would serve as an incitement to the living towards the highest scientific achievements. Serving the past is thus transformed into serving the common welfare of present days. The speech addresses the princes directly as patrons of science; it falls to them and their royal father to further the welfare of the home country in this capacity.

---

28 Gottsched Johann Christoph, "Gedächtnisrede auf den unsterblich verdienten Domherrn in Frauenburg Nicolaus Copernicus, als den Erfinder des wahren Weltenbaues", in: Kühne A. – Kirschner S. (eds.), *Biographia Copernicana. Die Copernicus-Biographien des 16. bis 18. Jahrhunderts* (Berlin: 2004) 240–267.

29 Herder Johann Gottfried, "Etwas von Nikolaus Kopernikus Leben, zu seinem Bilde" in Kühne A. – Kirschner S. (eds.), *Biographia Copernicana. Die Copernicus-Biographien des 16. bis 18. Jahrhunderts* (Berlin: 2004) 292–299.

The proposition ties on to this argumentation, which is at the same time general and situationally specific, announcing that Copernicus's life and achievements would be honoured in an act of exemplary commemoration. The text then introduces Copernicus as a 'great ornament of the 16th century', as 'sole originator of the new astronomy and science'[30] and as 'first inventor and instructor of the true world edifice'.[31] Taking up the title of the speech, this formulation concludes the proposition.

Touching on Copernicus's place and date of birth and family relations, Gottsched's argumentation mentions his education that brought Copernicus to Cracow University via the grammar school in Toruń. The mention of mathematics and astronomy leads the text into a long digression dealing with the history and dignity of celestial observation; the digression names Anaxagoras and Thales, Georg Peurbach, Johannes Regiomontanus and Dominicus di Novara.[32]

Arguing in the Renaissance style of artists' biographies, the speech makes a point in stating that this Dominicus immediately accepted his highly gifted pupil Copernicus 'as an assistant in his astronomical work. Verily, not a mean honour! if you consider that Copernicus then was not quite four and twenty years old'.[33]

Copernicus's mathematics professorship in Rome is mentioned next; it supplied him with such success that he 'even attracted artists and craftsmen to his lecture hall'.[34] The merits Copernicus gained in Warmia as a canon and physician in long decades after his time in Rome are dealt with in one short paragraph; after all, he then only did what many others had done as well.

The argumentation instead turns to two achievements that are considered outstanding: firstly, Copernicus serves as an example to refute the prejudice that men with a great gift for theory are inept at practical matters. Copernicus the politician and Copernicus the economist are the centre points of the speech's explanations.[35] Secondly, the text postulates that Copernicus as a scientist by far surpassed his achievements in those fields.

30 'große Zierde des sechszehnten Jahrhunderts [...] einziger Urheber der neuen Astronomie und Naturkunde'. Gottsched, "Gedächtnisrede" 245.

31 'erster Erfinder und Lehrer des wahren Weltbaues'. Ibid. 246.

32 Cf. ibid. 247.

33 'zum Gehülfen seiner astronomischen Arbeiten [...]. Fürwahr, keine geringe Ehre! wenn man erwäget, daß Copernicus damals noch nicht vollkommen vier und zwanzig Jahre alt gewesen'. Ibid.

34 'so gar Künstler und Handwerker in seinen Hörsaal locket'. Ibid. 248.

35 Cf. Ibid. 248–250.

> God's great works in nature moved his wits to such a degree and made him fall in love with them so strongly that in comparison he considered even the most important feats of mankind as mere child's play. Oh thrones! oh dominions! oh bishoprics! oh war and court offices! and what else human delusion would consider to be great: you cannot blind a true philosopher. The cognition of the hidden truth holds much stronger appeals for him; and the miracles of divine omnipotence pervade his soul so much that in contrast he regards all the world's treasures and splendours as trifles. Thus thought our Copernicus. And this especially shall render me the most important reason for true fame.[36]

Here, Copernicus the destroyer of illusions emerges, albeit quite in keeping with the idealised image of the great philosopher who thinks nothing of mundane honours and strives for the truth exclusively. Yet, Gottsched does not see the search for truth in itself as the distinguishing criterion for this kind of fame but rather the novelty of the results that follow from the search. Consequently, the text pits Columbus as a discoverer against Vespucci and names from more recent times the scientific innovators Simon Marius, the discoverer of the moons of Jupiter, Otto Guericke, the inventor of the air pump, Leibniz as a mathematician and Leonhard Christoph Sturm, the mathematician and theoretician of architecture.[37]

Like those, Copernicus was dissatisfied with the tradition of his discipline.

> An edifice, he said as it were to himself, erected by a prudent master builder, must be regular, suitable and beautiful; but where disorder, unpleasantness and confusion rule, certainly an ignorant person must have supervised the work. Now there was, according to the common teachings of the stargazers who adopted the Ptolemaic world edifice and placed the Earth in the centre of the world, nothing but confusion and disorder in it. [...] They built circle above circle and aggravated them with

36 'Die großen Werke Gottes in der Natur hatten seinen Verstand dermaßen gerühret, und so stark in sich verliebt gemacht, daß er auch die wichtigsten Werke der Menschen nur für ein Kinderspiel dagegen ansah. Ihr Thronen! ihr Herrschaften! ihr Bischofthümer! ihr Kriegs- und Hofämter! und was sonst der menschliche Wahn für groß zu achten pflegt: ihr alle könnt einen wahren Weltweisen nicht blenden. Die Erkenntniß der verborgenen Wahrheit hat viel stärkere Reizungen für ihn; und die Wunder der göttlichen Allmacht erfüllen seine Seele so sehr, daß er aller Welt Schätze und Herrlichkeiten darüber für Kleinigkeiten hält. So dachte unser Copernicus. Und dieses eben soll mir den wichtigsten Grund zu einem wahren Ruhme abgeben'. Ibid. 251.

37 Cf. Ibid.

> yet more circles to explain the course of the planets. They confabulated crystal skies, a fiery sphere, spiral lines and other phantasms as much as they liked. Copernicus, however, [...] as a philosopher understood: that the blame for this disorder had to be put on men much rather than on the sagacious creator of the world; yay that the reason for such confusions had to be based only on the imperfection of human reason.[38]

In this way, Gottsched's speech indeed does not save the text of the Bible but at least it saves the Christian god and science, which can continue postulating this god as the rational reason for the existence of all things. An embedded sentence reveals the emergence of the lone hero of the intellect: Copernicus makes his axiomatic claim 'as it were to himself'.

Quite obviously, Gottsched partakes in Blumenberg's later insight that Copernicus exposed the condition of the possibility of man's dazzlement, i.e. sensual evidence. In a historical review from antiquity onwards, the speech subsequently turns towards this sensual evidence and its potentially false implications. Copernicus demonstrated the correctness of his model on the basis of observations but in terms of and by 'good reasoning and highly probable proofs'.[39] Evidence, on the one hand, is the condition of possibility for fallacy, yet, at the same time, at the hands of a great mind that will not be dazzled, it is the condition of possibility for 'good reasoning and highly probable proofs' that transcend it.

Gottsched's speech here almost comes to an end; all that remains is to sum up and evaluate some historical reactions to the new model. Even Luther, although not mentioned by name but still identifiable, is branded a deluded spirit for his disdainful remarks about Copernicus.[40] All progress in the science

38 'Ein Gebäude, sprach er gleichsam zu sich selbst, das von einem klugen Baumeister aufgeführet worden, muß ordentlich, bequem und schön seyn: wo aber Unordnung, Widerwärtigkeit und Verwirrung herrschen, da hat gewiß ein Unverständiger die Aufsicht über das Werk gehabt. Nun war aber, nach der gemeinen Lehre der Sternseher, die den prolemäischen Weltbau annahmen, und die Erde in den Mittelpunkt der Welt setzten, nichts als Verwirrung und Unordnung in derselben. [...] Man bauete Zirkel über Zirkel, und beschwerete diese noch mit neuen Kreisen, den Lauf der Planeten zu erklären. Man erdichtete kristallene Himmel, eine Feuersphäre, Schraubenlinien, und andre Hirngespinste, so viel man wollte. Copernicus aber [...] sah als ein Weltweiser wohl ein: daß die Schuld dieser Unordnungen weit eher auf die Menschen, als auf den weisen Schöpfer der Welt fallen mußte; ja daß der Grund solcher Verwirrungen, bloß auf der Unvollkommenheit des menschlichen Verstandes beruhen müsse.' Ibid. 252.

39 'gute Vernunftschlüsse, und höchstwahrscheinliche Beweise [...]'. Ibid. 253.

40 Cf. ibid. 254, footnote 14.

of astronomy is made manifest through the names of—amongst others—Tycho Brahe, Kepler, Galilei, Gassendi, Halley and Newton.

With an apostrophe the text formulates the conclusion of its argumentation: Copernicus himself is called upon and thanked: 'You have shown us a specimen of true freedom in philosophy!'[41] Nevertheless, the speech only ends after one more, rather longish apostrophe that addresses the two princes; as patrons of the sciences they enhance the fame and welfare of Saxony.

By comparison, Herder's argumentation follows a different path, focussing on Copernicus the man of practice rather than on the philosopher. The text starts with an extensive depiction of the facts upon which Copernicus built: the antecedent heliocentric theories. 'All pieces of the Copernican opinion were, in other words, old: he himself did not deny having built his edifice upon those very ruins. But he was the man of such power to build it'.[42] Copernicus is thus introduced as a genius—albeit without a mention of the term. Here, Herder's argumentation coincides with Gottsched's on a deeper level: both texts emphasise the lone hero.

Copernicus's early, conventional education in inept places stands out in sharp contrast to his swift lustre und growing renown. 'Nicolaus Copernicus was born in a country which is almost regarded a literary desert [...] and was educated in a country which is almost even more so regarded, at Cracow in Poland'.[43]

The subsequent argumentation is intrinsically focussed on aesthetical criteria; aesthetics[44] is its subject and medium of proof at the same time. Ptolemy's model, says Herder's text, lacks order and consequently disturbed Copernicus's 'sense for symmetry and relation towards the whole', his 'sense for drawing'. 'So, imagination, painting, poetry beckoned up to the greatest discoveries which we think there are, and held the ladder'.[45] Subsequently, the text indeed touches on the biography of Copernicus and recalls with good reason that the new

41 'Du hast uns ein Muster einer wahren Freyheit im Philosophiren gewiesen!' Ibid. 257.

42 'Alle Stücke der Kopernikanischen Meynung waren also schon alt: er selbst leugnete es nicht, daß er eben auf diesen Trümmern zu seinem Gebäude gekommen. Er aber war der Mann von Kraft, ders baute [....]' Herder, "Etwas von Nikolaus Kopernikus Leben" 294.

43 'Nikolaus Kopernikus ward in einem Lande geboren, das fast für eine literarische Wüste gilt [...] und ward in einem Lande erzogen, das fast noch mehr dafür gilt, zu Krakau in Pohlen'. Ibid.

44 Cf. also Jäger M., *Die Ästhetik als Antwort auf das kopernikanische Weltbild. Die Beziehungen zwischen den Naturwissenschaften und der Ästhetik* (Hildesheim – Zürich – New York: 1984).

45 'Sinn für Symmetrie und Verhältniß zum Ganzen [...] Zeichnungsgefühl [...] Zu den größten Entdeckungen also, die wir dafür halten, winkte Einbildung, Malerey, Poesie herauf und hielt die Leiter'. Herder, "Etwas von Nikolaus Kopernikus Leben" 294.

theory first enjoyed the highest ecclesiastic esteem; the text's focus, however, is on Copernicus's portrait (in the version by Jean Jacques Boissard), which serves as an artistic object and physiognomic medium.[46] As early as at the beginning of the second paragraph, the text says: 'At the image of Copernicus we want to tell the reader something about the man'.[47] At the beginning of its last third the argumentation reverts to the image.

> What Copernicus alone, in other words, kept so completely and long to himself was what we also read in his face: the unbiased quiet, the juvenile gazing ahead without insolence and pretensions combined with the strength, with the retention to himself which the figure of the noble Sarmatian shows. One sees, the man clearly looks out of himself; [...] Earth is just as little the centre of his being as of his world edifice.[48]

Finally, the relatively ample discussion of the mathematical gain through Copernicus's book also hinges on this aesthetic-psychological operation of the text. With good reason Herder states that the calculations by Tycho Brahe and other successors were more exact. This is explained by calling Copernicus a great mind; 'pettiness he abhorred heartily'.[49] The text concludes with another referral to the tradition of Copernicus portraits: 'I do neither know if the image by Boissard is equal to the Strasburg one which Bernegger had brought from Prussia? nor if the one painted by Copernicus himself for Tycho exists somewhere?'[50]

In my view, Gottsched's and Herder's differing horizons of exegesis do not so much betray their respective functional contexts but rather more so a historical caesura. Gottsched's Copernicus is an enlightenment philosopher who elucidated contexts of delusion, which in a sense puts him in line with Blumenberg's point. Herder's Copernicus, on the other hand, is a great mind

46 Cf. also Claus Zittel's essay in this volume.

47 'Wir wollen bey Kopernikus Bilde dem Leser etwas vom Manne sagen.' Herder, "Etwas von Nikolaus Kopernikus Leben", 294.

48 'Was also Kopernikus allein so ganz und lange in sich hielt, war, was wir auch in seinem Gesicht lesen, die unbefangne Ruhe, das jugendliche Vorsichblicken ohn' Anmaaßung und Prätensionen, verbunden mit der Stärke, mit der Haltbarkeit auf sich selbst, die die Gestalt des edlen Sarmaten weiset. Man siehet, der Mann blickt rein aus sich heraus; [...] die Erde ist so wenig der Mittelpunkt seines Daseyns, als seines Weltgebäudes.' Ibid. 296f.

49 'Kleinkrämerey war er von Herzen Gram'. Ibid. 296.

50 'Ich weiß nicht, ob es [sc. das Bild von Boissard] dem Strasburgischen gleich ist, das Bernegger aus Preussen kommen ließ? noch ob das von Kopernikus selbst für Tycho gemahlte irgendwo existire?' Ibid. 298.

operating in and from the spirit of aesthetics; his world-historical achievement does not need to be substantiated any more, only Copernicus the man is of interest here.[51] Both texts, however, coincide in one point: marking and commemorating Copernicus as a self-confident and—almost by implication—lonely individual: a true bourgeois hero.

Selective Bibliography

Blumenberg H., "Kopernikus im Selbstverständnis der Neuzeit", in *Akademie der Wissenschaften und der Literatur Mainz, Abhandlungen der geistes- und sozialwissenschaftlichen Klasse* (Wiesbaden: 1964) 337–368.

Copernicus Nicolaus, *De revolvtionibvs orbium cœlestium, libri VI.* (Nuremberg, Johann Petreius: 1543).

Freud S., "Eine Schwierigkeit der Psychoanalyse", *Imago. Zeitschrift für Anwendung der Psychoanalyse auf die Geisteswissenschaften* V (1917). 1–7.

Freud S., "Vorlesungen zur Einführung in die Psychoanalyse [1916–17], chapter 18: Die Fixierung an das Trauma, das Unbewußte", in idem, *Gesammelte Schriften* (Leipzig – Vienna – Zurich: 1924) vol. VII, 282–295.

Gottsched Johann Christoph, "Gedächtnisrede auf den unsterblich verdienten Domherrn in Frauenburg Nicolaus Copernicus, als den Erfinder des wahren Weltenbaues", in: Kühne A. – Kirschner S. (eds.), *Biographia Copernicana. Die Copernicus-Biographien des 16. bis 18. Jahrhunderts* (Berlin: 2004) 240–267.

Günther F.F., "Die kopernikanische Wende als anthropologische Denkfigur bei Nietzsche und Elias", in Günther F. – Holzer A. – Müller E. (eds.), *Zur Genealogie des Zivilisationsprozesses. Friedrich Nietzsche und Norbert Elias* (Berlin: 2010) 175–191.

Heinse Daniel, *Ardinghello* (Leipzig: 1961).

Herder Johann Gottfried, "Etwas von Nikolaus Kopernikus Leben, zu seinem Bilde" in Kühne A. – Kirschner S. (eds.), *Biographia Copernicana. Die Copernicus-Biographien des 16. bis 18. Jahrhunderts* (Berlin: 2004) 292–299.

Jäger M., *Die Ästhetik als Antwort auf das kopernikanische Weltbild. Die Beziehungen zwischen den Naturwissenschaften und der Ästhetik* (Hildesheim – Zürich – New York: 1984).

Kant Immanuel, *Critique of Pure Reason*, trans. N. Kemp-Smith (London: 1950).

Kluwe S., "Trauma und Triumph: Die kopernikanische Wende in Dichtung und Philosophie. Stationen der Rezeptionsgeschichte von Gryphius bis Nietzsche", in:

---

51 A wider framework would be required to show to what extent Herder's position provides the condition of possibility for later literary representations of Copernicus.

Brodersen K. – Gebhardt H. (eds.), *Weltbilder*, Heidelberger Jahrbücher 47 (Berlin – Heidelberg – New York: 2003) 179–220,

Kühlmann W., "Neuzeitliche Wissenschaft in der Lyrik des 17. Jahrhunderts. Die Kopernikus-Gedichte des Andreas Gryphius und Caspar Barlaeus im Argumentationszusammenhang des frühbarocken Modernismus", *Jahrbuch der Deutschen Schillergesellschaft* 23 (1979) 124–153.

Nietzsche F., *Jenseits von Gut und Böse*, in idem, *Sämtliche Werke, Kritische Studienausgabe in 15 Bänden* (KSA), eds. G. Colli – M. Montinari (Munich – New York: 1980) vol. V.

Russell H.N., "Copernicus' Role in Kant's Revolution" *Journal of the History of Ideas* 20 (1959) 274–281.

Schröder R., "War die copernicanische Reform der Astronomie ein Weltbildwandel?", in Markschies C. – Zachhuber J. (eds.), *Die Welt als Bild : interdisziplinäre Beiträge zur Visualität von Weltbildern*, Arbeiten zur Kirchengeschichte 107 (Berlin: 2008) 91–112.

Ziegler und Kliphausen Heinrich Anshelm von, *Täglicher Schau-Platz der Zeit* [...] (Leipzig, Gleditsch: 1695).

Zittel C., "Boscovich", in Ottmann H. (ed.), *Nietzsche-Handbuch* (Stuttgart: 2000) 404–409.

CHAPTER 12

Max Brod: *Tycho Brahes Weg zu Gott*

Jörg Jungmayr

On the occasion of Max Brod's 100th birthday (24 May 1984), Peter Demetz published an article in the *Frankfurter Allgemeine Zeitung*, stating with much regret that Brod had almost completely vanished from the perception of the literary public. Demetz says there:

> The many Prague authors of German tongue and pen have turned into background actors without history in the literary perception of our times, gathering together in the backdrop around Franz Kafka and Rainer Maria Rilke. [...] It has fallen to Max Brod [...] to be praised as a faithful friend and mediator, the works of whom to read no one bothers. He is rather famous for what he did *not* do on one occasion (i.e. destroy Kafka's manuscripts, as his friend had requested) than for everything he thought and wrote during a long, productive life in his two homes, in Prague and Tel Aviv, as an author and critic [...].[1]

These findings can be corroborated by the status of research well into the new millennium. Looking into the secondary material concerning Max Brod up to around 2004, one quickly establishes that Brod—if at all—is mainly dealt with in a context with Franz Kafka.[2] A certain interest was focused on Brod the Zionist; else, one only finds a few essays regarding single literary or

1 'Die vielen Prager Schriftsteller deutscher Zunge und Feder haben sich im literarischen Bewußtsein unserer Epoche zu geschichtslosen Statisten verwandelt, die sich im Hintergrund um Franz Kafka und Rainer Maria Rilke scharen. [...] Max Brod [...] fällt dabei das besondere Schicksal zu, als treuer Freund und Mittler gerühmt zu werden, ohne daß man sich die Mühe nähme, seine eigenen Arbeiten zu lesen. Er ist eher berühmt für das, was er einmal nicht getan hat (nämlich Kafkas Manuskripte zu vernichten, wie es sein Freund wünschte), als für all das, was er in einem langen und produktiven Leben, in seiner doppelten Heimat in Prag und Tel Aviv, als Schriftsteller und Kritiker dachte und schrieb [...].' Demetz P., "Das Genie von der Post. Zum hundertsten Geburtstag: Max Brod, der Schriftsteller", *Frankfurter Allgemeine Zeitung* 119 (1984) 25.

2 Cf. for example on the friendship between Brod and Kafka, which was quite problematic and determined by Brod's arrogance: Koch H.G., "Kafkas Max und Brods Franz. Vexierbild einer Freundschaft", in Plachta B. (ed.), *Literarische Zusammenarbeit* (Tübingen: 2001) 245–256.

 | DOI 10.1163/9789004281127_014

philosophical texts.[3] E.g., only one study appeared in the mentioned period, that is in 1992, with reference to *Tycho Brahe*: "Angoisse et créativité dans *Tycho Brahes Weg zu Gott*" by Marie-Odile Blum.[4]

More comprehensive collections of essays, monographs or biographies are almost totally missing; on Brod's 100th birthday, a collected volume was published by Margarita Pazi,[5] as well as a biographical sketch, especially refurbished for the occasion, by Berndt W. Wessling;[6] a catalogue of an exhibition in Vienna in 1987[7] and an essay collection edited by Klaus-Eckehard Bärsch, "Max Brod im Kampf um das Judentum"[8] of 1992 can be mentioned—but apart from these: Brod appears as an appendix to Kafka, and in this context less so as his discoverer and patron but rather as a quantité négligeable.

The acknowledgement of Brod as an immensely prolific author, a blessed networker and a phenomenal headhunter with an infallible instinct for new talents has evaporated almost completely. Who indeed still knows that Brod not only gave the initial spark to the unending Kafka boom but that he also discovered Franz Werfel and Robert Walser,[9] who were to become icons of 20th-century German-language literature, and beside them lesser-known authors like the expressionist Franz Janowitz (getting himself involved in a fierce

3 Cf. exemplarily here: Schillemeit J., "Chancen und Grenzen der Willensfreiheit. Fragen der Ethik und Religionsphilosophie bei Max Brod, Felix Weltsch und Kafka (1920)", in *Prager deutschsprachige Literatur zur Zeit Kafkas* (Vienna: 1989) vol. I, 165–181; printed again in Schillemeit J. (ed.), "Kafka-Studien" (Göttingen: 2004) 327–347.

4 Blum M.O., "Angoisse et créativité dans 'Tyhos Brahes Weg zu Gott' de Max Brod", in Paul J.M. (ed.), *Le pessimisme. Idée féconde, idée dangereuse* (Nancy: 1992) 227–238.

5 Pazi M., *Max Brod 1884–1984. Untersuchungen zu Max Brods literarischen und philosophischen Schriften* (New York et al.: 1987).

6 Wessling B.W., *Max Brod. Ein Porträt zum 100. Geburtstag* (Gerlingen: 1984).

7 Lerperger R., *Max Brod. Talent nach vielen Seiten* (Vienna et al.: 1987).

8 Bärsch C.E., *Max Brod im Kampf um das Judentum Zum Leben und Werk eines deutsch-jüdischen Dichters aus Prag* (Vienna: 1992).

9 In *Arkadia. Ein Jahrbuch für Dichtkunst*, the only issue of which appeared Leipzig in 1913 at Kurt Wolff publishers, the editor Max Brod published texts among others by Franz Kafka, Robert Walser, Franz Werfel und Franz Janowitz. Regarding Brod and Werfel cf. Brod M., *Streitbares Leben (1884–1968)*, 2nd edn. (Frankfurt a.M.: 1979) 11–47; Brod M., *Der Prager Kreis* (Frankfurt a.M.: 1979) 169–172; cf. also Vassogne G., *Max Brod in Prag. Identität und Vermittlung* (Tübingen: 2009) 31–35, 179–180. Regarding Brod and Walser cf. Brod, *Streitbares Leben* 74–75, 244–253; also Vassogne, *Brod in Prag* 179–181. Cf. also Wagner K., " 'Direktoriell'. Robert Walser liest Max Brod", in Fattori A. – Gigerl M. (eds.), *Bildersprache, Klangfiguren. Spielformen der Intermedialität bei Robert Walser* (Munich: 2008) 155–167.

dispute with Karl Kraus about the 'right of primogeniture' to Janowitz)?[10] Who still remembers that it was Brod who made his Czech compatriots recognise Jaroslav Hašek's *Soldat Schweijk*, which had been published in cheap booklet instalments, as an epochal occurrence?[11] Who bothers about the aesthetician, the philosopher of religion and socialist, the physicist and mathematician Max Brod?[12] Who has a notion of the fact that with his commitment he helped one of the most important composers of the 20th century, Leoš Janáček, achieve

10 In 1913, Brod published 16 poems by Franz Janowitz in *Arkadia*. Karl Kraus in his obituary *In memoriam Franz Janowitz* in Mai 1918 (*Die Fackel* 474–483 [Mai 1918] 69–70) reproached the editor of *Arkadia* to have abused the inexperience of a young author. Karl Kraus himself had published collected poems by Franz Janowitz in 1919 under the title *Auf der Erde* and furnished them with introductory verses. Cf. on this Brod, *Streitbares Leben* 76–80; also Davian D.G., "Max Brod and Karl Kraus", in Pazi, *Max Brod 1884–1984* 207–231, here 223. Cf. also Janowitz F., *Auf der Erde und andere Dichtungen. Werke, Briefe, Dokumente*, ed. D. Sudhoff (Innsbruck: 1992) 145–159, 183–186, 255f.

11 Cf. on this Brod, *Streitbares Leben* 264–267; Šrámková B., *Max Brod und die tschechische Kultur* (Wuppertal: 2010) esp. 237–249; Vassogne, *Brod in Prag* 205–213. Cf. also Völker K., "Hašeks 'Schweijk'-Roman auf der Bühne. Die Piscator Inszenierung von 1928. Von Brod zu Brecht und die Folgen", in Pazi M. – Zimmermann H.D. (eds.), *Berlin und der Prager Kreis* (Würzburg: 1991) 225–241.

12 Cf. on this Brod, *Streitbares Leben*, esp. "Begegnungen an der Hochschule. Philipp Frank, Einstein, Alfred Weber, Christian Ehrenfels, Hugo Bergmann" 197–241. The knowledge of physics that Max Brod acquired with Ernst Lecher, the originator of metrology in the field of high frequency technology, came in handy for his Tycho Brahe and the Galilei. Cf. Brod, *Streitbares Leben* 199: 'But since a scientific department also belonged to medicine there was one among them whose student I became full of joy: Professor Lecher who propounded "experimental physics" in the most ingenious and witty way. For two semesters, I did not miss a single of his lectures and still remember the amphitheatrically built auditorium with the many apparatuses exactly as a space of the solemn muse; [...]. The room with the scales, mercury columns, retorts, contrivances for measuring the height of fall, with all the working devices the memory images of which then helped me when I wrote my "Tycho Brahe" and "Galileo".' – 'Doch da zur Medizin auch die naturwissenschaftliche Abteilung gehörte, war doch einer unter ihnen, dessen Hörer ich mit wahrer Freude wurde: Professor Lecher, der auf die ingeniöseste und witzigste Art 'Experimental-Physik' vortrug. Ich versäumte zwei Semester lang keine seiner Stunden und habe den amphitheatralisch gebauten Hörsaal mit den vielen Apparaten noch genau als einen Raum der ernsten Muse in Erinnerung; [...]. Den Raum mit den Waagen, Quecksilbersäulen, Retorten, Apparaten zur Messung der Fallhöhe, mit alle dem Arbeitsgerät, dessen Erinnerungsbilder mir dann halfen, als ich meinen 'Tycho Brahe' und den 'Galilei' schrieb."

his international breakthrough?[13] Who indeed still reads his numerous novels, which often reminisce Prague, with their historical, philosophical, erotic content—who would even publish them anew?

On the other hand, however, the fact that Brod has fallen so much out of time may present a chance to rediscover and re-read him beyond all current compulsion to utilise him—many things point to this if one looks at the Brod research of the last ten years. In addition to studies on single aspects, especially the monographs by Pavel Doležal, Gaëlle Vassogne and Barbora Šrámková deserve mention, which deal with Brod as a mediator and bridge builder between nationalities, cultures and religions.[14]

Brod's novel *Tycho Brahes Weg zu Gott* has found attention again: Arno A. Gassmann in his study on the father-son conflict in authors of the Prague Circle also deals with *Tycho Brahe*[15] and since 2007 a new edition of the English translation by Felix Warren Crosse (1928) is available again, with an introduction by Peter Fenves.[16]

A big problem, however, presents itself to the study of this novel and moreover of Brod's complete oeuvre: his bequest is inaccessibly locked away in banks in Tel Aviv and Zurich about which a rancorous dispute has flared up between the National Library of Israel on the one hand and Eva Hoffe and Ruth Wiesler on the other. The two are the daughters and heiresses to Ilse Ester Hoffe, to whom Brod bequeathed his legacy in 1961.[17] What makes this fierce battle so depressing is that the sole interest in it focuses on still-unknown Kafka manuscripts and the expectation of a colossal profit that comes with

13 Cf. on this Brod, *Streitbares Leben* 267–278; Šrámková, *Tschechische Kultur*, esp. "Die große Entdeckung. Leoš Janáček" 76–93; Vassogne, *Max Brod in Prag* 191–193, 196–197, 214–222.

14 Doležal P., *Tomáš G. Masaryk, Max Brod und das Prager Tagblatt (1918–1938). Deutsch-tschechische Annäherung als publizistische Aufgabe* (Frankfurt a.M. et al.: 2004), Vassogne, *Max Brod in Prag*; Šrámková, *Tschechische Kultur*. The correspondence between Max Brod and Hans-Joachim Schoeps was published in 2nd edn.: Schoeps J.H. (ed.), *Im Streit und das Judentum. Max Brod – Hans-Joachim Schoeps. Briefwechsel* (Hildesheim et al.: 2011).

15 Gassmann A.A., *Lieber Vater, Lieber Gott? Der Vater-Sohn-Konflikt bei den Autoren des engeren Prager Kreises (Max Brod – Franz Kafka – Oskar Baum – Ludwig Winder)* (Oldenburg: 2002) 48–58.

16 Brod M., *Tycho Brahe's path to God. A novel*, trans. F.W. Crosse (Evanston, Ill.: 2007).

17 Cf. on this Schult C., "Die Erbschaft", *Der Spiegel* 40 (2009) 152–154. Also Koldehoff S., "Kafka, Brod, Israel und ein Prozess in alle Ewigkeit", *Die Welt* (22.07.2010): URL = <http//www.welt.de/kultur/article8576264/Kafka-Brod-Israel-und-ein-Prozess-in-alle-Ewigkeit.html>; Müller A.L., "Der Prozess. Posse, Drama oder Krimi?", *Jüdische Zeitung* (November 2009): URL = <http://j-zeit.de/archiv/artikel.2123.html>; cf. also the interview by Lothar Müller with Andreas Kilcher in the *Süddeutsche Zeitung* (26.09.2010): URL = <http://www.sueddeutsche.de/kultur/2.220/nachlass-von-franz-kafka-wem-gehoert-kafka-1.004762>.

them, while the voluminous correspondence by Brod and his diary are at best of marginal interest. As long as the bequest is accessible neither in the National Library in Jerusalem nor in the German Literary Archive in Marbach or elsewhere, important questions concerning the *Tycho* novel cannot be answered at all or only rudimentarily: with whom did Brod exchange ideas on his novel, what literature did he use for his preparatory work,[18] how did work on the text proceed, in what timeframe was it accomplished?

Tycho Brahe is arguably Brod's most successful novel, marking his change of position from 'indifferentism'[19] to 'transcendental realism'[20] and positioned

18 Did Brod know the first Brahe biography that the French philosopher and scientist Pierre Gassendi propounded in 1654 [*Tychonis Brahei vita. Acessit Nicolai Copernici, Georgii Peurbachii, & Johannis Reiomontani, astronomorum celebrium, vita* (The Hague, Adrian Vlacq: 1654)]? Did Brod consult the astronomical literature, especially the one published in Prague (Dvorský F., *Neues über J. Kepler* [1880]; Hasner J., *Tycho Brahe und J. Kepler in Prag* [1872]; *Brevissimum planimetriae compendium. Sua manu exaravit Tycho Brahe*, ed. F.J. Studnicka [1903])? What additional (popular) science literature of his time (e.g. Guenther L., *Kepler und die Theologie. Ein Stück Religions- und Sittengeschichte aus dem 16. und 17. Jahrh* (Gießen: 1905); Herz N., *Keplers Astrologie* (Vienna: 1895); Dreyer J.L.E., *Tycho Brahe. Ein Bild eines wissenschaftlichen Lebens und Arbeitens im 16. Jahrhundert*, transl. M. Bruhns (Karlsruhe: 1894) did Brod use?

19 Cf. on this Brod, *Prager Kreis* 155: 'Everything, however, was overshadowed by the question: "Is the human will free? Or is it, as Schopenhauer puts it, 'necessitated', brought to its decisions with necessity?" I did not find a gap anywhere in the proofs of my god Schopenhauer and had to resign to it. This "laying down of the arms of ethics" turned into indifferentism for me within which I found good and bad equally commendable ("omnia admirari") and on an equal footing because they were caused in the same compulsive way. We have no possibility to choose. We are slave to to the causal chain, the cruel device that Kafka calls the "killer chain".'—'Doch alles wurde von der Frage überschattet: 'Ist der menschliche Wille frei? Oder wird er, wie Schopenhauer sagt, "nezessitiert", mit Notwendigkeit zu seinen Entschließungen gebracht?' Ich fand nirgends eine Lücke in den Beweisen meines Gottes Schopenhauer und mußte mich dreinschicken. Diese "Waffenstreckung der Ethik" wurde mir zum "Indifferentismus", in dem ich das Böse wie das Gute gleich preisenswert ("omnia admirari") und gleichberechtigt, weil auf gleiche Weise zwanghaft kausiert fand. Wir haben keine Möglichkeit zu wählen. Wir sind der Kausalkette, dem grausamen Apparat verfallen, den Kafka die "Totschlägerkette" nennt.' On indifferentism cf. also Voigts M., " 'Tod den Toten!' Indifferentismus und Utopie in den frühen Novellen Max Brods", in Grave J. – Sprengel P. – Vandevoorde H. (eds.), *Anarchismus und Utopie in der Literatur um 1900. Deutschland, Flandern und die Niederlande* (Würzburg: 2005) 108–119; Vassogne, *Brod in Prag*, esp. "Eine persönliche Antwort: der Indifferentismus" 25–31.

20 Cf. on this Brod, *Prager Kreis* 203: 'Or, as Goethe expresses it in a single sentence (in a conversation with Riemer in 1827): "The spirit of the real is the true ideational." Plato's doctrine in *one* sentence! If one tried to represent *this* real and if one succeeded and

between *Schloß Nornepygge. Der Roman des Indifferenten*[21] (1908; celebrated by Kurt Hiller as one of the key texts of modernity, i.e. expressionism)[22] and *Heidentum, Christentum, Judentum* (1921).

Brod's detachment from indifferentism was put into effect under the influence of Martin Buber. Impressed by reading Buber's *Daniel. Gespräche von der Verwirklichung*, Brod writes to the author on 5 June 1913:

> I have finished your book with extraordinary excitement. Conversations 3 and 5 appear to me to really give the most meaningful. "The danger of loving"—with this expression, I believe, you have even deeper penetrated the bottom of these things that are so hard to represent. [...] Thus, I appreciated that on page 143ff. you referred to my theory of "indifferentism" if I am not all mistaken.—To what extent I now regard this theory, which I advocated in my first books, only as an immature preliminary stage and misunderstanding of my present "ecstasy of redemption", I am

---

if one could also express it in one's own existence, in life—would not this be the true, the *absolute realism*, which would even and in fact above all and chiefly comprise the transcendent, which would represent the divine and the human, the social, at a time, in one and the the same pure draught of a line?'—'Oder, wie Goethe es in einem einzigen Satz (im Gespräch mit Riemer 1827) ausdrückt: "Der Geist des Wirklichen ist das wahre Ideelle."—Platons Lehre in *einem* Satz! Wenn man nun *dieses* Wirkliche darzustellen versuchte und wenn es gelänge und wenn man es auch im eigenen Existieren, im Leben ausdrücken könnte—wäre das nicht der eigentliche, der *absolute Realismus*, der auch und sogar vor allem und in erster Linie das Transzendente einbezöge, der das Göttliche darstellte und das Humane, Soziale zugleich, in ein und demselben reinen Linienzug?'

21 Cf. on this Pazi M., "Max Brods Roman 'Schloß Nornepygge' (1908), in idem (ed.), *Max Brod (1884–1984)* 115–132.

22 Hiller K., "Max Brod (Stadien einer Enttäuschung)", in *Die Weisheit der Langeweile. Eine Zeit- und Streitschrift*, vol. 1 (Leipzig: 1913) 156: '"Nornepygge"—that was a thunderclap, a rage, a stupefaction; to me, the strongest, most essential, holiest experience, 1909; [...] One rages afterwards, does not know how to contain oneself, one runs about overheated and is mad with pain, with joy.'—'"Nornepygge"—das war ein Donner, eine Raserei, eine Betäubung; war mir ein stärkstes, wesentlichstes, heiliges Erlebnis, 1909; [...] Man tobt danach, man weiß sich nicht zu halten, man rennt überheiß herum und ist verrückt vor Schmerz, vor Freude.' Cf. on this additionally: idem, *Leben gegen die Zeit* (Hamburg: 1969) 82: '[...] Max Brod the poet from Prague whom we valued enormously for his consequently scepticist fantasto-novel "Schloß Nornepygge" [...].'—'[...] Max Brod, der von uns wegen seines konsequent skeptizistischen Phantasto-Romans 'Schloß Nornepygge' enorm geschätzte Dichter aus Prag [...].'

> afraid I cannot explain in brief. Perhaps I may once confide myself to you on this matter orally.[23]

In a letter from 26 November 1913, Brod interprets Buber's concept of realisation as 'ecstasy of redemption', seeing it as a social act in which a reconciliation of the rational (empirical world) and the irrational (ecstasy) is performed. To this idea he wants to give shape in his novel, *Tycho*:

> Actually, since the beginning of my development I have had in mind some kind of reconciliation of reason and the irrational. No conflagration, of course; but a concourse in a mutual, completed ideal! Hence my Tycho.—To put it concretely: Your "realisation" interests me less in its epistemological side (as it does my friend, Dr Felix Weltsch) but rather in its ethical aspect.—Realisation has an ethical element in the unconditional, the unrelated.—Indeed, even more: I believe to interpret your "Daniel" correctly when I pair it up with your introductory essay to the "Ecstatic Confessions".—(Orientation = bustle – realisation = ecstasy).[24]

In his reply from 6 December 1913, Buber firmly contradicts the equation of ecstasy and realisation: ecstasy as an escapist individual experience is 'episodic-isolating' while realisation as a social act is 'lasting-conjoining'.

---

23 'Ich habe Ihr Buch mit ungewöhnlicher Spannung ausgelesen. Das 3. und 5. Gespräch scheint mir nun wirklich das Bedeutendste zu geben. "Die Gefahr zu lieben"—mit diesem Ausdruck, glaube ich, sind Sie noch tiefer in den Grund dieser so schwer darstellbaren Dinge eingedrungen. [...] So war es mir auch lieb, daß Sie auf S. 143ff. [sich auf] meine Lehre des "Indifferentismus" bezogen, wenn ich nicht irre.—Inwiefern ich jetzt diese ganze Theorie, die ich in meinen ersten Büchern verfocht, eigentlich nur als unreife Vorstufe und Mißverständnis meiner heutigen "Erlösungs-Ekstase" betrachte, kann ich leider nicht in kurzem auseinandersetzen. Vielleicht darf ich mich Ihnen einmal mündlich in dieser Sache anvertrauen.' Buber M., *Briefwechsel aus sieben Jahrzehnten. 1897–1918* (Heidelberg: 1972) 335–336.

24 'Mir schwebt eigentlich seit Beginn meiner Entwicklung irgend eine Versöhnung von Ratio und Irrationalem vor. Natürlich keine Vermengung; wohl aber ein Zusammentreffen im beiderseitigen, zu Ende geführten Ideal! Daher mein Tycho.—Um konkret zu reden: Mich interessiert Ihre "Verwirklichung" weniger von der erkenntniskritischen Seite aus (wie meinen Freund Dr. [Felix] Weltsch) als von der ethischen her.—Die Verwirklichung hat im Unbedingten, Beziehungslosen ein ethisches Element.—Ja, noch mehr: ich glaube, Ihren "Daniel" richtig zu deuten, wenn ich ihn mit Ihrem Einleitungsessai zu den "Ekstatischen Konfessionen" zusammenbringe.—(Orientierung = Getriebe – Verwirklichung = Ekstase).'—Buber, *Briefwechsel* 348f.

> Today, instead of the exhaustive explanation I had in mind, I can give you but a few hints that should, however, satisfy you for now. [...] Realisation is by no means ecstasy. In ecstasy man experiences the soul (Confessions XII bottom to XIII middle), in realisation, the world (Daniel 71f. [I, 39f]). In ecstasy his experience is receiving (Conf. XV), in realisation, acting. Ecstasy is episodic-isloating (Conf. XIX top and XXIIIf.), realisation, acting-conjoining (Dan. 42, 44 i.a. [I, 27, 29]). In realisation, cognition and ethos are amalgamated: man can only realise the world by acting it [...].[25]

Work on Tycho Brahe was already concluded in the summer of 1914 but the outbreak of WWI delayed the printing.[26] First appearing in 1915 as a preprint in the white pages,[27] then as a book by Kurt Wolff publishers in Munich or Leipzig and Vienna respectively,[28] since 1927 furnished with a preface by Stefan Zweig,[29] the novel made it to a total circulation of 117,000 copies until

25 'Ich kann Ihnen heute statt der ausführlichen Darlegung, die ich vorhatte, nur einige Hinweise geben, die Ihnen aber zunächst genügen dürften. [...] Verwirklichung ist keineswegs Ekstase. In der Ekstase erlebt der Mensch die Seele (Konfessionen XII unten bis XIII Mitte), in der Verwirklichung die Welt (Daniel 71f. [I, 39f]. In der Ekstase erlebt er empfangend (Konf. XV), in der Verwirklichung tuend (Dan. 74ff. [I, 41ff]. Die Ekstase ist episodisch-isolierend (Konf. XIX oben und XXIIIf.), die Verwirklichung dauernd verbindend (Dan. 42, 44 u.a. [I, 27, 29]). In der Verwirklichung sind Erkenntnis und Ethos verschmolzen: der Mensch kann die Welt nur erkennen, indem er sie tut [...].' Ibid. 350f. Cf. on this also Vassogne, *Brod in Prag* 67.

26 Brod, *Prager Kreis* 160: 'My book *Tycho Brahes Weg zu Gott* (finished in 1914 just before the war, published only in 1916 [*recte 1915*], delayed by the outbreak of the war, marks the decision [*breakthrough into freedom; the author*].'—'Mein Buch *Tycho Brahes Weg zu Gott* (beendet 1914 knapp vor dem Krieg, erschienen erst 1916 [*recte 1915*], durch den Kriegsausbruch verzögert, bezeichnet die Entscheidung [*Durchbruch in die Freiheit; the author*].'

27 Brod M., "Tycho Brahes Weg zu Gott. Roman", *Die weißen Blätter* 2 (1915) 73–120, 216–241, 349–374, 506–517, 628–674, 732–804. Cf. Kayser W. – Gronemeyer K., *Max Brod: mit unveröffentlichten Briefen Max Brods an Hugo und Olga Salus und an Richard Dehmel* (Hamburg: 1972) nr. 237.

28 Brod M., "Tycho Brahes Weg zu Gott. Roman". Cf. Kayser – Gronemeyer, *Max Brod* nr. 236 and 238. Brod dedicated the novel to Franz Kafka: 'To my friend Franz Kafka'—'Meinem Freunde Franz Kafka'.

29 Also Zweig sees in *Brahe* the triumph over indifferentism and interpretes the novel in the context of Brod's *Reubeni* novel that had been published in 1925. Cf. on this the reprint of the afterword from the edn. 1955, 265–270. In the context with *Brahe*, Zweig recalls the first encounter with Max Brod in Prague. Brod for his part has the meeting take place in Vienna. Cf. on his Brod, *Streitbares Leben* 308; cf. on this also: Matuschek O., *Stefan Zweig. Drei Leben—Eine Biographie* (Frankfurt a.M.: 2006) 61: 'Curiously enough and probably

1955. In 1978 und 1984 Suhrkamp delivered another two editions but later on there were no further prints. The Max Brod Bibliography by Werner Kayser and Horst Gronemeyer lists translations into English (1928), Italian (1933), Hebrew (1935) and Danish (1950),[30] there are, however, also renderings in Czech (1917), Yiddish (1921), Polish (1922) and French (1932).

Tycho Brahe met an unrestrictedly positive resonance when it first came out. Of the numerous reviews,[31] I would like to mention four: Hugo Bergmann classes *Brahe* in a fruitful field between reflection and intuition, between Otto Weininger and Saul Tchernichovski, stating that Max Brod with his novel had found the path to himself, his Jewishness and his social obligations.[32] In the view of Rudolf Fuchs, *Brahe* is less a historical but rather a utopian novel because it calls for the democracy of the intellect.[33] Max Hermann-Neiße sees Brod as a diagnostician of his times in a direct succession to Flaubert: 'Here is the artist-creatures' chalice and ordure, asylum, barbarisation, salacious circling on an embattled disc. [...] In Brod's epic, behind symbols and garbs which fit the body brilliantly, contemporary man is being redeemed.' Hermann-Neiße closes his review with a quote from Jakob Böhme: 'Solemnity has never been more necessary than now that all carts are pushed over and in great confusion.'[34] In view of an increasing irrationalism, through which the individual is liable to '[...] lose its meaning [...]' and to turn '[...] utmost solemnity into foolish dalliance [...]', Felix Weltsch characterises *Tycho Brahe* as a champion of reason in the battle against irrationality, instinct and intuition, whereby reason '[...]

from friendliness and for dramaturgical reasons, Zweig as well as Brod have transferred in their texts their first meeting to the native cities of the other respectively. When and where they actually met for the first time cannot be established with certainty but it was probably in Prague.'—'Kurioserweise und wohl aus Freundlichkeit und dramaturgischen Gründen haben sowohl Zweig als auch Brod in ihren Texten ihr erstes Zusammentreffen in die Geburtsstadt des jeweils anderen verlegt. Wann und wo sie sich wirklich erstmals sahen, ist nicht sicher festzustellen, doch war es wahrscheinlich in Prag.'

30 On the individual titles see Kayser – Gronemeyer, *Max Brod* nr. 280ff.

31 The reviews of *Brahe* are listed in: Vassogne, *Max Brod* 342; additionally: Kayser – Gronemeyer, *Max Brod* nr. 238.

32 Bergmann H., "Max Brods neuer Roman", *Der Jude. Eine Monatsschrift* 1 (1916/17) 134–136.

33 Fuchs R., "Tycho Brahes Weg zu Gott", *Die Aktion* 6 (1916) 656f.

34 'Hier ist der Künstler-Kreaturen Kelch und Kot, Asyl, Verwilderung, schlüpfriges Kreisen auf bedrängter Scheibe. [...] In Brods Epos wird, hinter Symbolen und Trachten, die genial auf den Leib passend Symbol und Tracht sind, der Mensch unserer Tage erlöst.'—'Der Ernst ist niemals nötiger gewesen als eben itzunder, da alle Wagen umgestoßen und in großer Verwirrung sind.' Hermann–Neiße M., "Tycho Brahes Weg zu Gott", *Zeit-Echo* (1915/16) 125f.

establishes itself all the more firmly where her opponents believed to be safe from her: in warm emotion and in passionately pulsating life.'[35]

In his characterisation of the novel's two protagonists, Tycho Brahe and Johannes Kepler, Brod uses formative encounters in Prague before WWI. The figure of Brahe with his massive corporeality bears features of the composer Max Reger, whom Brod met in Prague in December 1910:

> It was in one of the strained years before WWI. We were standing on the platform of Prague State Station and saw in the window of the express train from Germany, which was just arriving, a vigorously waving giant figure whose dimensions and rhythms of movement seemed to burst the framed opening. We: that is, the gentlemen of the "Bohemian Quartet" out of whom I especially liked the second violinist Josef Suk. He had invited me to come along and welcome the famous guest. The waving giant was Max Reger. What that man meant to me is hard to explain or inexplicable at all. [...] So now this divine man stood before me, powerfully gross, a colossus rooted in the ground, a hero to look up to whom also in the flesh, fate and awe commanded me. [...] On the next day, well towards noon, we picked him up from the hotel "Blue Star", showed him the Prague castle. Now he was solemn and grand. Never again did I have so strong a feeling that around a person of genius electric sparks were spitting frantically towards the ground. I was then working on my novel "Tycho Brahe". The sanguineous irascible figure of the great astronomer—and the ungovernable composer, fully dedicated to his genius—they conflated within me into a single living person. Thus my Tycho bears many traits of Reger and those two days which I was permitted to spend in Reger's proximity.[36]

35 '[...] durch ihr Daseinsrecht erhärte[t], wo sich ihre Gegner so sicher vor ihr glaubten: im warmen Gefühl und im leidenschaftlich pulsierenden Leben.' Weltsch F., "Tycho Brahes Weg zu Gott", *Die Schaubühne* 13 (1917) vol. I, 474–478.

36 'Es war in einem der spannungsreichen Jahre vor dem Ersten Weltkrieg. Wir standen auf dem Perron des Prager Staatsbahnhofs und sahen im Fenster des eben einfahrenden D-Zuges aus Deutschland eine lebhaft winkende Riesengestalt, deren Ausmaße und Bewegungsrhythmen die gerahmte Öffnung zu sprengen schienen. Wir: das heißt die Herren des "Böhmischen Quartetts", von denen mir der Sekondgeiger Josef Suk besonders lieb war. Er hatte mich eingeladen, mitzukommen und den berühmten Gast mit abzuholen. Der winkende Riese war Max Reger. Was dieser Mann für mich bedeutete, läßt sich schwer oder gar nicht klarmachen. [...] Und nun stand also dieser göttliche Mann vor mir, machtvoll dick, ein Koloß, in die Erde eingewurzelt, ein Heros, zum dem auch leiblich emporzublicken Schicksal und Ehrfurcht mir geboten. [...] Am nächsten Tag, so ziemlich gegen Mittag, holten wir ihn im Hotel "Blauer Stern" ab, zeigten ihm die Prager

To Brahe's opponent, Johannes Kepler, Brod lends traits of Albert Einstein:

> We meet Prof Frank also in the Fanta-Circle, we debated there with him and with Einstein whose keenest interpreter Frank was. While Einstein himself was not such an orthodox "Einsteinian" and thrilled me ever anew into astonishment, even exaltation when I observed the ease with which he changed his point of view in the discussion, experimenting, tentatively taking the opposing point of view and seeing the whole now from a totally changed angle. [...] This characteristic of his scientific courage and ever-newly commencing, I have emulated in my Tycho Brahe novel in the figure of my Kepler while wishing to draw Tycho Brahe himself rather as the rigid scientist who becomes set on his system.[37]

In the constellation Brahe – Kepler, which is characterised by the reversal of the initial teacher-pupil relationship (Brahe refers to Kepler in the beginning as 'Benjamin', as 'Johannes')[38] into a pupil-teacher relationship ('But I follow

---

Burg. Jetzt war er ernst und großartig. Nie wieder habe ich so stark das Gefühl gehabt, daß um eine geniale Person die elektrischen Funken wirbelnd toll zur Erde knisterten. Ich arbeitete damals gerade an meinem Roman "Tycho Brahe". Die blutvolle jähzornige Gestalt des großen Astronomen – und der unbändige, seinem Genius völlig hingegebene Komponist—sie verschmolzen in mir zu einem einzigen lebendigen Menschen. So kommt es, daß mein Tycho in vielem die Züge Regers und jener zwei Tage trägt, die ich in Regers Nähe zubringen durfte.' Brod, *Streitbares Leben* 256–258.

Brod dates the encounter even more precisely (257): 'It is not time yet, however—at least a year is missing until 1914. We sit and drink. Esepecially Reger is drinking eagerly.'—'Doch noch ist es nicht so weit—mindestens noch ein Jahr fehlt bis 1914. Wir sitzen und trinken. Besonders eifrig trinkt Reger.' Here, Brod is deceived by his memory for Reger gave only one concert in Prague, on 20 December 1911. Cf. on this: *Max Reger in seinen Konzerten. Programme der Konzerte Regers*, ed. Schreiber I. (Bonn: 1981) 363.

37 'Wir trafen ja Prof. Frank auch im Fanta-Kreis, wir debattierten dort mit ihm und mit Einstein, dessen eifrigster Interpret Philipp Frank war. Während Einstein selbst kein so orthodoxer "Einsteinianer" war und mich immer aufs neue zur Verwunderung, ja Begeisterung hinriß, wenn ich beobachtete, mit welcher Leichtigkeit er in der Diskussion experimentierend seinen Standpunkt änderte, sich einmal probeweise auch auf den entgegengesetzten Standpunkt zu stellen wußte und das Ganze nun unter verändertem Winkel völlig neu betrachtete. [...] Diese seine Eigenschaft des wissenschaftlichen Mutes und Immer-Neubeginnens habe ich in meinem Tycho-Brahe-Roman in der Figur meines Keplers nachgebildet, während ich in Tycho Brahe selber eher den starren, auf sein System sich versteifenden Gelehrten zeichnen wollte.' Brod, *Streitbares Leben* 201.

38 Brod M., *Tycho Brahes Weg zu Gott. Roman* (Munich: 1915) 32, 137. Subsequent quotes from this edition.

you, Kepler, you are now my paragon, my teacher, I am the pupil.'[39]), Brod reflects his complex and contradictory relationship to the young Franz Werfel:

> In 1910 the essentially more unpretentious and, as I believe, more personal "Diary In Verses" had followed. Werfel ties on to those verses with his Friend Of The World poems. [...] Thus, the rare case came to pass that the teacher was then on his knees before his pupil. A couple of years later I used this strange relationship as a basic stimulus in my novel "Tycho Brahes Weg zu Gott". [...] If truth be told, as I have hinted before, my friend Werfel contributed to the figure of Kepler things much more essential and painful than Einstein, who was but a fleeting guest on my path of fate. In Kepler I wished to depict the young genius who is favoured by luck and is urged by his dispositions and by external circumstances without being much influenced by the obligations of friendship and gratitude, while Tycho represents the older man [...]. In historical truth, Kepler resembled Tycho much more than I have shown; [...] albeit with characteristic divergences that one day lighted up to me as the idea of the whole novel and clarified the basis of this fiction as it were momentarily, in a single flash as an encounter with myself and my fate.[40]

The novel, whose plot—interspersed with retrospective insertions—concentrates on only the few months from February till October 1600 (on

39 'Ich aber folge dir, Kepler, Du bist jetzt mein Vorbild, mein Lehrer, ich bin der Schüler.' Ibid. 206.

40 '1910 war das wesentlich schlichtere, und wie ich glaube, persönlichere "Tagebuch in Versen" gefolgt. An diese Verse knüpfte Werfel mit seinen Weltfreund-Gedichten an [...]. Es ereignete sich also der seltene Fall, daß damals der Lehrer vor seinem Schüler auf den Knien lag. Einige Jahre später habe ich diese merkwürdige Beziehung in meinem Roman "Tycho Brahes Weg zu Gott" zur Grundanregung genommen. [...] In Wahrheit hat, wie schon angedeutet, zur Figur Keplers mein Freund Werfel viel Wesentlicheres und Schmerzlicheres beigetragen als Einstein, der doch nur ein flüchtiger Gast auf meinem Schicksalswege war. In Kepler wollte ich das vom Glück begünstigte junge Genie darstellen, das sich von seinen Anlagen treiben und von äußeren Umständen, daher auch von Freundschaftsverpflichtungen und Dank wenig beeinflussen läßt, während Tycho den älteren Mann darstellt, [...]. In der geschichtlichen Wahrheit war Kepler dem Tycho weit ähnlicher, als ich ihn dargestellt habe; [...]—allerdings mit charakteristischen Abweichungen, die mir eben als Idee des ganzen Romans eines Tages aufgeleuchtet sind und die Basis dieser Dichtung in mir gleichsam momenthaft, in einem einzigen Blitz als Begegnung mit mir selbst und meinem Schicksal klargelegt haben.' Brod, *Streitbares Leben* 13, 202f. Cf. on this also: Dorn A.M., *Leiden als Gottesproblem. Eine Untersuchung zum Werk von Max Brod* (Freiburg et al.: 1981) esp. controversy with Franz Werfel 35–39.

24 October that year Tycho Brahe died), and it is set in two places, Benatek castle and Prague. There are three discernible thematic complexes that overlap and relate to each other:

1. the *Commentariolus* manuscript—in this text, which originates from Heilsberg in 1509, Copernicus develops his theory of a heliocentric world system—and its reception by Tycho Brahe and Johannes Kepler
2. Brahe and Kepler as scientific and human antipodes
3. Brahe's path to god as a parable to Brahe's path to himself

The *Commentariolus* manuscript functions as a plot ferment here. In the beginning, the reader is informed that Brahe got in touch with Kepler by letter on order to discuss Copernicus's theory with him.

> [...] and thus he also did not hold back that he regretted Kepler's commendation of the Copernican world system, hoping, however, to convert him eventually to his own, Tychonian constellation. This he wrote immediately in his first letter [...].[41]

Since his position in Graz has become untenable, Kepler accepts the invitation and lets his host guide him through the instruments' hall in Benatek castle. Beside his instruments for the observation of the stars and his astronomical records—'A treasure worthy to be kept next to the gems of the princes [...]', Kepler remarks on it—, Brahe shows him the *Commentariolus* manuscript.

> It was only in the next apartment, which contained his library, that he found his contentment. He showed rare books, with especial pride an original manuscript by Copernicus, the unprinted "Commentariolus", which had been distributed to contemporaneous friends of the scientist in only few copies and reached Tycho through the grace of Hagecius.[42] Kepler reached for it. Without releasing the script, Tycho looked him in the face meaningfully: "You see how unjust it is to pass me off as an

41 '[...] und so hielt er auch damit nicht zurück, daß er die Lobesworte Keplers auf das Kopernikanische Weltsystem bedaure, daß er aber hoffe, ihn einmal zur eigenen, zur Tychonischen Konstellation zu bringen. Dies schrieb er gleich im ersten Brief [...].' Brod, Tycho Brahe (1915) 3.

42 Tadéaš Hajek, private physician to Rudolf II., plays an ambiguous and scheming role in the novel.

> opponent of this sage. No one can hold him and his works in higher esteem than I." Herewith he put the leaves back in their place.[43]

Readily Brahe lends his own records to his guest but refuses to hand him the *Commentariolus*: ' "This work is outdated", he said.'[44] Why the *Commentariolus* should be outdated and what the differing views of Brahe and Kepler would consist in, we do not get to know; it is only on one occasion that Brahe makes a brief remark on Copernicus:

> [...] I am indeed not as modest as Copernicus. He once said he would consider himself happy if the true course of the stars would not deviate from his calendar more than ten arc minutes. This, however, appears too undemanding to me.[45]

It becomes clear only later wherein the difference between Brahe and Kepler lies with regard to Copernicus. Brod describes their conflicting positions, which are the novel's focal point, as follows:

> In sinful reticence they had only debated the minor problems of their science. From now on Tycho wanted to have a go at the main point, which of course did not present itself as unambiguously as he would have liked to persuade himself.
>
> Kepler had indeed pronounced himself in his first book against the doctrine of Claudius Ptolemy, which then still prevailed, according to which the earth stands immoveable in the centre of the universe and the planets, among which numbered the sun and the moon, would circle her. But Tycho, who prided himself on his agreement with Kepler, could assert

43 'Erst im nächsten Gemach, das seine Bibliothek enthielt, ließ er's sich behagen. Er zeigte seltene Bücher vor, mit besonderem Stolz eine Original-Handschrift des Kopernikus, den ungedruckten "Commentariolus", der nur in wenigen Exemplaren an die zeitgenössischen Freunde des Gelehrten und durch des Hagecius Güte [...] an Tycho gelangt war. Kepler griff danach. Ohne die Schrift loszulassen, sah ihm Tycho bedeutungsvoll ins Gesicht: "Du siehst, wie unrecht man tut, wenn man mich als Gegner dieses Weisen ausgibt. Niemand kann ihn und seine Werke höher in Ehren halten als ich." Damit stellte er die Blätter wieder an seinen Ort.' Brod, *Tycho Brahe* (1915) 46f.

44 'Dieses Werk ist veraltet'. Ibid. 51.

45 '[...] So bescheiden wie Kopernikus bin ich nun freilich nicht. Der sagte einmal, er wolle sich glücklich schätzen, wenn der wirkliche Lauf der Sterne von seinem Kalender nicht mehr als zehn Bogenminuten abweiche. Das scheint mir allerdings allzu genügsam.' Ibid. 124f.

> that he, too, did not adhere to Ptolemy but rather like Kepler found it suspicious that one had to take refuge in all-too contrived constructions of ancillary circles, epicycles, in order to harmonise the truly observed planetary orbits with that theory.—About fifty years ago now Kepler had appeared with the view that the sun stands in the centre, circled by the earth and the planets. Also this theory, however,—and Kepler as well as Tycho had to know this—could not dispense with a similar albeit not quite as convoluted supplementation with epicycles and did not fit reality exactly.[46]

Brahe seeks to overcome the contradiction between his geocentric point of view and his own observations, which were admirable for the time—admirable because a telescope was not at hand then—by placing the earth in the centre and having her orbited by the moon and the sun on the one hand, and on the other hand assuming with Copernicus that the planets circle around the sun. With this hypothesis, Brahe wants to draw Kepler onto his side:

> 'Well, how is it then with the world system?', Tycho asked again as Kepler had not responded immediately. 'How have you made your choice between Copernicus and myself?' 'A difficult question which cannot receive a simple answer', Kepler replied, smiling subtly. 'Certainly, certainly', Tycho hurried to interpose, 'this cannot be said in a few words. I myself, as you know, am not an opponent of Copernicus. Only ignorant

46 'In sündhafter Zurückhaltung hatten sie bisher nur Nebenprobleme ihrer Wissenschaft erörtert. Von nun an wollte Tycho auf die Hauptsache losgehen, die freilich nicht so eindeutig lag, wie er sich gern eingeredet hätte.
Kepler hatte sich allerdings schon in seinem ersten Buch gegen die damals noch seit Alexandrinerzeit herrschende Lehre des Claudius Ptolemäus ausgesprochen, nach welcher die Erde unbeweglich im Mittelpunkt des Weltalls stand und die Planeten, zu denen auch Sonne und Mond zählten, sowie der Fixsternhimmel, an acht kristallenen Sphären befestigt, um sie kreisten. Aber Tycho, der so gern seine Übereinstimmung mit Kepler feststellte, konnte geltend machen, daß auch er am Ptolemäus nicht festhielt, vielmehr ebenso wie Kepler es verdächtig fand, daß man zu sehr gekünstelten Konstruktionen von Nebenkreisen, Epizyklen, seine Zuflucht nehmen mußte, um die wirklich beobachtete Bahn der Gestirne mit dieser Theorie in Einklang zu bringen.—Nun war vor etwa fünfzig Jahren Kopernikus mit der Ansicht aufgetreten, daß die Sonne im Mittelpunkt stehe, von der Erde und den Planeten umkreist. Aber auch diese Theorie (und das mußte doch Kepler ebensogut wie Tycho wissen) konnte einer ähnlichen, wenn auch nicht ganz so verwickelten Ergänzung durch Epizykel nicht entraten und paßte nicht genau auf die Wirklichkeit.' Ibid. 228f.

> people say that about me... Oh, I admire Copernicus's genius, I venerate this proud independent mind. Look at these three feeble worm-eaten wooden staves.' He pointed to a corner of the hall. [...] Here, touch them, palpate them piously. What else do you think you hold in your hands than the veritable *parallacticum* [*triquetrum*] with which our immortal Copernicus observed the stars? Yes, *his levibus baculis*! [...] Oh, it was one of my finest days when his instrument arrived at Hveen, a present from the cathedral chapter in Frauenburg. *O tanti monumenta viri*![47]

Brahe emphasises that his criticism of Copernicus is complex and not as simplistic as Luther's, who thought to refute Copernicus with the objection '[...] that according to the Scripture Joshua made the sun stand still and not the earth [...].'[48] All this, however, does not impress Kepler; he counters with the objection that the '[...] technical tools and experiences[...]' are not '[...] so comprehensive as to decide this question.'[49] Brahe still does not acknowledge defeat and he tries to convince Kepler of the correctness of his geocentric theory by pointing out the different material quality of the earth and the planets.

> As likely as not, the inert earth would not be able to run at all while the planets, sylph-like objects made of light and another fine matter, would by nature be made for flying and together with the sky of fixed stars—albeit not on the firm Ptolemaic spheres but especially in their dashing free revolution—testify to God's prudence and omnipotence.

47 '"Nun, wie steht es also mit deinem Weltsystem?", fragte Tycho nochmals, da Kepler nicht sofort geantwortet hatte. "Wie hast du dich zwischen Kopernikus und mir entschieden?" "Eine schwere Frage, die keine einfache Antwort erhalten kann," erwiderte Kepler, unmerklich lächelnd.
"Gewiß, gewiß," beeilte sich Tycho einzuwerfen, "das ist nicht mit wenigen Worten zu sagen. Ich selbst, wie du weißt, bin ja kein Gegner des Kopernikus. Nur unverständige Menschen sagen das von mir... O ich bewundere das Genie des Kopernikus, ich verehre diesen stolzen, unabhängigen Geist. Sieh hier diese drei schwachen, wurmstichigen Holzstäbe." Er wies in eine Ecke des Saales. [...] "Hier, fasse sie, befühle sie andächtig. Was anderes glaubst du in Händen zu halten als das wahrhaftige Parallacticum, mit dem unser unsterblicher Kopernikus die Sterne beobachtet hat? Ja, his levibus baculis! [...] Ach, es war einer meiner schönsten Tage, als dieses Instrument in Hveen anlangte, ein Geschenk des Frauenburger Domkapitels an mich. O tanti monumenta viri!"' Ibid. 232f.

48 '[...] daß nach der heiligen Schrift Josua die Sonne stillstehen ließ und nicht die Erde [...].' Ibid. 233.

49 '[...] technischen Hilfsmittel und Erfahrungen [...] so weit sind, um in dieser Frage entscheiden zu können.' Ibid. 234.

> Thus, Copernicus's opinion, which moreover had already been established in antiquity by Aristarchus of Samos and would therefore not have remained unknown to Ptolemy, would contradict if not the mathematical but all the more the physical laws [...].
> "That the planets are blended from ethereal matters, exactly this remains to be proven after all", Kepler answered with a certain harshness, while the others were hardly breathing any more. Tycho startled. [...] Now argument was weighed against argument. The egregious idea that the earth itself was but a planet like those remote small stars, Mercury and Venus, stood against the arbitrariness that Kepler tackled [...].[50]

Kepler's arguments do not only support the Copernican world system, they also consequently mean—and this really makes them revolutionary—that the universe is subject to the same physical conditions as the earth.

The discussion between Brahe and Kepler functions as a peripety. The crisis follows, a severe psychosomatic illness that befalls Brahe, a crisis which threatens to slip into a catastrophe if it was not for the catharsis at the end, i.e. Brahe's path to god. It is not without reason that I here operate with terms from the theory of tragedy: Max Brod himself in his essay *Von Sinn und Würde des historischen Romans*[51] established the proposition that the intensification and economy of the drama must find their equivalent in the historical novel if it does not want to be '[...] but a series of entertaining injections [...].'[52]

50 'Die träge Erde könne wohl überhaupt nicht laufen, während die Planeten, sylphenhafte Gebilde aus Licht und einem anderen feinen Stoffe, von Natur aus zum Fluge geschaffen seien und zusammen mit dem Fixsternhimmel freilich nicht auf den festen ptolemäischen Sphären, sondern gerade in ihrem rasenden freien Umschwung von der Weisheit und Allmacht Gottes Zeugnis ablegten. So widerspreche die Meinung des Kopernikus, die überdies schon im Altertum durch Aristarch von Samos aufgestellt worden sei, daher dem Ptolemäus nicht unbekannt geblieben sein könne, zwar nicht den mathematischen, aber um so mehr den physikalischen Gesetzen [...].
"Daß aber die Planeten aus ätherischem Stoffe gemischt sind, das steht doch eben noch zum Beweis," antwortete Kepler mit einer gewissen Strenge, während die anderen kaum mehr atmeten. Tycho erschrak. [...] Nun wurde Argument gegen Argument abgewogen. Die ungeheuerliche Idee, daß die Erde selbst nur ein Planet sei, wie jene fernen kleinen Gestirne Mercurius und Venus, stand gegen die Willkürlichkeit der von Kepler bekämpften Hilfskreise [...].' Ibid. 262f.

51 Brod M., "Von Sinn und Würde des historischen Romans", *Neue Rundschau* 67 (1952) 491–502, here 491.

52 '[...] nicht eine Folge von Unterhaltungs-Injektionen sein'. Ibid. 491.

The object of a historical novel is the elaboration of basic patterns,[53] not the scholarly documentation of sources; in order to achieve this, '[...] finding and invention [...]' are required, i.e. the conscious styling and fictional transformation of the historical topic.[54]

Let us, however, revert to the discussion between Brahe and Kepler: their differing argumentative approach to the central problem, geocentric vs. heliocentric system, makes it clear that we are dealing with two contrarious types of scientists: Kepler is the impartial and unideological scientist, concerned only with abstract cognition which expresses itself through the proportionality of numbers. His strictly analytical, impassive approach to the object of research on the one hand makes him much less prone to misjudgements; on the other hand, he is prepared to subject himself to any power if only it enables him to work uninhibited. Brahe is totally different: he is a whole-blood person searching for certainty and happiness. He understands astronomy as a social obligation. In the interest of his task, he must use all political contacts available to him and at the same time be cautious of his economic independence, which he achieves by making a profitable enterprise out of his science.

Between Brahe and Kepler there initially exists, as has been pointed out in the context of the young Franz Werfel, a teacher-pupil relationship; the more Brahe attempts, however, to rope Kepler in for his own ends, the more Kepler eludes him. Brahe suspects Kepler eventually of making common cause with his scientific and political opponents. His illness helps Brahe to the realisation that his personal relationship with Kepler can only be saved if he releases him unconditionally and lets him find his own scientific path. As a farewell present he delivers his astronomic records and Copernicus's *Commentariolus* up to him:

> "Well, I want to go now", said Kepler and shuffled his foot.
> Tycho embraced him violently. "My Benjamin, what else could I do to you, what joy, fulfil what wish? The manuscripts, yes, take them. I am indeed so happy that they may help you somehow ... But wait, another thing!" He hurried into the library and returned with the "Commentariolus" by Copernicus, which he had refused to Kepler on the day of his arrival:

53 Cf. ibid. 495.

54 '[...] Findung und Erfindung [...]'. Ibid. 497.

> "Take this little work also. I give it to you. Maybe you find something useful in it."[55]

Indeed Kepler finds something useful in the little work because in the following period he delivers the proof that the Copernican system is not only a hypothetical model but also a physical fact.

With the separation from Brahe, Kepler regains his own freedom to act and is able to break free from the cul-de-sac into which he had manoeuvred himself. At the end of the novel there are his encounters with Rabbi Loew and Rudolf II.

The combination of a Talmudic scholar and a catholic potentate, of cabbala and astrology, of ghetto and cabinet of curiosities, is in itself the stuff which the Prague-German bestseller-dreams are made on—one needs only to recall the *Golem* by Gustav Meyrink and *Nachts unter steinernen Brücken* by Leo Perutz – Brod, however, refrains from turning his *Brahe* into a magical Panopticon and instead focuses on the question: how does man relate to something that is beyond him and what are the consequences for his intraworldly behaviour?

For Karl Marx, the case was clear: 'Where the belief in the here and now crumbles, the beyond becomes the fashion.'[56] Now Brod, who, one must not forget, was a confident socialist, argues to the exact contrary. By taking the beyond into this world and speaking of the 'miracle of the here and now', he makes the 'hic et nunc' the real epistemological benchmark. In his religio-philosophical treatise *Heidentum, Christentum, Judentum* from 1921, Brod

55 '"Nun will ich also gehen," sagte Kepler und scharrte mit dem Fuß.
Tycho umarmte ihn heftig: "Mein Benjamin, was könnte ich dir denn noch antun, welche Freude, welchen Wunsch erfüllen? Die Manuskripte, ja, nimm sie. Ich bin ja so glücklich, dass sie dir irgendwie helfen können . . . Doch warte, noch etwas!" Er eilte in die Bibliothek und kam mit dem "Commentariolus" des Kopernikus zurück, den er gleich am Ankunftstage Keplern verweigert hatte: "Nimm auch diese kleine Schrift. Ich schenke sie dir. Vielleicht findest du etwas Nützliches in ihr." ' Brod, *Tycho Brahe* (1915) 344.

56 'Wo der Glaube ans Diesseits zerfällt, kriegt das Jenseits Konjunktur.'—The quote is attributed to Karl Marx. Cf. however: Marx K., "Zur Kritik der Hegel'schen Rechts-Philosophie. Einleitung" (Deutsch-Französische Jahrbücher. Lfg. 1/2.1844), in Marx K. – Engels F., *Gesamtausgabe* (*Mega*), 1st section, vol. II, 1 (Berlin: 1982) 170–183, here 171: 'Therefore, the criticism of religion is in its *seed* the *criticism of the vale of tears* whose *halo* is religion. [. . .] Thus it is *history's task* to establish the *truth of the here and now* after *the beyond of truth* is gone.' – 'Die Kritik der Religion ist also im *Keim* die *Kritik des Jammerthales*, dessen *Heiligenschein* die Religion ist. [. . .] Es ist also die *Aufgabe der Geschichte*, nachdem das *Jenseits der Wahrheit* verschwunden ist, die *Wahrheit des Diesseits* zu etablieren.'

explains what he understands by the 'miracle of the here and now' on the basis of the legend of Rabbi Simeon bar Yochai: fleeing from the Romans, Rabbi Simeon and his son must hide in a cave. The escape turns into isolation and when they leave their cave again, the isolation has become the destruction of the here and now:

> There they saw the people tilling and sowing. And they spoke: Oh, these people neglect the matters of eternal life and occupy themselves with temporal life! And wherever Bar Joachi and his son directed their gaze, everything was scorched.[57]

It is only when Rabbi Simon is prepared to accept life with all its contradictions that he experiences the miracle of the here and now: 'Now that a miracle has occurred to us, I will establish good devices [...]'[58]—'good devices' in the context of the legend meaning: establishing markets, trading, building baths etc.

In the conversations that Brahe conducts with Loew and Rudolf, Brod has pre-formulated the miracle of the here and now. Brahe meets the rabbi on the Prague castle when the latter returns from an audience with the emperor. They engage in a conversation, in the course of which it becomes clear that the history of the Jewish people is a parable to his own restless life. He wants to experience the cause of '[...] all his failed ventures [...]', all his '[...] slights, insults, jeopardies within and without which he suffered on account of his teachings [...]'[59] and urges the rabbi:

> "Tell me now, how is it possible to suffer such a lot of affliction? How is it possible to endure all this? And all this for nothing, for a couple of letters?"
> "It is not at all the question how we endure it", said the old man in a soft voice which had mollified more and more throughout the conversation,

57 'Da sahen sie die Menschen pflügen und säen. Und sie sprachen: Ach, diese Menschen vernachlässigen die Dinge des ewigen Lebens und beschäftigen sich mit dem zeitlichen Leben! Und wohin Bar Joachi und sein Sohn ihre Blicke richteten, dort verbrannte alles.' Brod M., *Heidentum, Christentum, Judentum. Ein Bekenntisbuch*, 2 vols. (Munich: 1921) vol. I, 221.

58 'Da uns nun ein Wunder geschehen ist, will ich eine gute Einrichtung treffen [...].' Ibid. 222.

59 '[...] all seiner verfehlten Unternehmungen [...] Kränkungen, Beleidigungen, Gefahren innen und außen, die er seiner Lehre wegen erlitt [...].' Brod, *Tycho Brahe* (1915) 387.

> "we have a doctrine: God is not there for the just to serve him and support him, but the just is there to serve god and to support him."[60]

At the beginning of the novel, Brahe had been wondering about the reversed relationship of the above and the below, the beyond and the here and now; at a closer look, this prepares the topic that the rabbi in his answer consistently concludes:

> Do therefore not believe that I am only concerned about fathoming the harmony of the star world. That would still be patchwork. That superior harmony is only a reflected splendour of the one below; god has stretched a deep cohesion between my happiness and the laws of the sky, and the law of both of them, the terrestrial regiment of the world and the celestial alike, I must hold in my fist or else I have lived in vain.[61]

Brahe, after taking leave from Rabbi Loew, is admitted to the emperor but he does not meet him as the superior representative of power but as the needy melancholy private person who has entrenched himself behind the valuables of his cabinet of curiosities. The initial inhibition soon gives way to an animated conversation on the relationship between astrology and astronomy in the course of which the emperor remarks:

> "More terrible, however, than this assumption is that rumour has it, he [i.e. Kepler] had expressed himself filthily about astrology and does not pursue this art at all. [...] Look, I know from the most trustworthy persons that Kepler adheres more to the Copernican quirks than your own well turned-out Diataxis mundi." [...] The emperor had spoken rightly. Kepler was Tycho's scientific opponent, no doubt.[62]

60 '"Nun sagt mir aber, wie ist es möglich, soviel Leid zu erdulden? Wie ist es möglich, dies alles auszuhalten? Und all dies um ein Nichts, um ein paar Buchstaben?"
"Es ist gar nicht die Frage, wie wir es aushalten," sprach der Alte mit weicher Stimme, die sich während des Gesprächs mehr und mehr besänftigt hatte, "wir haben eine Lehre: Gott ist nicht um des Gerechten willen da, um ihm zu dienen und ihn zu stützen, sondern der Gerechte ist da, um Gott zu dienen und ihn zu stützen."' Ibid. 387f.

61 'Denn glaube ja nicht, daß es mir nur darum zu tun ist, die Harmonie der Sternenwelt zu ergründen. Das wäre immer noch Stückwerk. Diese obere Harmonie ist nur ein Abglanz der unteren; einen tiefen Zusammenhang hat Gott zwischen mein Glück und die Gesetze des Himmels gespannt, und das Gesetz dieser beiden, des irdischen Weltregiments wie des Himmels, muß ich in der Faust halten, sonst habe ich vergebens gelebt.' Ibid. 111f.

62 '"Doch ärger als diese Vermutungen ist, daß die Rede geht, er habe sich unflätig über die Astrologiam geäußert, treibe auch diese Kunst überhaupt nicht. [...] Seht, ich weiß von

Exactly at this point, however, the 'miracle of the here and now' begins to be effective: Brahe does not help Rudolf by asking him for support of his own stalling projects but by espousing his scientific opponent and making headway for the emperor's appointment of Kepler as his court mathematician with immediate effect.

•••

The problem of the freedom of action has been briefly addressed in the context of Brahe's decision to let Kepler take his own course. This problem has two sides for Brod, a religio-philosophical one and a scientific one.

Brod addresses the religio-philosophical aspect in *Heidentum, Christentum, Judentum*. There, he asks himself the question where freedom would be if man in his misery were only in a position to choose between happiness and unhappiness.

Brahe, too, finds himself confronted with this question: in addition to Kepler, he releases his daughters knowing full well that their failure is bound to occur and he is guilty of it[63] no matter if he keeps them under his guardianship or lets them follow their own paths. One, Brahe says, will perish of her old-maidishness, the other, of her sensuality, of the life she lives; and yet: they are responsible for their lives, not he.

The scientific aspect becomes clear when Brahe's system is confronted with Kepler's. Brahe's system still bears an ontological reference—*operari sequitur esse*, acting follows from being—, while Kepler's system is exclusively committed to the principle of causality. Causality, however, means that there is no contingency, no exception from the rule, and therefore in consequence: there is no freedom of action.

Brod readdresses the problem of determinism and indeterminism after WWII, i.e. in 1947 in *Diesseits und Jenseits* and in 1948 in his second Copernican novel, *Galilei in Gefangenschaft*.

With their empirical-experimental research, Kepler and Galilei collapsed Aristotle's deductive doctrinal system that was based on unproved hypotheses. In the *Galilei* novel, Brod makes the old and the new science collide: Galilei prompts a colleague to look through the telescope; the colleague refuses claiming that empirical observations would be nothing but misperceptions in the sense of the Aristotelian entelechy.

---

den vertrauenswürdigsten Personen, daß dieser Kepler den Kopernikanischen Schrullen mehr anhängt als Eurer eigenen wohlgeratenen Diataxis mundi." [...] Der Kaiser hatte richtig gesprochen. Kepler war Tychos wissenschaftlicher Gegner, kein Zweifel.' Ibid. 405.

63 Ibid. 297f.

In *Diesseits und Jenseits*, Brod holds the view that the freedom empiricism has gained by overcoming Aristotelianism is relinquished again with the totality of mono-causality. Heisenberg's uncertainty principle, too, does not change anything here, for—here, Brod quotes Werner Heisenberg—the '[...] proof of the invalidity of the law of causality presupposes (as proof) the validity of the law of causality [...].'[64] The constraint of causality—which deals with the functional relationship between things, not with the things themselves—can only be prised open if one bethinks oneself of the Platonic—not the Aristotelian!—dimension of being and comprehends freedom as the 'ultimate state of mind'. Consciousness and with it, freedom are determined by being and, consequently, cognition needs to '[...] admit being into the circle of its endeavours [...]'[65] beside its scientific orientation. 'If mankind', Max Brod asserts, 'had not forgotten this orientation towards being, [...] the shameful social injustices, the exploitation of enormous working social strata, master morality, Fascism and the gruesome years of the two world wars would never have happened.'[66]

It is this ontological relatedness that puts Brahe in a position to pursue his way to the end autonomously, in spite or rather because of all errors—exactly this is expressed by the novel's title: *Tycho Brahes Weg zu Gott*.

English translation: Wolfgang Neuber

Selective Bibliography

Bärsch C.E., *Max Brod im Kampf um das Judentum Zum Leben und Werk eines deutsch-jüdischen Dichters aus Prag* (Vienna: 1992).

Blum M.O., "Angoisse et créativité dans 'Tyhos Brahes Weg zu Gott' de Max Brod", in Paul J.M. (ed.), *Le pessimisme. Idée féconde, idée dangereuse* (Nancy: 1992) 227–238.

Brod M., *Diesseits und jenseits*, 2 vols. (Winterthur: 1947).

Brod M., *Der Prager Kreis* (Frankfurt a.M.: 1979).

Brod M., *Heidentum, Christentum, Judentum. Ein Bekenntisbuch*, 2 vols. (Munich: 1921).

Brod M., *Streitbares Leben (1884–1968)*, 2nd edn. (Frankfurt a.M.: 1979).

64 'Beweis der Ungültigkeit des Kausalgesetzes setzt (als Beweis) die Gültigkeit des Kausalgesetzes voraus'. Brod M., *Diesseits und jenseits*, 2 vols. (Winterthur: 1947) vol. I, 196f.

65 'Sein in den Kreis ihrer Bemühungen aufnehmen'. Brod, *Diesseits und jenseits*, vol. II, 143.

66 'Hätte die Menschheit diese Richtung aufs Sein nicht vergessen, [...] so wäre es nie zur beschämenden sozialen Ungerechtigkeit, zur Ausbeutung riesiger arbeitenden Volksschichten, zur Herrenmoral, zum Faschismus und den grausigen Jahren der zwei Weltkriege gekommen.' Ibid. 143f.

Brod M., *Tycho Brahes Weg zu Gott. Roman* (Munich: 1915).

Brod M., *Tycho Brahe's path to God. A novel*, trans. F.W. Crosse (Evanston, Ill.: 2007).

Buber M., *Briefwechsel aus sieben Jahrzehnten. 1897–1918* (Heidelberg: 1972).

Doležal P., *Tomáš G. Masaryk, Max Brod und das Prager Tagblatt (1918–1938). Deutsch-tschechische Annäherung als publizistische Aufgabe* (Frankfurt a.M. et al.: 2004).

Dorn A.M., *Leiden als Gottesproblem. Eine Untersuchung zum Werk von Max Brod* (Freiburg et al.: 1981).

Gassmann A.A., *Lieber Vater, Lieber Gott? Der Vater-Sohn-Konflikt bei den Autoren des engeren Prager Kreises (Max Brod – Franz Kafka – Oskar Baum – Ludwig Winder)* (Oldenburg: 2002).

Janowitz F., *Auf der Erde und andere Dichtungen. Werke, Briefe, Dokumente*, ed. D. Sudhoff (Innsbruck: 1992).

Koch H.G., "Kafkas Max und Brods Franz. Vexierbild einer Freundschaft", in Plachta B. (ed.), *Literarische Zusammenarbeit* (Tübingen: 2001) 245–256.

Lerperger R., *Max Brod. Talent nach vielen Seiten* (Vienna et al.: 1987).

Matuschek O., *Stefan Zweig. Drei Leben—Eine Biographie* (Frankfurt a.M.: 2006).

Pazi M., *Max Brod 1884–1984. Untersuchungen zu Max Brods literarischen und philosophischen Schriften* (New York et al.: 1987).

Schillemeit J.: "Chancen und Grenzen der Willensfreiheit. Fragen der Ethik und Religionsphilosophie bei Max Brod, Felix Weltsch und Kafka (1920)", in *Prager deutschsprachige Literatur zur Zeit Kafkas* (Vienna: 1989) vol. 1, 165–181.

Schoeps J.H. (ed.), *Im Streit und das Judentum. Max Brod – Hans-Joachim Schoeps. Briefwechsel* (Hildesheim et al.: 2011).

Šrámková B., *Max Brod und die tschechische Kultur* (Wuppertal: 2010).

Vassogne G., *Max Brod in Prag. Identität und Vermittlung* (Tübingen: 2009)

Völker K., "Hašeks 'Schweijk'-Roman auf der Bühne. Die Piscator Inszenierung von 1928. Von Brod zu Brecht und die Folgen", in Pazi M. – Zimmermann H.D. (eds.), *Berlin und der Prager Kreis* (Würzburg: 1991) 225–241.

Voigts M., " 'Tod den Toten!' Indifferentismus und Utopie in den frühen Novellen Max Brods", in Grave J. – Sprengel P. – Vandevoorde H. (eds.), *Anarchismus und Utopie in der Literatur um 1900. Deutschland, Flandern und die Niederlande* (Würzburg: 2005) 108–119.

Index Nominum

www.ingramcontent.com/pod-product-compliance
Lightning Source LLC
LaVergne TN
LVHW010542300926
849162LV00002B/214